AF361986

Small-Scale Hydropower and Energy Recovery Interventions: Management, Optimization Processes and Hydraulic Machines Applications

Small-Scale Hydropower and Energy Recovery Interventions: Management, Optimization Processes and Hydraulic Machines Applications

Editors

Mosè Rossi
Massimiliano Renzi
David Štefan
Sebastian Muntean

MDPI • Basel • Beijing • Wuhan • Barcelona • Belgrade • Manchester • Tokyo • Cluj • Tianjin

Editors
Mosè Rossi
Department of Industrial
Engineering and
Mathematical Sciences
Marche Polytechnic
University
Ancona
Italy

Massimiliano Renzi
Faculty of Science and
Technology
Free University of
Bozen-Bolzano
Bolzano
Italy

David Štefan
Viktor Kaplan Department of
Fluid Engineering
Brno University of
Technology
Brno
Czech Republic

Sebastian Muntean
Center for Fundamental and
Advanced Technical Research
Romanian Academy -
Timisoara Branch
Timisoara
Romania

Editorial Office
MDPI
St. Alban-Anlage 66
4052 Basel, Switzerland

This is a reprint of articles from the Special Issue published online in the open access journal *Sustainability* (ISSN 2071-1050) (available at: www.mdpi.com/journal/sustainability/special_issues/ hydropower_hydraulic).

For citation purposes, cite each article independently as indicated on the article page online and as indicated below:

LastName, A.A.; LastName, B.B.; LastName, C.C. Article Title. *Journal Name* **Year**, *Volume Number*, Page Range.

ISBN 978-3-0365-5554-6 (Hbk)
ISBN 978-3-0365-5553-9 (PDF)

Contents

Preface to "Small-Scale Hydropower and Energy Recovery Interventions: Management, Optimization Processes and Hydraulic Machines Applications"

Several topics in the small-scale hydropower sector are of great interest for pursuing the goal of a more sustainable relationship with the environment.

The goal of this Special Issue entitled "Small-Scale Hydropower and Energy Recovery Interventions: Management, Optimization Processes and Hydraulic Machines Applications"was to collect the most important contributions from experts in this research field and to arouse interest in the scientific community towards a better understanding of what might be the main key aspects of the future hydropower sector.

Indeed, the Guest Editors are confident that the Special Issue will have an important impact on the entire scientific community working in this research field that is currently facing important changes in paradigm to achieve the goal of net-zero emissions in both the energy and water sectors.

Mosè Rossi, Massimiliano Renzi, David Štefan, and Sebastian Muntean

Editors

sustainability

MDPI

Editorial

Small-Scale Hydropower and Energy Recovery Interventions: Management, Optimization Processes and Hydraulic Machines Applications

Mosè Rossi [1,*] , **Massimiliano Renzi** [2] , **David Štefan** [3] and **Sebastian Muntean** [4]

1   Department of Industrial Engineering and Mathematical Sciences, Marche Polytechnic University, Via Brecce Bianche 12, 60131 Ancona, Italy

2   Faculty of Science and Technology, Free University of Bozen-Bolzano, Piazza Università 1, 39100 Bolzano, Italy

3   Viktor Kaplan Department of Fluid Engineering, Brno University of Technology, Technická 2896/2, 616 69 Brno, Czech Republic

4   Center for Fundamental and Advanced Technical Research, Romanian Academy—Timisoara Branch, Bulevardul Mihai Viteazu 24, 300223 Timisoara, Romania

*   Correspondence: mose.rossi@staff.univpm.it

Citation: Rossi, M.; Renzi, M.; Štefan, D.; Muntean, S. Small-Scale Hydropower and Energy Recovery Interventions: Management, Optimization Processes and Hydraulic Machines Applications. *Sustainability* **2022**, *14*, 11645. https://doi.org/10.3390/su141811645

Received: 13 September 2022
Accepted: 15 September 2022
Published: 16 September 2022

Publisher's Note: MDPI stays neutral with regard to jurisdictional claims in published maps and institutional affiliations.

The overuse of fossil fuels has brought considerable climate change to our planet, affecting not only human life, but also the ecosystem of the Earth (e.g., the melting of glaciers, drought in warm/hot seasons, etc.). Indeed, an effect of raising CO_2 concentration in the atmosphere is the steep gradient of the global temperature increase (almost +0.08 °C per decade, starting from 1900), which can be slowed down only by rethinking the current paradigms of energy production and use.

The introduction and promotion of technologies capable of exploiting cleaner and renewable sources, as well as delivering core energy-related services such as electricity and drinkable water, is currently one of the main paths to be followed for tackling the global warming issue. Water, as one of these sources, is an essential input for life and energy production; conversely, a significant amount of energy is required for the extraction, treatment, distribution, and disposal of water for civil and industrial uses. Thus, these resources must be managed properly to grant a secure supply to the population while meeting sustainability goals. This tradeoff constitutes the so-called water–energy nexus that motivates scientists to further investigate, optimize, and invest in this important research topic dealing with the correct and smart use of water.

Renewables are considered the key factor to proceeding with the decarbonization path. According to the International Energy Agency (IEA), renewables have increased electricity production by 7% in one year, namely, from 2019 to 2020 [1]. In particular, wind and photovoltaic (PV) technologies together accounted for almost 60% of this increase. The share of renewables in the global electricity market reached almost 29% in 2020. Renewable power deployment still needs to expand significantly to achieve net-zero emissions by 2050. In the medium term, the target is to reach more than 60% of the electricity generation using renewables by 2030 [2]. Hydropower is the largest source of electricity worldwide; indeed, its generation increased by 124 TWh (+3%) in 2020, reaching 4418 TWh and generating more than all other renewable technologies combined. Hydropower is supposed to have a 3% average annual generation growth until 2030 to provide 5870 TWh/year of electricity; thus, an average of 48 GW of new capacity must be installed to achieve the net-zero emissions target [3].

Currently, most of the hydropower plants worldwide are large-scale systems; these systems were widely installed in the twentieth century, exploiting the most convenient locations in terms of high height differences and flow rates, as these sources were the ones that granted the highest payback in energetic and economic terms. The technologies used in these installations are now considered to be mature, even though there are some recent studies regarding the further revamping and digitalization of large-scale hydraulic turbines, such as Francis turbines or pump turbines, in the case of storage power plants. For instance, Chen et al. (Contribution 1) proposed an analytical model of a Francis turbine operating at the best efficiency point (BEP). This shows a much faster response in predicting the internal flow field and working performance compared to experimental campaigns and computational fluid dynamics (CFD), which both require high investments in equipment and supercomputers. On the other hand, the problematic of pressure pulsations during the operation sequences of pump-storage power plant has been investigated by Himr et al. (Contribution 2), which addressed this energy dissipation phenomenon to the second (bulk) viscosity of water.

As an alternative option to further foster the exploitation of hydropower potential, the small-scale hydropower sector is currently being considered for future hydropower scenarios, for example, as a downsized alternative to traditional plants, and as a solution to exploit the so-called "hidden hydropower potential", which involves energy recovery solutions and the exploitation of alternative hydraulic geodetic potentials, such as pressurized water networks and irrigation systems. These solutions are particularly attractive in regions where traditional hydropower sites have been extensively exploited, such as in Europe and North America. Generally, the water pressure in distribution networks is lowered through pressure-reducing valves (PRVs) down to 3–4 bar from higher values that depend either on the height of the water source or the outlet pressure of pumping stations used to distribute the water to the end-users. During this process, the energy of the water is wasted in the forms of heat, noise, and vibration [4]. For this reason, a kind of energy recovery intervention in water distribution networks can be the replacement of PRVs with hydraulic turbines capable of exploiting the water potential available in these networks. Biner et al. (Contribution 3) proposed a new micro-hydroelectric system whose core element is a counter-rotating micro turbine, focusing on a gross capacity between 5 kW and 25 kW. This kind of hydraulic machine requires low capital expenditure with an economic return within 10 years.

However, other technologies that have been used in several traditional hydraulic applications can be adopted for energy generation and recovery purposes, such as pump as turbines (PATs), which are considered an effective and much cheaper solution than traditional hydraulic turbines. PATs are hydraulic pumps (e.g., axial or radial type) that operate in reverse mode, namely, as a turbine, to produce power. The main advantages of this kind of technology are (i) their large availability of different scales in the market, (ii) low investment cost, i.e., 10–20 times lower than a traditional hydraulic turbine, and (iii) easiness of installation, while one drawback might be their lower efficiency compared to conventional hydraulic turbines. PATs are usually applied in a power range between 5 and 500 kW, and their producibility can be varied by modifying their rotational speed according to the operating conditions onsite, leading to a payback periods (PBPs) of approximately 6 years. In recent years, several studies have been carried out by analyzing this technology to better assess its pros and cons, as well as room for possible improvement. Binama et al. (Contribution 4) investigated the behavior of a PAT from a fluid dynamic point of view, focusing on the flow structure formation mechanism when the impeller rotational speed changes. The results showed a worsening of both flow and pressure fields by increasing the flow rate exploited by the hydraulic machine, but a slight improvement in pressure pulsation levels by increasing the rotational speed of the impeller. Contrarily, the increase in the rotational speed when dealing with part-load operating conditions led to worse results. Kan et al. (Contribution 5) focused on the hydrodynamics of PATs, which have been shown to cause progressive deterioration in inner surface smoothness. This phenomenon leads to an increase in friction losses that affect the overall performance of the hydraulic machine. CFD simulations in an axial-flow PAT have been carried out to address this problem, while the hydraulic machine operates at different operating conditions. The results showed that the wall roughness gradually decreases the head, the mechanical power, and the efficiency of the analyzed PAT due to the axial velocity distribution uniformity and the increase in velocity-weighted average swirl angles. PATs can be used not only for producing/recovering energy in water systems, but also as storage technology. Lugauer et al. (Contribution 6) focused on the operation of a micro-pump storage system from an economic point of view. A custom-built simulation model was based on pump and turbine maps that were either given by manufacturers/other works in the literature, or obtained through similarity laws. In particular, 11 PATs controlled by a frequency converter for various generation and load scenarios were evaluated. The results showed that systems with 22 kW power output and heads greater than 70 m are the most profitable since a levelized cost of electricity (LCOE) of 0.292 €/kWh and total storage efficiency of 42% have been reached.

In addition to energy recovery interventions within distribution networks, the residual water energy content can also be recovered in tap-water systems within buildings (e.g., faucets). In this regard, Chen et al. (Contribution 7) developed a suitable micro-pipe mixed-flow turbine with a 15 mm diameter. This turbine has been designed firstly with CFD, and then the obtained results have been validated with experiments, showing a power output and an efficiency of 6.40 W and 87.13%, respectively, after using a multi-objective orthogonal optimization method to enhance its performance. Such a small device can be used, for example, to power some microelectronics adopted in water distribution lines as digital devices to monitor the performance and the reliability of the network.

Moving to the irrigation systems, there are a lot of weir structures that are used to control the water elevation, water velocities, etc., thus representing an energy waste in terms of unexploited potential. The installation of a hydraulic turbine in this kind of system, in addition to performing the same task of the weir structures previously mentioned, allows recovering energy, thus contributing to distributed energy production. Therefore, a two-fold advantage could be achieved by both controlling the natural water flow and also recovering unexploited hydraulic energy potential. Cassan et al. (Contribution 8) studied a hydrostatic pressure wheel capable of regulating flow discharge and water height in open channels, and elaborated a model of the water depth–discharge–rotational speed relationship that considers the different energy losses present in the turbine. Experimental tests were made to calibrate the head loss coefficients before the hydraulic machine was installed in a real application. Finally, the wheel was able to achieve the goals of both water level regulation and energy recovery, and can be described as simple, robust, and environmentally friendly.

Nevertheless, the use of turbines is not the only solution for improving water systems; indeed, besides the optimal management of water supply systems through variable-speed pumping stations that has been studied by Drăghici et al. (Contribution 9), there are water networks all over the world that are subjected to considerable water leakages due to the limited maintenance of both supply and distribution pipelines. Just to give an idea, the World Bank (WB) estimates the average value of non-revenue water at 25–30%, which is due to water losses [5]. In addition to the loss of an important life source, water leakages also lead to economic losses: to limit them as much as possible, smart meters are a suitable solution for achieving satisfactory results in terms of both water, energy, and economic savings. A smart meter records the water consumption and provides information to both suppliers and end-users, such as water quality and temperature. Furthermore, smart meters monitor the supply network in real-time, and are highly responsive in detecting issues in the network such as leaks. Water distribution networks can be then divided into sub-branches where a fixed number of smart meters are installed, thus facilitating the control of these sub-networks. This division is called district metered areas (DMAs), and allows companies to check zones where water leakages, both physical (e.g., inside the water network and on the residential side) and apparent (e.g., authorized and unauthorized consumptions), might appear. Bonthuys et al. (Contribution 10) focused on the study of leakage reduction in a water distribution system located in Stellenbosch (South Africa), and a genetic algorithm was used to identify and optimize the location and size of hydro-turbine installations for energy recovery purposes. Both energy recovery installations and pipe replacements showed a reduction in leakage of up to or more than 6% by replacing 10-year-old pipes.

As deeply analyzed and discussed in this Editorial, several topics in the small-scale hydropower sector are of great interest for pursuing the goal of a more sustainable relationship with the environment. The goal of this Special Issue entitled *"Small-Scale Hydropower and Energy Recovery Interventions: Management, Optimization Processes and Hydraulic Machines Applications"* was to collect the most important contributions from experts in this research field and to arouse interest in the scientific community towards a better understanding of what might be the main key aspects of the future hydropower sector. Indeed, the Guest Editors are confident that the Special Issue will have an important impact on the entire scientific community working in this research field that is currently facing important changes in paradigm to achieve the goal of net-zero emissions in both the energy and water sectors.

List of Contributions:

1. Chen, Y.; Zhou, J.; Karney, B.; Guo, Q.; Zhang, J. Analytical Implementation and Prediction of Hydraulic Characteristics for a Francis Turbine Runner Operated at BEP. *Sustainability* **2022**, *14*, 1965, https://doi.org/10.3390/su14041965.
2. Himr, D.; Habán, V.; Štefan, D. Inner Damping of Water in Conduit of Hydraulic Power Plant. *Sustainability* **2021**, *13*, 7125, https://doi.org/10.3390/su13137125.
3. Biner, D.; Hasmatuchi, V.; Rapillard, L.; Chevailler, S.; Avellan, F.; Münch-Alligné, C. DuoTurbo: Implementation of a Counter-Rotating Hydroturbine for Energy Recovery in Drinking Water Networks. *Sustainability* **2021**, *13*, 10717, https://doi.org/10.3390/su131910717.
4. Binama, M.; Kan, K.; Chen, H.-X.; Zheng, Y.; Zhou, D.-Q.; Su, W.-T.; Ge, X.-F.; Ndayizigiye, J. A Numerical Investigation into the PAT Hydrodynamic Response to Impeller Rotational Speed Variation. *Sustainability* **2021**, *13*, 7998, https://doi.org/10.3390/su13147998.
5. Kan, K.; Zhang, Q.; Zheng, Y.; Xu, H.; Xu, Z.; Zhai, J.; Muhirwa, A. Investigation into Influence of Wall Roughness on the Hydraulic Characteristics of an Axial Flow Pump as Turbine. *Sustainability* **2022**, *14*, 8459, https://doi.org/10.3390/su14148459.
6. Lugauer, F.J.; Kainz, J.; Gehlich, E.; Gaderer, M. Roadmap to Profitability for a Speed-Controlled Micro-Hydro Storage System Using Pumps as Turbines. *Sustainability* **2022**, *14*, 653, https://doi.org/10.3390/su14020653.
7. Chen, H.; Kan, K.; Wang, H.; Binama, M.; Zheng, Y.; Xu, H. Development and Numerical Performance Analysis of a Micro Turbine in a Tap-Water Pipeline. *Sustainability* **2021**, *13*, 10755, https://doi.org/10.3390/su131910755.
8. Cassan, L.; Dellinger, G.; Maussion, P.; Dellinger, N. Hydrostatic Pressure Wheel for Regulation of Open Channel Networks and for the Energy Supply of Isolated Sites. *Sustainability* **2021**, *13*, 9532, https://doi.org/10.3390/su13179532.
9. Drăghici, I.A.; Rus, I.-D.; Cococeanu, A.; Muntean, S. Improved Operation Strategy of the Pumping System Implemented in Timisoara Municipal Water Treatment Station. *Sustainability* **2022**, *14*, 9130, https://doi.org/10.3390/su14159130.
10. Bonthuys, G.J.; van Dijk, M.; Cavazzini, G. Optimizing the Potential Impact of Energy Recovery and Pipe Replacement on Leakage Reduction in a Medium Sized District Metered Area. *Sustainability* **2021**, *13*, 12929, https://doi.org/10.3390/su132212929.

Funding: This research received no external funding.

Conflicts of Interest: The authors declare no conflict of interest.

References

1. International Energy Agency (IEA). Renewables 2020: Analysis and Forecast to 2025. Available online: https://www.iea.org/reports/renewables-2020 (accessed on 8 August 2022).
2. International Energy Agency (IEA). Renewable Power. Available online: https://www.iea.org/reports/renewable-power (accessed on 8 August 2022).

3. International Energy Agency (IEA). Hydropower. Available online: https://www.iea.org/reports/hydropower (accessed on 8 August 2022).
4. Majidi, M.; Etzadi-Amoli, M. Recapturing wasted energy in water pressure reducing valves via in-conduit hydropower generators. *Measurement* **2018**, *123*, 62–68. [CrossRef]
5. El-Zahab, S.; Zayed, T. Leak detection in water distribution networks: An introductory overview. *Smart Water* **2019**, *4*, 5. [CrossRef]

 *sustainability*

Article

Analytical Implementation and Prediction of Hydraulic Characteristics for a Francis Turbine Runner Operated at BEP

Yu Chen [1,2,*], Jianxu Zhou [3], Bryan Karney [2], Qiang Guo [3] and Jian Zhang [3]

1   School of Mechanical Engineering, Nanjing Institute of Technology, Nanjing 211167, China
2   Department of Civil Engineering, University of Toronto, Toronto, ON M5S 1A4, Canada; karney@ecf.utoronto.ca
3   College of Water Conservancy and Hydropower Engineering, Hohai University, Nanjing 210098, China; jianxuzhou@163.com (J.Z.); guoq1228@126.com (Q.G.); jzhang@hhu.edu.cn (J.Z.)
*   Correspondence: yuchen@njit.edu.cn

Abstract: The extensive investigation and profound understanding of the hydraulic characteristics of the Francis turbine are crucial to ensure a safe and stable hydraulic system. Especially, predicting the runner's hydraulic efficiency with high fidelity is mandatory at the early stage of a new hydropower project. For these purposes, the current technologies mainly include experimentation and CFD simulation. Both methods generally have the demerits of a long period, massive investment and high requirements for supercomputers. In this work, an analytical solution is therefore introduced in order to predict the internal flow field and working performance of the runner while the Francis turbine operates at the best efficiency point (BEP). This approach, based on differential-geometry theory and the kinematics of ideal fluid, discretizes the blade channel by several spatial streamlines. Then, the dynamic parameters of these streamlines are determined in a curved-surface coordinate system, including velocity components, flow angles, Eulerian energy and pressure differences across the blade. Additionally, velocity components are converted from the spatial-velocity triangle to the Cartesian coordinate system, and the absolute-velocity vectors as well as the streamlines are subsequently derived. A validation of this approach is then presented. The analytical solution of hydraulic efficiency shows good agreement with the experimental value and simulation result. Additionally, the distributions of pressure differences over the blade, velocity and Eulerian energy are well predicted with respect to the CFD results. Finally, the discrepancy and distribution of the dynamic parameters are discussed.

Keywords: analytical solution; mathematic analysis; Francis turbine; hydraulic characteristics; efficiency prediction; BEP

Citation: Chen, Y.; Zhou, J.; Karney, B.; Guo, Q.; Zhang, J. Analytical Implementation and Prediction of Hydraulic Characteristics for a Francis Turbine Runner Operated at BEP. *Sustainability* **2022**, *14*, 1965. https://doi.org/10.3390/su14041965

Academic Editors: Mosè Rossi, Massimiliano Renzi, David Štefan and Sebastian Muntean

Received: 4 December 2021
Accepted: 2 February 2022
Published: 9 February 2022

Publisher's Note: MDPI stays neutral with regard to jurisdictional claims in published maps and institutional affiliations.

1. Introduction

The current global energy system is experiencing an optimization process and aims to increase its use of clean energy, including solar, wind, nuclear, hydro and other renewables. However, this system consisting of diverse energy sources is apt to trigger swings in the power grid. In this case, hydropower, which is superior in terms of rapid startup and shutdown as well as in its effective regulation of power quality, has long had a crucial role in stabilizing the power system.

The runner of a turbine serves as the core of the hydro-energy conversion. It directly determines the stability of the hydraulic system and further obliquely affects the robustness of the electric system. It is thereby essential to optimize the runner-design process and predict its working performance, particularly at the early stage of a hydropower project. For this purpose, experiments [1,2] and CFD simulations [3–5] are the two widely used technologies. However, satisfactory results require high-end instruments such as Particle Image Velocimetry [6,7], Laser Doppler Velocimetry [8,9] or Remote Sensing Survey. Likewise, accurate CFD results impose high demands on supercomputers. Consequently, the mathematical-analysis method seems a preferable alternative for overcoming these shortcomings. However, the relevant analytical methods to date have mainly been used for the runner's inverse design [10–13] and seldom for solving the runner's internal characteristics [14].

In this ambit, a few studies focused on the direct design of centrifugal pump and its flow analysis [15,16]. By extension, the flow on the stream surface (S1) was converted to a 2D Cartesian coordinate system by conformal transformation. The flow around the cascade was further described by equations of integral form, based on which the relative velocity on the blade surface was derived [17]. In their work, a discrepancy arose between calculations and measurements, presumably attributable to both ignoring the effects of blade shape on the flow and to the unsatisfied momentum equation in the blade channel [18]. This method was later improved with a direct and inverse iteration, and distributions of the relative and meridian velocities were obtained by solving continuity and momentum equations with the streamline-curvature method [19,20].

As for the analytical method in turbines, most studies have only focused on the flow patterns at the runner's inlet and outlet, particularly on their velocity triangles [21,22]. On the meridional plane, the blade of the pump-turbine was discretized and the distributions of Eulerian energy were deduced by a direct and inverse method [23]. Therein, a surface coordinate was first established on the S1 stream surface of a pump turbine. This decisive step implies the importance of describing flow patterns in the S1-surface coordinate system. Additionally, some relevant technologies have been reported. For instance, spherical coordinates associated with the conformal-transformation method were adopted to investigate the flow in a tubular turbine. The results revealed the relationship between guide-vane opening and the flow at the vane outlet [24]. Then, the runner blade was discretized in spherical coordinates and further designed with the help of cylindrical coordinates [25]. Special attention has also been dedicated to the mathematical models describing swirling flow in the turbomachinery [26] and downstream of the runner [27,28]. Then, the helical flow pattern was described by the partial differential function (PDF) of the stream function. This PDF can be solved by converting it to constrained variational problems or under simplified conditions, and then validated by the finite-element method [29,30].

Most of these studies tried to determine the direct and inverse design of reaction runners. Few of them obtained the hydraulic characteristics and internal flow field under ideal cases or by the virtue of measurements. Herein, an analytical approach based on the classic runner-design method is introduced, named the analytical method of characteristics (AMOC). This approach focuses on predicting the internal and external characteristics of a Francis turbine without performing extra tests or finite-element simulations; only the runner configuration and its operational parameters are required. In this study, a calculation procedure is performed at the BEP under incompressible and inviscid-flow conditions. This analytical work aims to accurately and effectively predict the optimal hydraulic efficiency and flow field in an existing runner, to quantitatively analyze the energy conversion on the blade and to reveal the pressure difference across the blade.

The paper is organized as follows: Section 2 presents the main procedures of the AMOC approach, with detailed steps in Appendix A. Section 3 takes the form of a case study of a Francis turbine, describing the corresponding results and its validation by the help of the finite-volume method. Section 4 contains conclusions and outlooks.

2. AMOC Methodology Implementation

2.1. Basic Assumptions

A mathematical model describing the flow in the blade channel is justifiably crucial to analytically determine the runner characteristics. Indeed, real flow is complicated due to its kinematics, dynamics and thermodynamics or even their interactions. In order to simplify this issue and effectively implement the AMOC, some hypotheses are listed as follows:

- The liquid is inviscid, single-phase and incompressible.

- The flow field in the runner is only subject to kinematics and described based on monistic theory, namely the velocity at a certain point is determined only by the streamwise distance from the origin of a streamline.
- The fluid flows smoothly in the blade channel, i.e., the relative velocity component is aligned with the local tangential direction of the relative spatial streamline.

2.2. Implementation Procedures

The implementation procedure of the AMOC approach is illustrated in Figure 1. This approach is based on the well-known classical-design theory of radial-axial turbine runners under incompressible and inviscid-flow conditions. The three-dimensional runner configuration, particularly the spatial blade channel, is discretized and projected onto the meridional plane. Closely following this, spatial streamlines are formed by the interaction of the blade and the stream surface. Then fluid kinematic relationships are constructed for each streamline and later the kinematic parameters are solved.

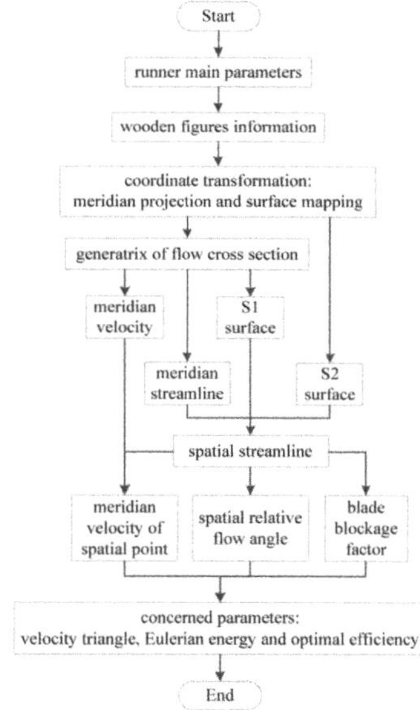

Figure 1. Calculation flow diagram of AMOC.

Furthermore, the flow at the blade-channel inlet is irrotational, with uniform radial and tangential velocity components. However, at the outlet of the channel, a uniform axial-velocity profile is considered. A blade-blockage factor is introduced in order to account for the effects of blade thickness. Some parameters are initially input in order to perform this approach, including:

- A model of the runner and the main geometry of the blade, such as wooden figures.
- Meridional channel geometry, such as the crown, band, and blade leading-edge and trailing-edge curves.
- The operational parameters at the BEP, such as available unit speed, head, and discharge.

Finally, the approach outputs several spatial streamlines and their relevant parameters including the velocity components, Eulerian energy, absolute flow angle, relative flow angle, runner's hydraulic efficiency and pressure difference over the blade.

The general descriptions of Figure 1 are presented as follows with additional details on the derivation shown in Appendix A.

2.2.1. Coordinate Transformation

Meridian projection and surface mapping are performed in this section. Geometries described in a Cartesian system are converted into a curved coordinate system by virtue of the stream surface (S1) and the blade-chamber surface (S2). Detailed deviations can be found in Appendix A.1. The coordination operations are defined by Equations (A1)–(A4).

2.2.2. Generatrix of Flow Section and Meridional Velocity

The cross section of the flow in the blade channel is perpendicular to the S1 stream surface by the classical design theory of radial-axial turbine runners. Its intersection with the meridian plane generates the so-called generatrix, which can be geometrically acquired by inserting inscribed circles in the blade channel. As shown in Figure A3, one generatrix, the arc $\widehat{AB}$ derived from Equation (A5), rotates round the Z-axis to form the flow cross section. Then the meridional velocity is computed in Appendix A.2.

2.2.3. Streamline Cluster Generation

Meridional streamlines discretize the flow channel into several passages, with each passage conveying the same discharge. These meridional streamlines (SL) and ten relative spatial streamlines (SSL) are derived by the steps in Appendix A.3.

Appendix A.4 elaborates the steps to determine the relative flow angle at each node on the spatial streamlines. Then, with the known meridional velocity of the SL, the corresponding meridional velocity on the SSL is derived in Appendix A.5. Blockage effects due to the blade thickness are taken into account by Equations (A15) and (A16).

Using the above procedures, all of the parameters with which to further implement the AMOC are obtained, including the revised meridional velocity $V_m{'}$, the relative flow angle β_i and the blockage factor $1/k_i$ at each node on the SSL.

By recalling the Law of sines and Law of cosines, the velocity components are derived from Equations (A17) to (A21).

Afterwards, the Eulerian energy at each node on the SSL is quantified as:

$$E_{Ui} = U_i \cdot V_{Ui} \tag{1}$$

where E_{Ui} denotes the Eulerian energy at node i.

Recalling the fundamental equation of a hydraulic turbine defined as the integral form [28]:

$$\int_{S_1} (\omega r_{b1} V_{Ub1}) \rho V_{mb1} dS_1 - \int_{S_2} (\omega r_{b2} V_{Ub2}) \rho V_{mb2} dS_2 = (\rho Q)(gH)\eta \tag{2}$$

where S_1 and S_2, respectively, denote the cross sections upstream and downstream of the runner.

This equation defines the change in flux of the moment of momentum along both the streamwise and spanwise directions. According to the second assumption in Section 2.1, this equation is integrated as:

$$\frac{\gamma}{g} Q_0 \cdot (E_{Ub1} - E_{Ub2}) = \gamma Q_0 H_0 \eta_{r0} \tag{3}$$

then, the runner hydraulic efficiency can be computed where, γ is the liquid unit weight and g is the gravity acceleration. E_{Ub1} and E_{Ub2} separately denote Eulerian energies at the blade inlet and outlet. H_0 is the rated head and η_{r0} is the runner's hydraulic efficiency at the BEP as computed by the AMOC.

Figure A5 shows the discretization of the blade inlet and outlet, as well as the notation of each subsection. With the length-weighted method, the Eulerian energy in Equation (3) is therefore replaced by the average value, quantified as:

$$\overline{E_{Ub1}} = \sum_{J=1}^{9} \zeta_{b1J} \cdot \frac{E_{Ub1J} + E_{Ub1(J+1)}}{2} \tag{4}$$

where

$$\zeta_{b1J} = \frac{L_{b1J}}{\sum\limits_{J=1}^{9} L_{b1J}}$$

$$\overline{E_{Ub2}} = \sum_{J=1}^{9} \zeta_{b2J} \cdot \frac{E_{Ub2J} + E_{Ub2(J+1)}}{2} \tag{5}$$

where

$$\zeta_{b2J} = \frac{L_{b2J}}{\sum\limits_{J=1}^{9} L_{b2J}}$$

where J denotes the ordinal of the SSL. $J = 1, 2, \ldots , 9$ due to ten SSLs derived here as shown in Figure A4. The subscript $b1$ and $b2$ denote blade inlet and outlet separately. $\overline{E_U}$ denotes the length-averaged Eulerian energy. E_{UJ} denotes the Eulerian energy on the Jth SSL. L_J denotes the length of the subsection. ζ_J denotes the weighted factor of each segment.

2.3. Transformation of Velocity Component

The velocity components are converted from velocity triangles to the Cartesian system by vector operation and coordinate operation in Appendix B. Then, the absolute streamlines are derived by interpolation and extrapolation.

3. Case Study and Results Discussion

3.1. Basic Parameters of Runner Model

This AMOC procedure was performed on a Francis turbine model (A858a-36.6) with an X-shaped blade that was conceptualized and developed by Prof. Hermod [31,32]. A prototype of this model serves at the Three Gorges right bank station. Corresponding runner and blade illustrations are separately presented in Figures 2 and 3. The main dimensions of this model and its operational parameters at the BEP are tabulated in Table 1. The X-blade runner features a twisted blade and a negative angle at the inlet in order to reduce the cross flow in the blade channel [32]. These features make the second and third assumptions more reasonable.

Figure 2. Runner of Francis turbine.

Figure 3. X-shaped blade.

Table 1. Main specifications of the Francis turbine model.

Geometric Dimensions	D1 mm	D2 mm	N	Blade shape
	372.9	366.0	15	X
BEP Parameters	n_0 r/min	Q_0 m^3/s	H_0 m	ρ kg/m^3
	1122	0.492	30	997

3.2. Validation and Result Analyses

With respect to the same turbine model, the results predicted by the AMOC approach were evaluated by experimental data and using the CFD software ANSYS Fluent 19.2. In the numerical simulation, the three-dimensional model and flow passage are illustrated in Figure 4. This system consisted of the spiral case, guide vanes, runner and draft tube. A grid with two million elements was employed to the whole domain. Then, the Navier–Stokes equations coupled with the k-ε turbulence model were solved in the steady state. A total head of 30 m was prescribed at the system inlet, and an average static pressure of 0 Pa was used at the system outlet. This simulation configuration has been widely used in the field of hydraulic turbomachines. Simulation results have shown good agreement with experiments [10,33], at least at the BEP. A ThinkPad laptop T460p was used to perform the simulation, and it took almost twelve hours to get satisfactory results for one operation point.

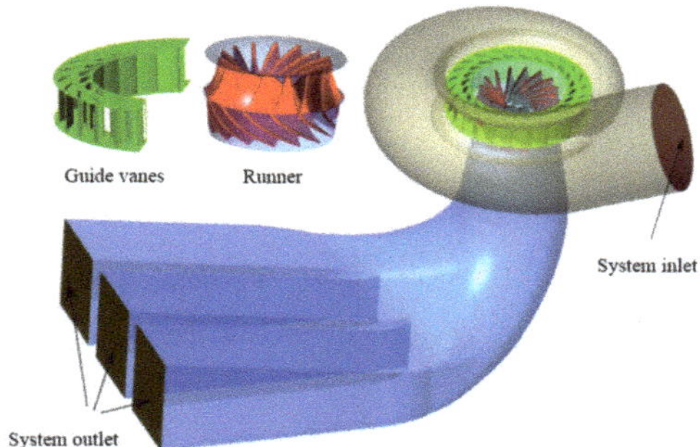

Figure 4. Geometric model of the whole flow system for the turbine.

3.2.1. Time Consumption and Turbine Hydraulic Efficiency

By performing the AMOC, the whole implementation process consumed around two hours to obtain all of the desired results using an identical laptop. From this point, the time consumption of AMOC was 1/6 of the CFD calculation. Here, the efforts of 3D modeling were not taken into account, let alone the various cases throughout the whole operation range.

Part of the data report obtained at the blade inlet and outlet after running the AMOC is listed in Table 2, including various components of the velocity triangles, Eulerian energy, the length of each subsection and the optimal hydraulic efficiency computed from Equation (1) to Equation (5). That is,

$$\eta_{r0} = \left(\overline{E_{Ub1}} - \overline{E_{Ub2}}\right)/g/H_0 = (305.507 - 26.846)/9.81/30 = 0.9469 \tag{6}$$

Table 2. Analytical results of AMOC implementation.

	SSL	V m/s	W m/s	U m/s	β °	α °	E_U m²/s²	L mm	$\overline{E_U}$ m²/s²
Blade inlet	SSL-0	16.229	4.195	17.085	71.236	14.169	268.829		
	SSL-0.0625	16.361	4.179	17.100	72.828	14.124	271.325	7.905	
	SSL-0.125	16.445	4.186	17.133	73.538	14.131	273.225	7.960	
	SSL-0.25	16.614	4.230	17.274	74.005	14.166	278.274	15.770	
	SSL-0.375	16.852	4.290	17.534	73.849	14.153	286.505	15.577	
	SSL-0.5	17.219	4.343	17.934	73.603	14.004	299.634	15.288	305.507
	SSL-0.625	17.602	4.436	18.446	72.174	13.881	315.193	15.122	
	SSL-0.75	17.932	4.594	19.067	68.891	13.826	332.012	14.758	
	SSL-0.875	18.273	4.776	19.853	64.022	13.591	352.623	14.080	
	SSL-1	18.511	5.081	20.726	57.639	13.406	373.192	13.066	
Blade outlet	SSL-0	5.385	8.943	5.921	35.663	104.470	−7.966		
	SSL-0.0625	5.533	10.272	6.761	29.733	112.968	−14.597	23.288	
	SSL-0.125	5.160	10.012	7.741	30.516	99.864	−6.843	21.686	
	SSL-0.25	5.392	9.451	9.767	32.539	70.507	17.574	35.446	
	SSL-0.375	5.853	10.144	11.594	30.310	61.004	32.895	26.720	
	SSL-0.5	6.031	11.290	13.249	26.973	58.104	42.221	20.423	26.846
	SSL-0.625	6.450	12.061	14.771	25.331	53.139	57.150	17.256	
	SSL-0.75	6.983	12.593	16.175	24.245	47.777	75.907	14.977	
	SSL-0.875	6.720	14.220	17.475	21.495	50.842	74.151	13.097	
	SSL-1	7.104	14.881	18.681	20.737	47.879	89.001	12.220	
η_{r0} %					94.69				

This turbine model was tested by the Harbin electrical-machine factory in China. The main results used to validate the analytical approach are shown in Table 3 [34]. The optimal efficiency derived from the test was 94.63%. This value in the runner hill chart is 94.7% as illustrated in Figure 5 [35]. Moreover, the optimal hydraulic efficiency of the runner predicted by the Navier–Stokes solver is 94.57%. The efficiency value computed by the different methods are compared in Table 4. The relative errors are calculated, and these minor discrepancies quantitatively validate that the AMOC enables the accurate prediction of the runner's optimal hydraulic efficiency. Admittedly, the experimental value of the efficiency was calculated between the cross sections upstream of the runner and downstream of the runner, and this value takes into account the hydraulic efficiency, the mechanical efficiency and the volumetric efficiency, respectively, whereas the analytical approach in this paper predicts only the hydraulic efficiency from the blade inlet towards the outlet. At the BEP, the mechanical and volumetric loss is negligible, which leads to a good agreement between the experimental and the analytical approaches.

Table 3. Main test results of the model runner.

Case and Variable	n_{11} r/min	Q_{11} m^3/s	η %	Cavitation Coefficient
BEP	73.7	670	94.63	/
Limitation case	/	992	86.60	0.08

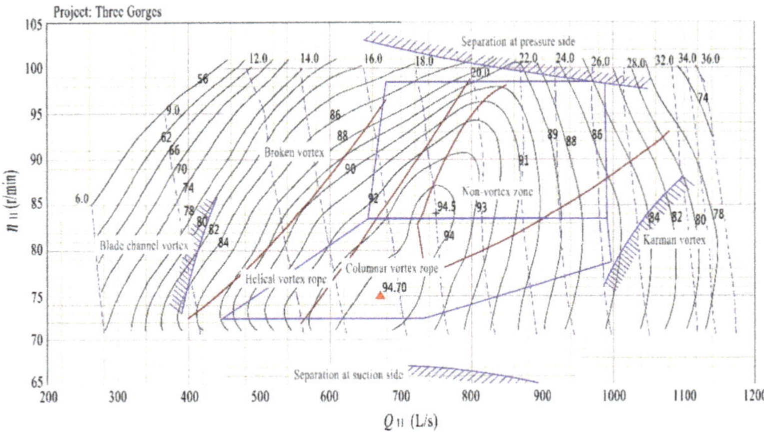

Figure 5. Hill chart for model A858a-36.6 under 30 m test head.

Table 4. Efficiency comparison with different methods.

Methods	Efficiency Value %		
AMOC	94.69		94.69
Test	94.63	94.63	
CFD		94.57	94.57
Relative error %	0.06	0.06	0.13

3.2.2. Velocity Distribution in Blade Channel

By data processing and velocity conversion, many more visual illustrations are presented. Figure 6 shows both the vector and scalar of the absolute velocity along the SSLs. Additionally, the absolute streamlines are partially plotted in the left figure by blue lines. It is evident that the radial flow gradually transfers to the axial direction due to the blade configuration. The meridional velocity correspondingly tends to perform in the same manner along the SSL. The runner additionally rotates clockwise. Therefore, the direction of absolute velocity evolves, namely being horizontal to the right at the blade inlet and inclined downwards at the outlet, as demonstrated by the blue lines in Figure 6a. This pattern greatly depends on the runner configuration and indeed, the flow features here coincide with the turbine at medium specific speed. Meanwhile, the length of the red arrows represents the magnitude of absolute velocity. Its streamwise evolution implies that the velocity smoothly decreases, which agrees well with the distribution of the velocity scalar in Figure 6b. Due to the effects of circumferential velocity, the maximum and minimum values separately emerge at the inlet near the band and at the outlet adjacent to the crown. These locations would possibly suffer from large stress, thereby inducing flow separation and other elusive hydraulic turbulence. Overall, in this section, the velocity distributions from the AMOC feature smooth transitions, which is justifiably rational at the BEP.

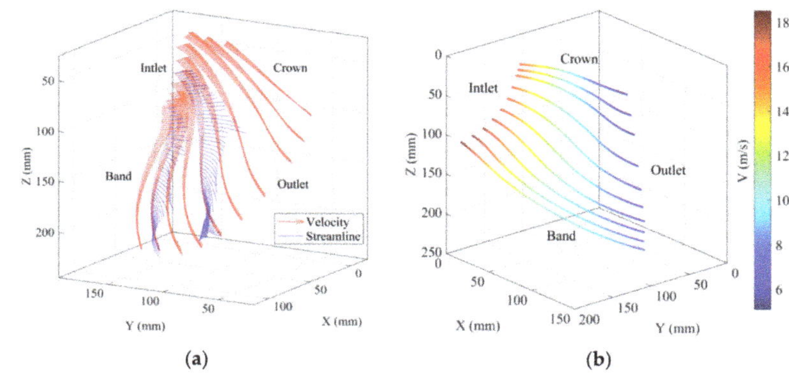

Figure 6. Distribution of absolute velocity on the SSL by AMOC: (**a**) Velocity vector and streamline; (**b**) Velocity scalar.

Additionally, the Navier–Stokes solver provides the absolute velocity and Eulerian energy distributing on a certain S2 surface, as shown in Figure 7. To further aid in the assessment of the results, the runner crown and the band are superimposed onto the flow fields. There is good agreement in the velocity distribution, as evidenced in Figures 6b and 7a. The absolute velocity in Figure 7a smoothly evolves from the blade's leading edge towards the trailing edge in the same manner as shown in Figure 6. Therefore, the main flow features and velocity magnitude are well approximated by the AMOC approach, and the S2 surface is basically divided into three parts: red, green and blue. Their boundaries, to some extent, coincide with the blade's leading and trailing edges. This implies the feature of the X-shaped blade, namely less cross flow from the crown to the band.

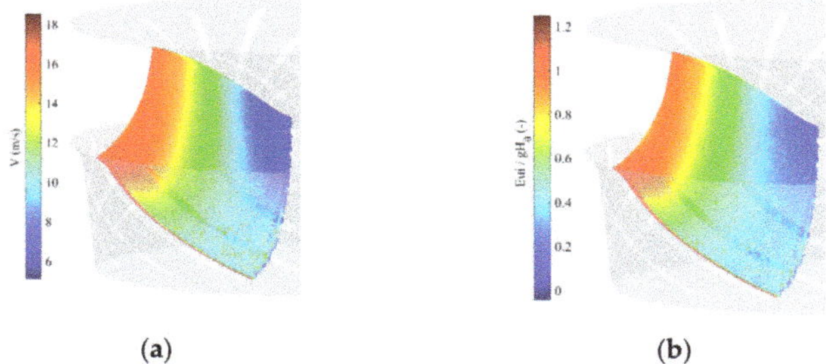

Figure 7. Distribution of results obtained by the Navier–Stokes solver on S2 surface: (**a**) Absolute velocity; (**b**) Dimensionless Eulerian energy.

3.2.3. Eulerian Energy in Blade Channel

In the (m, θ) curved coordinate system, the length of the SSL projection on the m-axis and gH_0 are used to normalize the coordinate and Eulerian energy, respectively, of each point on the corresponding SSL. Figure 8a displays the resultant distribution of the dimensionless Eulerian energy on five spatial streamlines. The contour of this energy on the S2 surface is illustrated in Figure 8b. Obviously, Eulerian energy shows both a spanwise ascent and a streamwise descent trend. On the other hand, the fluid potential energy is converted to mechanical energy by pushing the runner. This process is reflected in the figures as the Eulerian energy declines from the blade inlet towards the outlet. Hence, the slope of the curves in Figure 8a can be regarded as a criterion to define the local ability of energy conversion. This ability is proportional to the absolute value of the specific slope. By extension, a larger absolute value corresponds to the steep decline of the curve, triggering more energy conversion. Therefore, the corresponding location on the blade works more efficiently. As for the five curves in the figure, the gradient varies with the streamwise position. In a similar manner, it can be inferred that the whole blade surface converts energy unevenly, as seen in the color map in Figure 8b. This non-uniform energy transition affects the runner performance from a micro perspective and, in turn, provides a decisive step in the runner optimization, design and manufacturing process. Additionally, the right figure schematically highlights two special locations of maximal and minimal Eulerian energy, showing the same behavior as the absolute velocity in Figure 6b, and the energy value is observed to be close to zero on the outlet curve. This happens to coincide with the absence of residual momentum downstream of the runner in the optimal case. Similar distribution patterns can be found between Figures 7b and 8b. Due to the high dimensional interpolation required to obtain Figure 8b, the shape of the S2 surface is slightly changed. The general contour distribution is still in accord with the Fluent results.

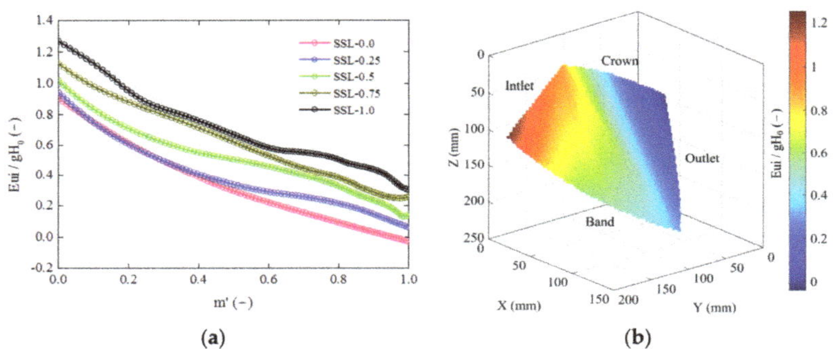

Figure 8. Distribution of Eulerian energy: (**a**) On three SSLs; (**b**) On the stream surface.

3.2.4. Pressure Difference over the Blade

The pressure difference over the blade is related to the meridional velocity and the meridional derivative of the velocity moment, expressed as [10,36]:

$$p^+ - p^- = \frac{2\pi\rho}{N} \cdot V'_m \cdot \frac{\partial(rV_u)}{\partial m} \tag{7}$$

where p^+ and p^- separately denote the static pressure on the pressure surface and the suction surface of the blade.

The coefficient of pressure difference is defined as:

$$C_p = \frac{p^+ - p^-}{\rho g H_0} \tag{8}$$

Consequently, the pressure difference at each node can be calculated using the corresponding revised meridional velocity and the circumferential component of absolute velocity, especially when the real blade thickness is taken into account. As a result, the coefficient of pressure difference across the blade is presented in Figure 9. As expressed by Equation (7), the pressure difference is proportional to the revised meridional velocity and to the derivative of the tangential velocity moment. These two parameters are larger towards the band. Therefore, the pressure difference theoretically increases in the spanwise direction. An overall agreement is observed between the AMOC approach and the Fluent results, though some noticeable discrepancies arise. This may be attributed to the hypotheses, errors during interpolation and the simplifications of this model. In this ambit, the treatment by degenerating the real blade into an S2 surface perhaps had the largest impact, since the AMOC calculates the pressure difference along a spatial streamline, while this value is obtained from the pressure surface and suction surface in the Navier–Stokes solver.

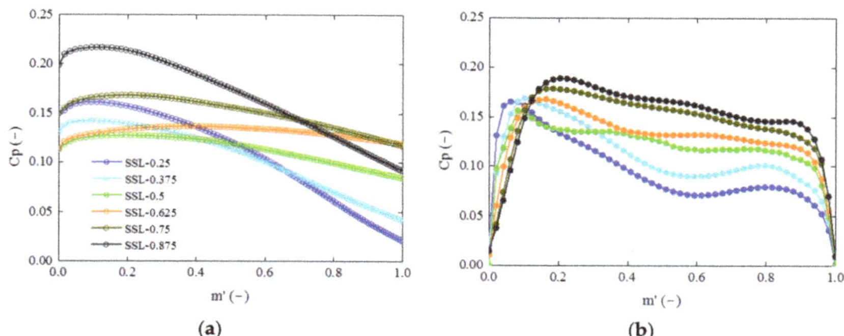

Figure 9. Distribution of C_p along spatial streamlines: (**a**) AMOC implementation; (**b**) ANSYS Fluent Navier–Stokes solver.

In Figure 9a, the spanwise C_p distribution seems to be chaotic due to the shape of the spatially twisted blade, while the streamwise distribution basically declines. During the first half chord, namely $0 < m' < 0.5$, most of the load was applied to the region adjacent to both the crown and the band, whereas the medium region was subjected to less load. On the contrary, the liquid load mainly worked on the downstream region near the band for $0.5 < m' < 1$. This coincides with the higher Eulerian energy near the lower part at the blade outlet, as shown in Figures 7b and 8b. It is notable that on the edge of the blade outlet ($m' = 1$) the pressure difference is slightly away from zero. This is probably attributed to the enlarged turbulence by the blade-channel vortex and to the outlet residual circulation, which ensured a preferable wake flow and further, a higher runner efficiency. In such a case, the flow pattern leaving the blade fails to meet the Kutta–Joukowski condition. In addition, when C_p reaches the peak value, the corresponding velocity difference over the blade also becomes the largest. The flow is likely to separate from the blade surface, and this separation can be arithmetically predicted by the definition of stagnation enthalpy in rotational machinery. This scenario will be quantitatively analyzed in the future work combining CFD technology. In summary, the C_p distribution mainly depends on the blade geometry and flow pattern in the blade channel. By calculating this value, the specific load distribution can be obtained, and the local energy conversion ability on the blade can be clearly evaluated.

4. Conclusions and Outlook

This paper predicted the characteristics of a Francis turbine operated at the BEP by an analytical approach, the AMOC. Implementations of this method were based on differential-geometry theory and the kinematics of ideal fluid, as well as the monistic theory, which was proposed decades ago but is relatively simple and can be efficiently performed. All of these merits make the monistic theory justifiably popular in mathematical models and analytical solutions.

By performing the AMOC procedures, the blade channel was discretized using the spatial relative streamlines, then the relevant hydraulic parameters were calculated, including velocity components, flow angles, Eulerian energy and the pressure difference across the blade. The runner's optimal hydraulic efficiency was further derived from the Eulerian energy. An arithmetical procedure was proposed to convert the velocity components from the spatial-velocity triangle to the Cartesian coordinate system. The resultant distributions revealed the dynamic evolution of the absolute-velocity vector, magnitude and streamline from the blade inlet to the outlet. Correspondingly, the Eulerian energy was distributed with a similar pattern, and its streamline gradient can represent the energy-conversion ability of the blade. Moreover, the pressure difference derived from flow dynamics elucidates the loading on a blade surface, especially for an X-shaped blade. Herein, most of the loading clusters were near both the crown and the band on the first half of blade, whereas regions near the band dominantly suffered from liquid loading on the downstream half.

The results from the AMOC were compared both to the model test and to the Navier–Stokes solution by the optimal hydraulic efficiency, the absolute velocity, the Eulerian energy and the pressure distribution across the blade. The analytical efficiency value was 94.69%, showing good agreement with the test and numerical simulation. A relevant minor discrepancy and similar flow patterns implied that the proposed AMOC can reliably predict the internal flow field and hydraulic efficiency of a Francis turbine at the BEP. The approach represents a key step towards runner design and selection with short duration and high efficiency.

The basic assumptions of this approach, i.e., inviscid fluid smoothly flowing through the blade channel, hardly deteriorated the precision of predicting the hydraulic efficiency. This satisfactory result is probably attributed to using the length-weighted-average method to calculate the Eulerian energy at the blade inlet and outlet. The uneven energy distribution was therefore cancelled out. Furthermore, the predicted pressure difference at the blade outlet generally coincided with the real conditions despite a slight deviation from the theoretical value. This phenomenon may benefit from a compensation for viscosity and backflow by introducing a blade-blockage factor, or from an induced pressure gradient by the residual circulation at the blade outlet.

One possible improvement to this AMOC approach would be to obtain a better prediction of the pressure difference, especially in the region near the blade inlet. In the AMOC approach, the blade degenerated to a smooth stream surface. However, the blade shape near the leading edge had a large curvature due to better dynamic characteristics. Therefore, it is in fact possible to carefully deal with the region adjacent to the blade's leading edge.

5. Patents

Yu Chen, Jianxu Zhou, Wenqing Jiang, et al. Calculation methodology and system of the best efficiency for a hydraulic reaction runner: China, ZL201810870576.1[P]. 21 April 2020.

Author Contributions: Conceptualization, methodology, computation, visualization, writing-original draft, Y.C.; methodology, resources, supervision, writing-review and editing, J.Z. (Jianxu Zhou) and B.K.; formal analysis, validation, writing-review, Q.G.; funding acquisition, J.Z. (Jianxu Zhou) and J.Z. (Jian Zhang). All authors have read and agreed to the published version of the manuscript.

Funding: This research was funded by the National Natural Science Foundation of China, grant number 51879087, 51839008, by the High-education Natural Science Foundation of Jiangsu Province, grant number 21KJB570001, and by the Research Foundation for High-level Talents of NJIT, grant number YKJ202131.

Institutional Review Board Statement: Not Applicable.

Informed Consent Statement: Not Applicable.

Data Availability Statement: Not Applicable.

Acknowledgments: The authors would like to thank the support by the China Scholarship Council and the Harbin electrical machine factory in China.

Conflicts of Interest: The authors declare no conflict of interest.

Appendix A

Appendix A.1. Coordinate Transformation

Two definitions are introduced firstly, the stream surface (S1) and the blade chamber surface (S2). S1 is a rotational symmetric surface such as the trumpet-shaped light-green face in Figure A1. It defines the flow trajectory in the channel without a blade from the radial to the axial direction. S2 is identical to the real blade surface when the runner has an infinite number of infinitely thin blades, such as the purple face in Figure A1. Meanwhile, S2 can be composed of the chamber lines of the spanwise blade cascade. Therefore, S2 is named the blade-chamber surface. In Figure A1, a curved coordinate system (m, θ) located on the S1 surface is established. The θ–axis denotes the azimuthal direction on the horizontal plane, while the m-axis denotes the generatrix line of the S1 surface on the vertical plane. The spatial relative streamline (SSL), which is the intersection of S1 and S2, is marked as the red curve in Figure A1. Node coordinates on this streamline are therefore converted from a Cartesian system to the (m, θ) system, expressed as:

$$r_i^2 = x_i^2 + y_i^2 \tag{A1}$$

$$z_i = z_i \tag{A2}$$

$$m_i = m_{i-1} + \left(\Delta z_i^2 + \Delta r_i^2\right)^{0.5} / r_i \tag{A3}$$

$$\theta_i = \mathrm{atan}\left(\frac{x_i}{y_i}\right) - \mathrm{atan}\left(\frac{x_0}{y_0}\right) \tag{A4}$$

where i denotes the ith node on the SSL. (x_i, y_i, z_i) denotes the coordinates of the ith node in the Cartesian system. (m_i, θ_i) denotes the coordinates of the ith node in the curved system. (x_0, y_0) are the coordinates of the SSL origin. $\Delta z_i = z_i - z_{i-1}$ is the axial interval of the adjacent nodes. $\Delta r_i = r_i - r_{i-1}$ is the radial interval of the adjacent nodes.

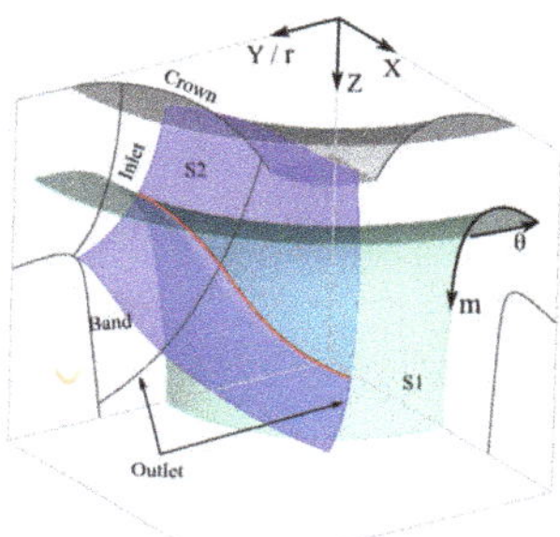

Figure A1. Blade-chamber surface, stream surface and their meridian projection.

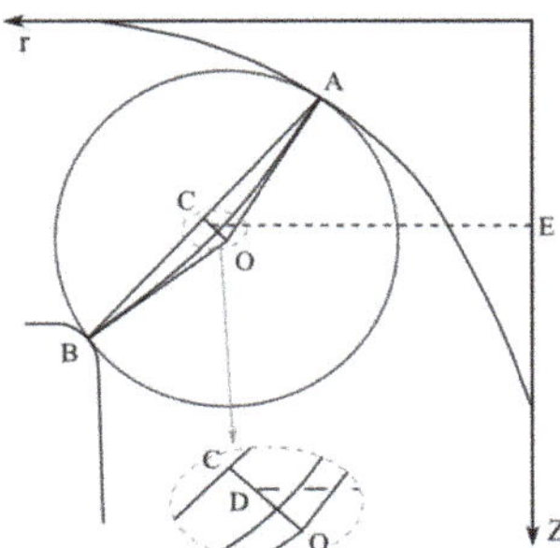

Figure A2. Generatrix of flow cross section in blade channel.

Appendix A.2. Generatrix of Flow Section and Meridional Velocity

In the meridional plane in Figure A2, an inscribed circle tangents to both the crown and the band with tangency points A and B. OA and OB denote the radius of the circle with the center O. Then, the generatrix can be equivalent to the arc $\hat{AB}$, which is tangent to the radius OA and OB, respectively, on points A and B. Its length can be calculated by the following empirical formula:

$$AB = \frac{2}{3}(\overline{AB} + \overline{OA}) \tag{A5}$$

where $\overline{OA}$ is the radius value and $\overline{AB}$ is the chord length.

Then, the arc $\hat{AB}$ rotates around the Z-axis to obtain the flow cross section. Its area S and meridian velocity V_m are subsequently computed:

$$S = 2\pi \cdot \overline{DE} \cdot AB \tag{A6}$$

$$V_m = Q_0/S \tag{A7}$$

where line OC is perpendicular to the chord AB. Point D is located on 1/3 of the line OC, and the length $\overline{CD} = 1/3\overline{OC}$. Q_0 denotes the discharge at the BEP.

Appendix A.3. Streamline Cluster Generation

The meridional channel is discretized by the meridional streamlines. An example of the medium streamline, named SL-0.5, is presented in Figure A3. It divides the channel into two parts. The coordinates of point F, which are the intersection of generatrix $\hat{AB}$ and SL-0.5, can be derived by the following equations. Equations (A8)–(A10) are established by geometrical features and Formula (A11) defines the same discharge passing arc $\hat{AF}$ and arc $\hat{FB}$.

$$(z_F - z_{O'})^2 + (r_F - r_{O'})^2 = R^2 \tag{A8}$$

$$(z_F - z_A)^2 + (r_F - r_A)^2 = R^2 + R^2 - 2R^2 \cos\gamma_1 \tag{A9}$$

$$(z_F - z_B)^2 + (r_F - r_B)^2 = R^2 + R^2 - 2R^2 \cos\gamma_2 \tag{A10}$$

$$R \cdot \gamma_1 \cdot 2\pi \cdot \frac{r_F + r_A}{2} = R \cdot \gamma_2 \cdot 2\pi \cdot \frac{r_F + r_B}{2} \tag{A11}$$

where (r, z) with different subscripts denotes the coordinates of corresponding points. O' is the center of arc $\hat{AB}$. R denotes its radius. γ_1 and γ_2 denote the radian angle of $\angle AO'F$ and $\angle BO'F$.

With various inscribed circles in the meridional channel, a series of intersection points are obtained, which then make up the streamline SL-0.5. In the same manner, SL-0.25 is produced, thereby evenly dividing the region between SL-0.5 and the crown. Likewise, SL-0.75 halves the channel between SL-0.5 and the band.

Subsequently, the S1 stream surface is established by rotating the meridional streamline around the Z-axis. Meanwhile, the S2 surface is defined by the centers of incircles and their tangency points in the wooden blade figures. The resultant relative spatial streamline is then generated by intersecting the S1 and S2 surfaces, named SSL-No., where the No. coincides with the number of the SL. Finally, ten SSLs including the crown and the band are illustrated in Figure A4. They represent ideal flow trajectories on the blade surface. Additionally, the origins and endpoints of these lines are separately located on the blade's leading edge and trailing edge.

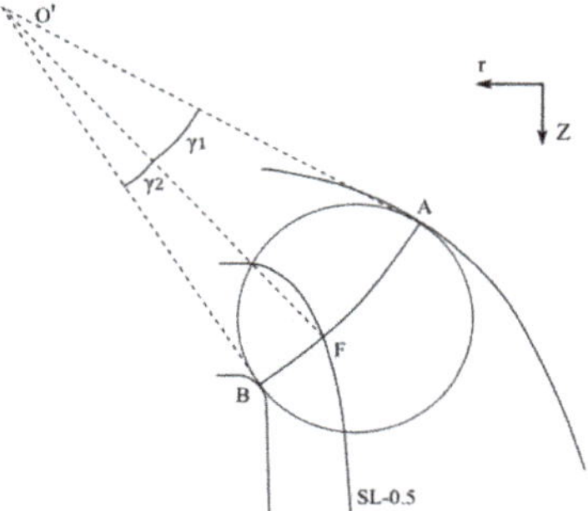

Figure A3. Sketch of meridional streamline.

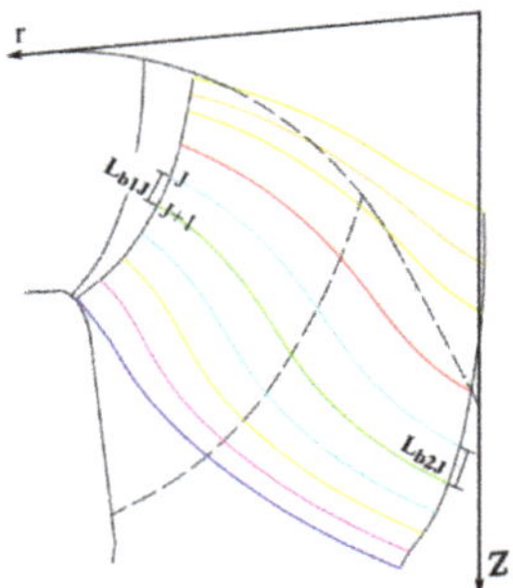

Figure A4. Sketch of spatial relative streamline.

Appendix A.4. Relative Flow Angle

Figure A5 shows the projection of the (m, θ) coordinates, velocity triangle and spatial streamline onto a plane. The spatial streamline is evenly divided into n segments in order to further discretize the spatial flow field. According to the third assumption, the relative flow angle (β) in Figure A5a at each node can be derived from interpolating the angles between the local tangent and the horizontal direction. Considering different positions of the node, three scenarios are presented as follows, and the corresponding schematic of the flow angle is shown in Figure A5b–d, respectively.

$i = 1$

$$\tan \beta_1 = \tau_2 \cdot \frac{m_2 - m_1}{\theta_2 - \theta_1} - \tau_3 \cdot \frac{m_3 - m_1}{\theta_3 - \theta_1} \tag{A12}$$

$$\tau_2 = \frac{\theta_3 - \theta_1}{\theta_2 - \theta_1}$$

$$\tau_3 = \frac{\theta_3 - \theta_2}{\theta_2 - \theta_1}$$

$2 \leq i \leq n$

$$\tan \beta_i = \tau_i \cdot \frac{m_i - m_{i-1}}{\theta_i - \theta_{i-1}} + \tau_{i+1} \cdot \frac{m_{i+1} - m_i}{\theta_{i+1} - \theta_i} \tag{A13}$$

$$\tau_i = \frac{\theta_{i+1} - \theta_i}{\theta_{i+1} - \theta_{i-1}}$$

$$\tau_{i+1} = \frac{\theta_i - \theta_{i-1}}{\theta_{i+1} - \theta_{i-1}}$$

$$i = n + 1$$

$$\tan \beta_{n+1} = \tau_n \cdot \frac{m_{n+1} - m_n}{\theta_{n+1} - \theta_n} - \tau_{n-1} \cdot \frac{m_{n+1} - m_{n-1}}{\theta_{n+1} - \theta_{n-1}} \tag{A14}$$

$$\tau_n = \frac{\theta_{n+1} - \theta_{n-1}}{\theta_n - \theta_{n-1}}$$

$$\tau_{n-1} = \frac{\theta_{n+1} - \theta_n}{\theta_n - \theta_{n-1}}$$

where β_i denotes the relative flow angle at the ith node and τ_i denotes the interpolating coefficient.

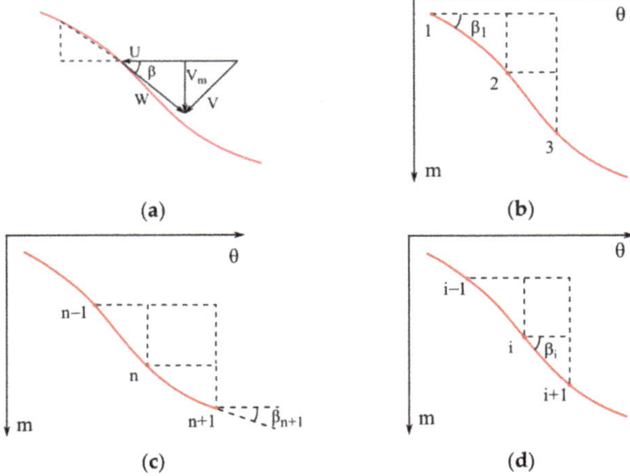

Figure A5. Velocity triangle and flow angle at different locations on SSL: (**a**) velocity triangle; (**b**) origin of streamline; (**c**) endpoint of streamline; (**d**) medium positions.

Appendix A.5. Meridional Velocity at SSL Nodes and Effects of Blade Thickness

It is notable that the SL is the meridional projection of the corresponding SSL, so they have identical m-axial components. Additionally, Equation (A7) defines the distribution of V_m on SL. Therefore, based on the (m, V_m) pairs on the SL, V_m at each node on the SSL is subsequently interpolated.

In addition, blade thickness, to some extent, blocks the flow passage, changing the velocity's magnitude and direction. Herein, the AMOC is only concerned with its effect on magnitude, while the flow direction is kept aligned with the SSL and the blocked flow passage is subjected to the unchanged discharge, which is expressed by:

$$S \cdot 2\pi r_i \cdot V_{mi} = S \cdot (2\pi r_i - Ne_i) \cdot V_{mi}' \tag{A15}$$

where r_i denotes the radius of the ith node. N denotes the blade number. e_i denotes the corresponding thickness derived from diameter of the incircles in the wooden blade figure. V_{mi}' denotes the revised meridional velocity at node i. Then the blockage factor, accounting for the thickness effect, is defined as:

$$\frac{1}{k_i} = \frac{V_{mi}}{V_{mi}'} = \frac{2\pi r_i - Ne_i}{2\pi r_i} \tag{A16}$$

Appendix A.6. Velocity Triangle

According to the resolved parameters, each component in the velocity triangle at every node can be computed. By extension, in the triangle sketched in Figure A5a, the circumferential velocity component is:

$$U_i = r_i \cdot \omega = r_i \cdot \frac{2\pi n_0}{60} \tag{A17}$$

where ω is the angular speed and n_0 is the rated rotation speed.

By recalling the Law of sines, the relative velocity is:

$$W_i = \frac{V_{mi}{'}}{\sin \beta_i} = \frac{k_i \cdot V_{mi}}{\sin \beta_i} \tag{A18}$$

With the Law of cosines in the velocity triangle, the velocity components and flow angles satisfy the following equations:

$$V_i^2 = U_i^2 + W_i^2 - 2 \cdot U_i \cdot W_i \cdot \cos \beta_i \tag{A19}$$

$$W_i^2 = U_i^2 + V_i^2 - 2 \cdot U_i \cdot V_i \cdot \cos \alpha_i \tag{A20}$$

Thus, the absolute velocity V_i and the corresponding absolute flow angle α_i can be calculated. Then, V_i is projected in the circumferential direction as:

$$V_{Ui} = V_i \cdot \cos \alpha_i \tag{A21}$$

where V_{Ui} denotes the tangential component of V_i at node i.

Appendix B

Transformation of Velocity Component

To obtain the absolute streamline through each node, the absolute velocity should be decomposed in the Cartesian coordinate system into three components, V_X, V_Y and V_Z. Specific decomposition is shown in Figure A6, and the positive direction of each component coincides with the coordinate axis.

Any two adjacent nodes on the SSL are designated as $F_i(x_i, y_i, z_i)$ and $F_{i+1}(x_{i+1}, y_{i+1}, z_{i+1})$. The projection of F_{i+1} onto the Z-axis is $F_{i+1}{'}(0, 0, z_{i+1})$. Then, the two vectors can be expressed by coordinates:

$$F_i F_{i+1} = (x_{i+1} - x_i, y_{i+1} - y_i, z_{i+1} - z_i) \tag{A22}$$

$$F_{i+1} F_{i+1}{'} = (-x_{i+1}, -y_{i+1}, 0) \tag{A23}$$

Due to the segment $F_{i+1}F_{i+1}{'}$ acting as the radius of the dotted circle line in Figure A6, it is known that $F_{i+1}F_{i+1}{'} \perp U$ and $F_{i+1}F_{i+1}{'} \perp V_Z$. That is, $F_{i+1}F_{i+1}{'}$ is the normal vector of the plane formed by U and V_Z. Recalling the vector operation:

$$F_i F_{i+1} \cdot F_{i+1} F_{i+1}{'} = |F_i F_{i+1}| \cdot |F_{i+1} F_{i+1}{'}| \cdot \cos \gamma_3 \tag{A24}$$

where γ_3 is the angle between vector $F_i F_{i+1}$ and $F_{i+1} F_{i+1}{'}$.

By introducing coordinate values of each vector, the Equation (A24) is elaborated as:

$$\begin{aligned} &-x_{i+1} \cdot (x_{i+1} - x_i) - y_{i+1} \cdot (y_{i+1} - y_i) = \\ &[(x_{i+1} - x_i)^2 + (y_{i+1} - y_i)^2 + (z_{i+1} - z_i)^2]^{0.5} \cdot (x_{i+1}^2 + y_{i+1}^2)^{0.5} \cdot \cos \gamma_3 \end{aligned} \tag{A25}$$

Then γ_3 is derived.

With the third assumption, the relative velocity W at node $\mathbf{F_{i+1}}$ ranges with the vector $\mathbf{F_i F_{i+1}}$ in Figure A6a, and the radial velocity V_r is the projection of W onto vector $\mathbf{F_{i+1} F_{i+1}}'$. Then the following two equations are deduced:

$$V_r = W \cdot \cos \gamma_3 \tag{A26}$$

$$V_Z^2 = V_m'^2 - V_r^2 \tag{A27}$$

The positive value of V_Z defines its vertically downward direction. The angle between the Y-axis and vector $\mathbf{F_{i+1}}'\mathbf{F_{i+1}}$ is set as γ_4 as shown in Figure A6b. Then

$$\mathbf{F_{i+1}}'\mathbf{F_{i+1}} \cdot \mathbf{j} = \left|\mathbf{F_{i+1}}'\mathbf{F_{i+1}}\right| \cdot \left|\mathbf{j}\right| \cdot \cos \gamma_4 \tag{A28}$$

where $\mathbf{j}$ is the positive unit vector of the Y-axis. The above equation is later elaborated by the coordinate operation and γ_4 is solely determined in $[0, \pi]$. That is:

$$y_{i+1} = \left(x_{i+1}^2 + y_{i+1}^2\right)^{0.5} \cdot 1 \cdot \cos \gamma_4 \tag{A29}$$

Therefore, V_X and V_Y can be obtained by decomposing V_r and V_U along the X-axis and Y-axis, respectively:

$$V_X = -V_r \cdot \sin \gamma_4 - V_U \cdot \cos \gamma_4 \tag{A30}$$

$$V_Y = -V_r \cdot \cos \gamma_4 + V_U \cdot \sin \gamma_4 \tag{A31}$$

The positive directions of V_r and V_U are presented in Figure A6b. In particular, at the origin of SSL, namely $i = 1$, the angle between W_1 and $\mathbf{F_1 F_1}'$ is approximated by the angle between $\mathbf{F_1 F_2}$ and $\mathbf{F_1 F_1}'$. Using the above procedures in this section, one can successfully convert the velocity components from the velocity triangle to the Cartesian coordinate system.

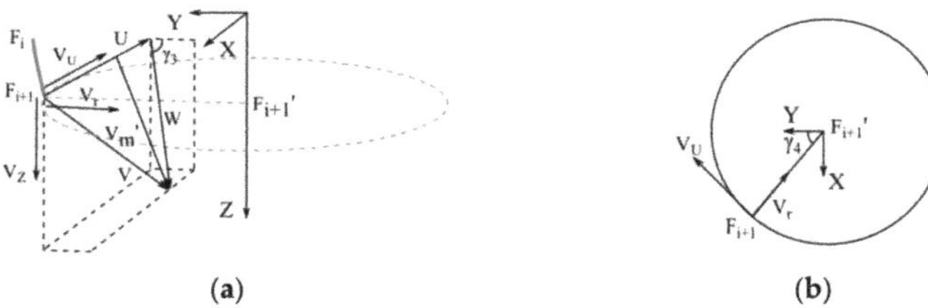

(a)   (b)

Figure A6. Diagram of the velocity triangle and decomposition: (**a**) spatial sketch; (**b**) top view.

References

1. Zhang, Y.; Zheng, X.; Li, J.; Du, X. Experimental study on the vibrational performance and its physical origins of a prototype reversible pump turbine in the pumped hydro energy storage power station. *Renew. Energy* **2019**, *130*, 667–676. [CrossRef]
2. Trivedi, C.; Gogstad, P.J.; Dahlhaug, O.G. Investigation of the unsteady pressure pulsations in the prototype Francis turbines—Part 1: Steady state operating conditions. *Mech. Syst. Sig. Process.* **2018**, *108*, 188–202. [CrossRef]
3. Zhang, L.; Wu, Q.; Ma, Z.; Wang, X. Transient vibration analysis of unit-plant structure for hydropower station in sudden load increasing process. *Mech. Syst. Sig. Process.* **2019**, *120*, 486–504. [CrossRef]
4. Li, D.; Sun, Y.; Zuo, Z.; Liu, S.; Wang, H.; Li, Z. Analysis of Pressure Fluctuations in a Prototype Pump-Turbine with Different Numbers of Runner Blades in Turbine Mode. *Energies* **2018**, *11*, 1474. [CrossRef]
5. Gohil, P.P.; Saini, R.P. Numerical Study of Cavitation in Francis Turbine of a Small Hydro Power Plant. *J. Appl. Fluid Mech.* **2016**, *9*, 357–365. [CrossRef]
6. Decaix, J.; Müller, A.; Favrel, A.; Avellan, F.; Münch, C. URANS Models for the Simulation of Full Load Pressure Surge in Francis Turbines Validated by Particle Image Velocimetry. *J. Fluids Eng.* **2017**, *139*, 121103. [CrossRef]
7. Favrel, A.; Müller, A.; Landry, C.; Yamamoto, K.; Avellan, F. Study of the vortex-induced pressure excitation source in a Francis turbine draft tube by particle image velocimetry. *Exp. Fluids* **2015**, *56*, 215. [CrossRef]
8. Skripkin, S.; Tsoy, M.; Kuibin, P.; Shtork, S. Swirling flow in a hydraulic turbine discharge cone at different speeds and discharge conditions. *Exp. Therm. Fluid Sci.* **2019**, *100*, 349–359. [CrossRef]

9.   Lai, X.; Liang, Q.; Ye, D.; Chen, X.; Xia, M. Experimental investigation of flows inside draft tube of a high-head pump-turbine. *Renew. Energy* **2019**, *133*, 731–742. [CrossRef]
10.   Leguizamón, S.; Avellan, F. Open-Source Implementation and Validation of a 3D Inverse Design Method for Francis Turbine Runners. *Energies* **2020**, *13*, 2020. [CrossRef]
11.   Li, W. Inverse design of impeller blade of centrifugal pump with a singularity method. *Jordan J. Mech. Ind. Eng.* **2011**, *5*, 119–128.
12.   Zangeneh, M.; Goto, A.; Takemura, T. Suppression of secondary flows in a mixed-flow pump impeller by application of 3D inverse design method. In *International Gas Turbine and Aeroengine Congress and Exposition*; ASME: Hague, The Netherlands, 1994.
13.   Hawthorne, W.R.; Wang, C.; Tan, C.S.; Mccune, J.E. Theory of Blade Design for Large Deflections: Part I—Two-Dimensional Cascade. *J. Eng. Gas Turbines Power* **1984**, *106*, 346–353. [CrossRef]
14.   Yin, J.; Wang, D. Review on applications of 3D inverse design method for pump. *Chin. J. Mech. Eng.* **2014**, *27*, 520–527. [CrossRef]
15.   Kan, K.; Chen, H.; Zheng, Y.; Zhou, D.; Binama, M.; Dai, J. Transient characteristics during power-off process in a shaft extension tubular pump by using a suitable numerical model. *Renew. Energy* **2021**, *164*, 109–121. [CrossRef]
16.   Kan, K.; Yang, Z.; Lyu, P.; Zheng, Y.; Shen, L. Numerical Study of Turbulent Flow past a Rotating Axial-Flow Pump Based on a Level-set Immersed Boundary Method. *Renew. Energy* **2021**, *168*, 960–971. [CrossRef]
17.   Qian, H.; Chen, N.; Lin, R.; Qu, L.; Pu, S. Design of mixed flow pump impeller by using direct-inverse problem iteration. *J. Hydraul. Eng.* **1997**, *1*, 25–30.
18.   Bing, H.; Zhang, Q.; Cao, S.; Tan, L.; Zhu, B. Iteration design of direct and inverse problems and flow analysis of mixed-flow pump impellers. *J. Tsinghua Univ.* **2013**, *53*, 179–183.
19.   Tan, L.; Cao, S.; Wang, Y.; Zhu, B. Direct and inverse iterative design method for centrifugal pump impellers. *J. Power Energy* **2012**, *226*, 764–775. [CrossRef]
20.   Bing, H.; Cao, S.; Tan, L. Iteration method of direct inverse problem of mixed-flow pump impeller design. *J. Drain. Irrig. Mach. Eng.* **2011**, *29*, 277–281.
21.   Ciocan, T.; Susan-Resiga, R.F.; Muntean, S. Modelling and optimization of the velocity profiles at the draft tube inlet of a Francis turbine within an operating range. *J. Hydraul. Res.* **2016**, *54*, 74–89. [CrossRef]
22.   Chen, Z.; Singh, P.M.; Choi, Y.-D. Francis Turbine Blade Design on the Basis of Port Area and Loss Analysis. *Energies* **2016**, *9*, 164. [CrossRef]
23.   Yin, J.; Li, J.; Wang, D.; Wei, X. A simple inverse design method for pump turbine. *IOP Conf. Ser. Earth Environ. Sci.* **2014**, *22*, 012030. [CrossRef]
24.   Han, F.; Huang, L.; Yang, L.; Kubota, T. Geometrical analysis of adjustable guide vanes for bulb turbine. *J. Eng. Thermophys.* **2010**, *31*, 263–266.
25.   Liu, P. *Channel Theory for High-Head Bulb Runner*; South China University of Technology: Guangzhou, China, 2010.
26.   Susan-Resiga, R.F.; Milos, T.; Baya, A.; Muntean, S.; Bernad, S. Mathematical and Numerical Models for Axisymmetric Swirling Flows for Turbomachinery Applications. *Sci. Bull. Univ. Politeh. Timis. Trans. Mech.* **2005**, *50*, 47–58.
27.   Susan-Resiga, R.; Ighişan, C.; Muntean, S. A Mathematical Model for the Swirling Flow Ingested by the Draft Tube of Francis Turbines. *WasserWirtschaft* **2015**, *13*, 23–27. [CrossRef]
28.   Susan-Resiga, R.F.; Muntean, S.; Avellan, F.; Anton, I. Mathematical modelling of swirling flow in hydraulic turbines for the full operating range. *Appl. Math. Modell.* **2011**, *35*, 4759–4773. [CrossRef]
29.   Susan-Resiga, R.F.; Avellan, F.; Ciocan, G.D.; Muntean, S.; Anton, I. Mathematical and Numerical Modelling of the Swirling Flow in Francis Turbine Draft Tube Cone. *Sci. Bull. Univ. Politeh. Timis. Trans. Mech.* **2005**, *50*, 1–16.
30.   Muntean, S.; Susan-Resiga, R.; Göde, E.; Baya, A.; Terzi, R.; Tîrşi, C. Scenarios for refurbishment of a hydropower plant equipped with Francis turbines. *Renew. Energy Environ. Sustain.* **2016**, *1*, 30. [CrossRef]
31.   Brekke, H. *Hydraulic Turbines. Design, Erection and Operation*; Norwegian University of Science and Technology: Trondheim, Norway, 2001.
32.   Brekke, H. Performance and safety of hydraulic turbines. In *25th IAHR Symposium on Hydraulic Machinery and Systems*; IOP Publishing: Bristol, UK, 2010.
33.   Leguizamón, S.; Ségoufin, C.; Phan, H.T.; Avellan, F. On the efficiency alteration mechanisms due to cavitation in Kaplan turbines. *J. Fluids Eng.* **2017**, *139*, 061301. [CrossRef]
34.   Li, W.; Gong, R. The Researching of Turbine Main Parameters Selection for Three Gorges Right Bank Power Station. *Large Electr. Mach. Hydraul. Turbine* **2009**, *2*, 37–42.
35.   Wang, G. Key Technologies for Huge Full Air-cooled Hydrogenerator Sets of Three Gorges Right Bank Power Station—Hydraulic Turbine Section. *Large Electr. Mach. Hydraul. Turbine* **2008**, *4*, 30–36.
36.   Daneshkah, K.; Zangeneh, M. Parametric design of a Francis turbine runner by means of a three-dimensional inverse design method. *IOP Conf. Earth Environ. Sci.* **2010**, *12*, 012058. [CrossRef]

sustainability

Article

Inner Damping of Water in Conduit of Hydraulic Power Plant

Daniel Himr *,† , Vladimír Habán † and David Štefan †

Faculty of Mechanical Engineering, Brno University of Technology, Technická 2896/2,
616 69 Brno, Czech Republic; haban@fme.vutbr.cz (V.H.); stefan@fme.vutbr.cz (D.Š.)
* Correspondence: himr@fme.vutbr.cz
† These authors contributed equally to this work.

Abstract: The operation of any hydraulic power plant is accompanied by pressure pulsations that are caused by vortex rope under the runner, rotor–stator interaction and various transitions during changes in operating conditions or start-ups and shut-downs. Water in the conduit undergoes volumetric changes due to these pulsations. Compression and expansion of the water are among the mechanisms by which energy is dissipated in the system, and this corresponds to the second viscosity of water. The better our knowledge of energy dissipation, the greater the possibility of a safer and more economic operation of the hydraulic power plant. This paper focuses on the determination of the second viscosity of water in a conduit. The mathematical apparatus, which is described in the article, is applied to data obtained during commissioning tests in a water storage power plant. The second viscosity is determined using measurements of pressure pulsations in the conduit induced with a ball valve. The result shows a dependency of second viscosity on the frequency of pulsations.

Keywords: second viscosity; bulk viscosity; measurement; hydraulic power plant; conduit; pressure pulsations; dissipation; damping

Citation: Himr, D.; Habáne, V.; Štefan, D. Inner Damping of Water in Conduit of Hydraulic Power Plant. *Sustainability* **2021**, *13*, 7125. https://doi.org/10.3390/su13137125

Academic Editor: Tomonobu Senjyu

Received: 23 April 2021
Accepted: 14 June 2021
Published: 25 June 2021

1. Introduction

The unsteady flow in a pipeline is a phenomenon of great interest. It is not possible to avoid an unsteady flow when operating a hydraulic system. For example, a water turbine generates instability when it starts or stops, when the power changes, when a black-out occurs etc. All of these events induce transients connected to pressure pulsations. The magnitude of the pulsations depends on many factors, and it is necessary to examine these factors carefully in advance to prevent accidents and problems regarding operation or regulation.

Even the steady operation of a turbine generates pressure pulsations, which travel through the system. The pulsations are caused by vortex rope [1] or rotor–stator interactions when the runner blade is passing the guide vane [2]. These pulsations can induce severe vibrations of the structural parts in the system.

The behavior of hydraulic systems is usually simulated as one-dimensional flow described by continuity and momentum Equations (1) and (2), where the wave speed and viscosity are important parameters. The wave speed influences the magnitude of a pressure surge and the speed of a response of the system to any event. Viscosity defines energy dissipation via friction and is usually included in the friction loss coefficient.

$$\frac{\partial p}{\partial t} + \frac{a^2 \rho}{S}\frac{\partial Q}{\partial x} = 0, \tag{1}$$

$$\frac{\partial Q}{\partial t} + \frac{S}{\rho}\frac{\partial p}{\partial x} + \frac{\lambda}{2DS}|Q|Q = g_p S. \tag{2}$$

Experience and experiments of various researchers showed that energy loss defined only by the friction loss coefficient can be insufficient. Moreover, when water changes phase or dissolved air is released, energy is dissipated by the following mechanisms: unsteady friction [3], friction in pipe material [4], fluid compression [5], when water changes phase or dissolved air is released [6,7]. Omitting these mechanisms can have a crucial impact on whether the analyzed system is deemed stable or unstable and on the selection of the method to control it.

The energy dissipation due to the compressibility of a liquid or gas can be expressed as a function of the bulk viscosity, which includes regular viscosity and the second coefficient of viscosity (second viscosity) (see Equation (3)).

$$b = \xi + \frac{2}{3}\eta.$$ (3)

Bulk viscosity is zero for dilute monoatomic gases according to Stokes' hypothesis. However, this assumption is not valid for other gases and liquids. An extensive review of bulk viscosity showed how the second viscosity influences equations describing fluid flow [5,8–12].

Thus far, many experiments have been performed to obtain the value of the bulk viscosity or the second viscosity. Karim [13] measured the second viscosity of water, methyl alcohol and ethyl alcohol in the 1950s. In recent work, Dukhin [14] compared three methods (Brillouin spectroscopy, laser transient grating spectroscopy and acoustic spectroscopy) for twelve liquids, including water. Holmes [15] investigated the dependence of the second viscosity of water on temperature using acoustic spectroscopy. He [16] performed similar experiments but used stimulated Brillouin scattering. All of the authors obtained similar results, but they did not describe any frequency dependence. However, all of the experiments were performed with very high pulsation frequencies (from 1 MHz to 1 GHz). The measurement of much lower pulsation frequencies (from 0.2 Hz to 10 kHz) is described in [17,18], where a hyperbolic dependence on frequency is described.

Determination of the second viscosity coefficient in a conduit of a hydraulic power plant is a main goal of the study. Measured data of the pressure pulsations are used. The mathematical background is briefly described in Section 2; a detailed description can be found in [17]. The measurement set-up is given in Section 3, and the following section discusses the obtained results.

2. Theory

A basic mathematical background is given in this section. One-dimensional flow is supposed in the pipe, which is a usual simplification in pipeline dynamics.

2.1. Continuity Equation

Generally, liquid flow can be described by the continuity equation and momentum equation. The continuity equation can be written in the following form:

$$\frac{d\rho}{dt}V + \rho\frac{dV}{dt} = 0.$$ (4)

Equation (5) provides a relationship between density and pressure.

$$\frac{dp}{dt} = a^2\frac{d\rho}{dt}.$$ (5)

For liquid in a pipe, Equation (4) can be adjusted into form (6) using Equation (5) when the Gauss–Ostrogradski formula is used.

$$\frac{1}{a^2}\frac{dp}{dt}S + \rho\frac{\partial Q}{\partial x} + \rho\frac{\int \vec{v}\vec{n}dP}{dx} = 0.$$ (6)

The integral in Equation (6) refers to the integration over the moving (deforming) pipe wall, where the inner surface is P, and the velocity is $\vec{v}$. The velocity can differ at different points. The integral equals zero when the pipe is rigid. An ideal contact between the liquid and the pipe wall is assumed.

When a thin walled pipe is assumed, Equations (7) and (8) can be used to describe the stress and deformation of the wall.

$$\sigma = \frac{D}{2e}p, \tag{7}$$

$$v = \frac{D}{2}\frac{\partial \varepsilon}{\partial t}. \tag{8}$$

The model of the body in Figure 1 helps to define a relationship between the stress and deformation of the wall.

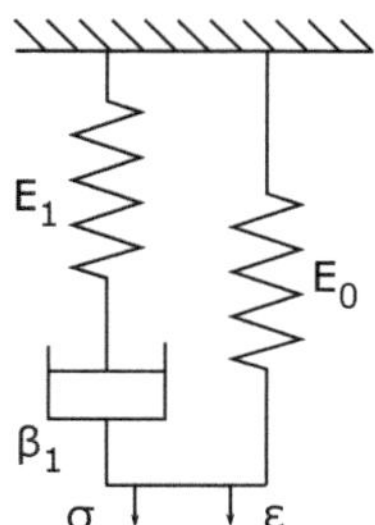

Figure 1. Model of the pipe wall.

The final shape of the continuity Equation (9) can be found after a Laplace transformation of Equations (6)–(8), where the complex wave speed is a function of the complex Young's modulus (see Equations (10) and (11)). A circular cross-section of the pipe is assumed.

$$0 = \frac{\rho a_c^2}{S}\frac{\partial \bar{Q}}{\partial x} + s\bar{p}, \tag{9}$$

$$a_c = a\sqrt{\frac{E_c e}{E_c e + a^2 \rho D}}, \tag{10}$$

$$E_c = E_0 + \frac{s E_1 \beta_1}{E_1 + s \beta_1}. \tag{11}$$

2.2. Momentum Equation

The momentum Equation for liquid (12) includes energy loss in the tensor Π.

$$\rho \frac{\partial \vec{v}}{\partial t} + \rho \vec{v}\,\text{grad}\,\vec{v} - \text{div}\Pi + \text{grad}\,p = 0. \tag{12}$$

Tensor Π describes the energy loss due to friction and compression (see Equation (13)).

$$\text{div}\Pi = \eta \Delta \vec{v} + (\eta + \xi)\text{grad div}\,\vec{v}. \tag{13}$$

It is possible to neglect the second term in Equation (12) when the liquid velocity is much lower than the wave speed. Then, one-dimensional flow in a pipe with a circular cross-section can be described by Equation (14).

$$\frac{\partial Q}{\partial t} - \frac{2\eta + \xi}{\rho}\frac{\partial^2 Q}{\partial x^2} - \left(\frac{4\eta}{\rho D^2}\right)^2 \int_0^t \Gamma(t - \tau)Q\mathrm{d}\tau + \frac{S}{\rho}\frac{\partial p}{\partial x} = 0. \tag{14}$$

Memory function Γ expresses the energy loss due to an unsteady velocity profile. It has a great impact on liquids with high viscosity (oil) or on flow in a capillary. Usually, it can be neglected, because its effect is much smaller than that of the second viscosity influence [19,20]. Thus, the final shape of the momentum equation can be written after a Laplace transformation when the flow rate derivation is substituted using Equation (9).

$$\frac{S}{\rho}\left[1 + (2\eta + \xi)\frac{s}{a_c^2}\right]\frac{\partial \bar{p}}{\partial x} + s\bar{Q} = 0. \tag{15}$$

2.3. Second Viscosity

The solution of Equations (9) and (15) can be written using a transfer matrix $\mathbf{T}(x,s)$ with a 2×2 size as follows:

$$\begin{pmatrix} \bar{Q} \\ \bar{p} \end{pmatrix} = \mathbf{T}(x,s)\begin{pmatrix} \bar{Q} \\ \bar{p} \end{pmatrix}_{x=0}. \tag{16}$$

Therefore, when the boundary conditions for $x = 0$ are known (beginning of the pipe), it is possible to obtain the Laplace transformations of the pressure and flow rate wherever in the pipe. The transfer matrix contains all of the properties and geometrical parameters of the pipe line and flowing fluid, including the wave speed and second viscosity.

When the boundary conditions and conditions in another place are known (for example, from an experiment), it is possible to use an appropriate optimization method to find some elements of the transfer matrix. This is a good way to compute second viscosity and wave speed.

These parameters can also be obtained using eigenvalues of the transfer matrix. The first step is to obtain a response of the system to an impulse. This is the experimental part. The impulse can be induced with the valve (when water flows, the valve closing causes pressure pulsations). The eigenvalue of the system can be determined from a Fourier transformation of the measured pressure pulsations, and this is a complex number. The imaginary part is the same as the eigen angular frequency of the pulsations. The real part can be computed from the corresponding amplitude of a system of a single degree of freedom (see Equation (17) and Figure 2).

$$s = \frac{\pi}{\sqrt{3}}\Delta f \pm i2\pi f. \tag{17}$$

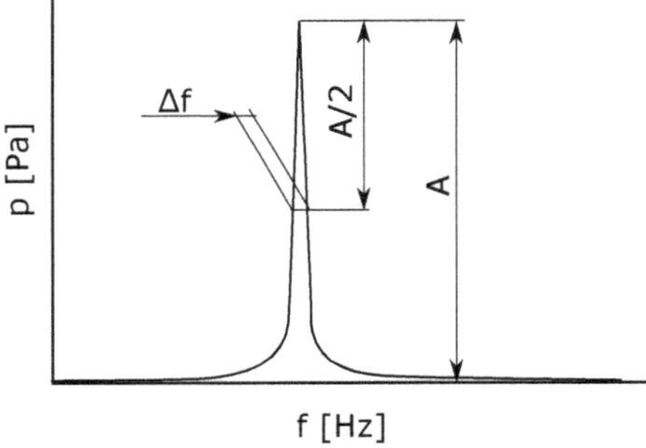

Figure 2. Determination of the real part of the eigenvalue.

The eigenvalue of the system equals the eigenvalue of the transfer matrix. The wave speed and second viscosity can be changed to meet this requirement. The transfer matrix can be written in various forms that include properties of the system (plain fluid, fluid + pipe, fluid + pipe with added mass). Detailed mathematical analysis is described in [17], and the chosen form of the transfer matrix with fluid properties has a form (18).

$$\mathbf{T}(x,s) = \begin{pmatrix} \cosh\left(\sqrt{\frac{s\cdot B}{A\cdot C}}\cdot x\right) & -\sqrt{\frac{s\cdot A}{B\cdot C}}\cdot\sinh\left(\sqrt{\frac{s\cdot B}{A\cdot C}}\cdot x\right) \\ -\sqrt{\frac{B\cdot C}{s\cdot A}}\cdot\sinh\left(\sqrt{\frac{s\cdot B}{A\cdot C}}\cdot x\right) & \cosh\left(\sqrt{\frac{s\cdot B}{A\cdot C}}\cdot x\right) \end{pmatrix}, \tag{18}$$

where coefficients A, B, C stand for:

$$A = \frac{S}{\rho} + \frac{2\cdot\eta+\zeta}{\rho^2}\cdot\frac{S\cdot s}{a^2}, \tag{19}$$

$$B = s - \frac{s\cdot J_1\left(\frac{D}{2}\cdot i\cdot\sqrt{\frac{s\cdot\rho}{\eta}}\right)}{s\cdot J_1\left(\frac{D}{2}\cdot i\cdot\sqrt{\frac{s\cdot\rho}{\eta}}\right) - \frac{D}{4}\cdot i\cdot\sqrt{\frac{s\cdot\rho}{\eta}}\cdot J_0\left(\frac{D}{2}\cdot i\cdot\sqrt{\frac{s\cdot\rho}{\eta}}\right)}, \tag{20}$$

$$C = \frac{a^2\cdot\rho}{S}. \tag{21}$$

Second viscosity and wave speed have to be optimized to obtain the zero determinant of the transfer matrix.

3. Measurement

The storage hydraulic power plant Dlouhé Stráně is the most powerful hydraulic power plant in the Czech Republic. It has two Francis turbines ($2 \times 325\,\mathrm{MW}$), which can also work as pumps. The turbines collect water from a bottom reservoir built on Dlouhá Desná river and pump it into a top reservoir on the top of Dlouhé Stráně mountain. The geodetic head is 510–$534\,\mathrm{m}$, and the flow rate during the turbine operation is $2 \times 68.5\,\mathrm{m}^3/\mathrm{s}$. Each turbine has its own conduit with a length of 1547 and 1495 m and a diameter of 3.6 m. A simplified scheme is shown in Figure 3.

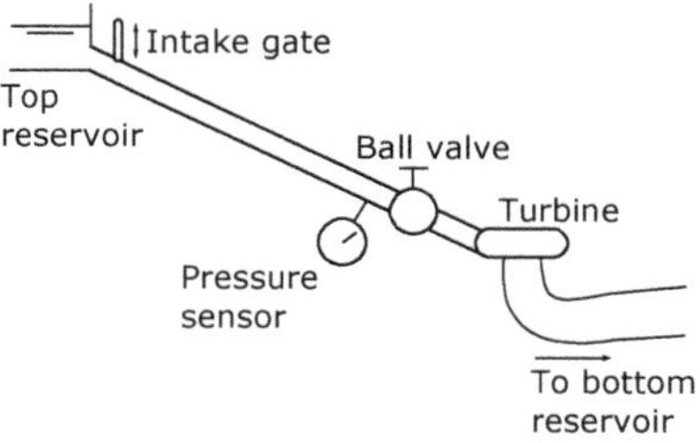

Figure 3. Scheme of one conduit with the turbine.

Measurement of the pressure pulsations was performed in November 2011 as a part of the commissioning tests in relation to turbine modernization. The obtained data primarily were used to determine the initial flow rate using the time-pressure method (Gibson's method). Twenty-eight sets of data were gathered, including the regular stop, emergency stop and disconnection of the electric network. Only data from the regular stop of the turbine are usable for second viscosity computations (see Figure 4).

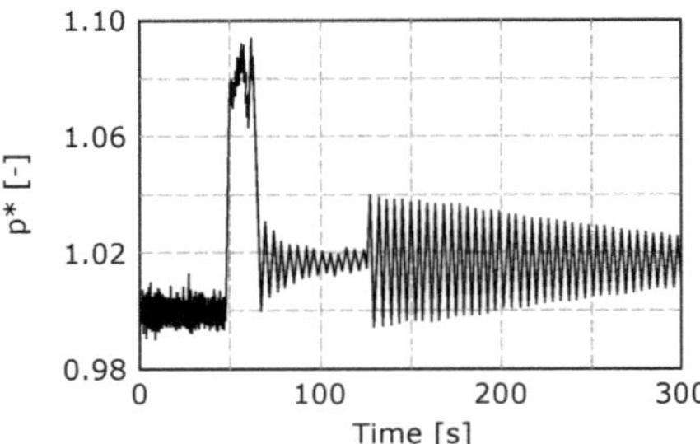

Figure 4. Pressure upstream of the ball valve during the regular stop.

Steady operation continued until the 50th second. Then, guide vanes started to close and caused a pressure surge that was 9% above the initial value. The guide vanes completely closed within 20 s. The ball valve upstream of the turbine also started to close but more slowly than the guide vanes. The pressure pulsations were damped until the 125th second. The ball valve finally closed at this time. The following pulsations correspond to an unsteady flow in a simple pipe with the boundary condition of constant pressure at the upstream end and a zero flow rate at the downstream end. These pulsations are mathematically well described and are usable for the evaluation of the wave speed and second viscosity (see Figure 5).

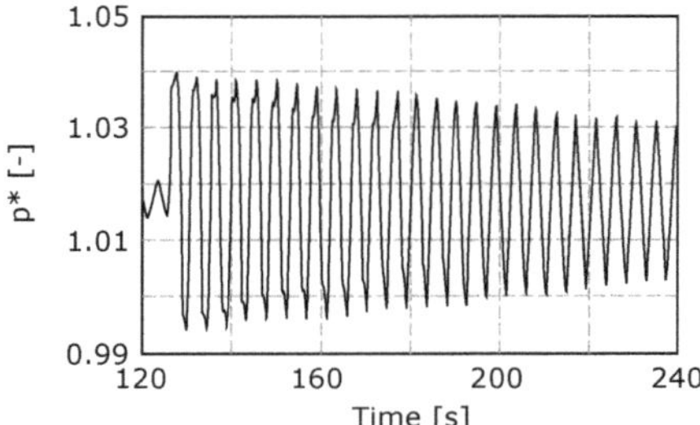

Figure 5. Pressure upstream of the ball valve during the regular stop after the ball valve fully closed.

The emergency stop is the same as the regular stop, but when the guide vanes finish closing, the intake gate also closes. The obtained pressure data are not appropriate for second viscosity evaluation, because the pulsations are strongly damped. Tests of closing after electricity disconnection also did not provide any usable data. Again, the system closed the intake gate, and the pulsations were damped too quickly.

4. Evaluation

Pressure oscillations after the 125th second were subject to discrete Fourier transformation, and then the eigen frequencies were determined. Table 1 shows the first seven eigen frequencies obtained from the measured pressure using Equation (17).

Table 1. Eigenvalues and corresponding shapes of pressure pulsations.

Real Part (α), Imaginary Part (ω), Frequency (f)	Shape of Pressure Amplitudes along the Conduit
$\alpha = 0.0121522 \, \text{rad/s}$ $\omega = 1.403 \, \text{rad/s}$ $f = 0.223 \, \text{Hz}$	
$\alpha = 0.01469911 \, \text{rad/s}$ $\omega = 4.235 \, \text{rad/s}$ $f = 0.674 \, \text{Hz}$	
$\alpha = 0.03323047 \, \text{rad/s}$ $\omega = 6.991 \, \text{rad/s}$ $f = 1.113 \, \text{Hz}$	
$\alpha = 0.03837075 \, \text{rad/s}$ $\omega = 9.792 \, \text{rad/s}$ $f = 1.558 \, \text{Hz}$	
$\alpha = 0.04869796 \, \text{rad/s}$ $\omega = 12.610 \, \text{rad/s}$ $f = 2.007 \, \text{Hz}$	
$\alpha = 0.08408768 \, \text{rad/s}$ $\omega = 15.355 \, \text{rad/s}$ $f = 2.444 \, \text{Hz}$	
$\alpha = 0.07548372 \, \text{rad/s}$ $\omega = 18.156 \, \text{rad/s}$ $f = 2.890 \, \text{Hz}$	

The shapes of the pressure amplitude along the conduit are shown in the right column. The left end is the location of the valve; thus, the pressure always has maximal amplitude there. The right end corresponds to the reservoir that maintains constant pressure there.

When the eigenvalues of the system are known, it is possible to determine the wave speed and second viscosity, which satisfy Equation (16). The values were obtained from eigenvalues in Table 1, and they are plotted in Figures 6 and 7.

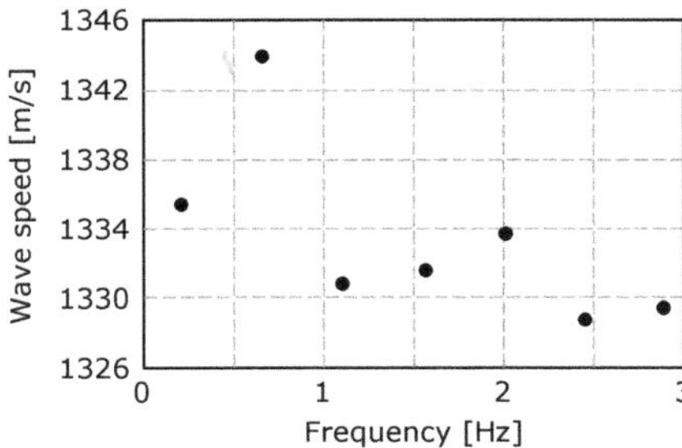

Figure 6. Wave speed with different frequencies.

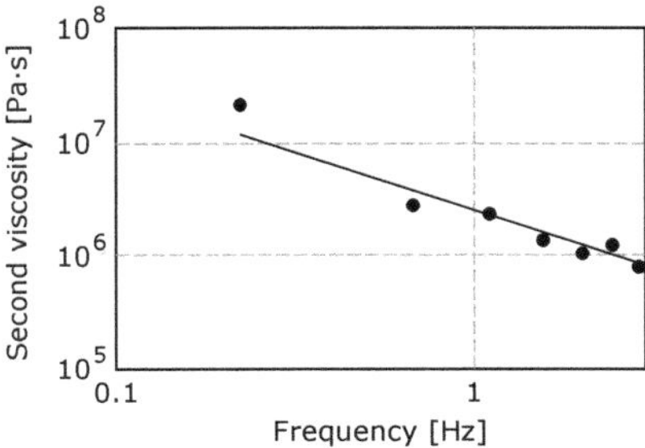

Figure 7. Second viscosity for different frequencies.

It is clear that the wave speed is independent of the frequency. The scattering is 1.1% from the mean value. The uncertainty was 0.2% of the sensor range, its time constant was 0.005 s and it was directly mounted on the pipe wall. Unlike the wave speed, the second viscosity strongly depends on the frequency. One can easily find a hyperbolic function that satisfactorily corresponds to the obtained values. Figure 7 is on a logarithmic scale; therefore, the hyperbola appears as a line (see following equation). The obtained result corresponds to [17].

$$\xi = \frac{C}{f}.\tag{22}$$

Bubbles of free air in the water have an impact on the wave speed and second viscosity [21]; however, in this case, the water had been pumped up and down several times before this measurement. Therefore, there was a very low level of air content, and it is possible to neglect its influence.

5. Summary

This paper describes an evaluation of the second viscosity of water. Pressure pulsations in a conduit of a storage power plant were measured at the ball valve. The measurement was performed as a part of commissioning tests and served for computations of the flow rate using the time-pressure method; however, it was also possible to use it for the evaluation of the second viscosity. The second viscosity was found from the eigenvalue of the conduit. The result showed a hyperbolic dependence of second viscosity on the frequency.

The presented methodology could be transferred to other hydraulic machines of any scale in order to verify the hydraulic system in which the machine will operate. This is necessary for economic and safer operations.

Author Contributions: Conceptualization, V.H.; methodology, V.H.; validation, D.H. and D.Š.; formal analysis, D.Š.; investigation, V.H. and D.H.; resources, D.Š.; data curation, V.H. and D.H.; writing—original draft preparation, D.H. and D.Š.; funding acquisition, D.H. and D.Š. All authors have read and agreed to the published version of the manuscript.

Funding: This paper was supported by the projects "Computer Simulations for Effective Low-Emission Energy" funded as project no. CZ.02.1.01/0.0/0.0/16_026/0008392 by Operational Programme Research, Development and Education, Priority axis 1: Strengthening capacity for high-quality research and project no. FSI-S-20-6235 "The Fluid Mechanics Principle Application as a Sustainable Development Tool."

Institutional Review Board Statement: Not applicable.

Informed Consent Statement: Not applicable.

Data Availability Statement: Restrictions apply to the availability of these data. Data were obtained from OSC, a.s. and are available from the authors only with the permission of OSC, a.s.

Conflicts of Interest: The authors declare no conflicts of interest.

Symbols

The following symbols are used in this manuscript:

a	wave speed
b	bulk viscosity
D	pipe diameter
E	Young's modulus
e	pipe wall thickness
f	frequency
g_p	gravity acceleration projection to pipe axis
i	imaginary unit

J_0 Bessel function
J_1 Bessel function
n normal vector oriented out of liquid
P pipe wall area
p pressure
Q flow rate
S pipe cross-section
s parameter of Laplace transformation
T transfer matrix
t time
V volume
v velocity
x longitudinal coordinate
β pipe wall damping
ε deformation
η dynamic viscosity
λ friction coefficient
ξ second viscosity
ρ density
σ stress
ω angular velocity

Subscripts:

c complex number
0 see Figure 1
1 see Figure 1

Superscripts:

$*$ value divided by steady value
$^-$ variable after Laplace transformation

References

1. Alligne, S.; Nicolet, C.; Tsujimoto, Y.; Avellan, F. Cavitation surge modelling in Francis turbine draft tube. *J. Hydraul. Res.* **2014**, *52*, 399–411. [CrossRef]
2. Liu, H.; Ouyang, H.; Wu, Y.; Tian, J.; Du, Z. Investigation of unsteady flows and noise in rotor-stator interaction with adjustable lean vane. *Eng. Appl. Comp. Fluid* **2014**, *8*, 299–307. [CrossRef]
3. Vítkovský, J. P.; Bergant, A.; Simpson, A.R.; Lambert, M.F. Systematic evaluation of one-dimensional unsteady friction models in simple pipelines. *J. Hydraul. Eng.* **2006**, *132*, 696–708. [CrossRef]
4. Keramat, A.; Kolahi, A.G.; Ahmadi, A. Waterhammer modelling of vicoelastic pipes with a time-dependent Poissons's ratio. *J. Fluid. Struct.* **2013**, *43*, 164–178. [CrossRef]
5. Pochylý, F.; Habán, V.; Fialová, S. Bulk viscosity—Constitutive equations. *Int. Rev. Mech. Eng.* **2011**, *5*, 1043–1051.
6. Cannizzaro, D.; Pezzinga, G. Energy dissipation in transient gaseous cavitation. *J. Hydraul. Eng.* **2005**, *131*, 724–732. [CrossRef]
7. Jablonská, J. Compressibility of the fluid. *EPJ Web. Conf.* **2014**, *67*, 322–327. [CrossRef]
8. Graves, R.E.; Argrow, B.M. Bulk viscosity: Past to present. *J. Thermophys. Heat Tr.* **1999**, *13*, 337–342. [CrossRef]
9. Dellar, P.J. Bulk and shear viscosities in lattice Boltzmann equations. *Phys. Rev. E* **2001**, *64*, 11. [CrossRef] [PubMed]
10. Zuckerwar, A.J.; Ash, R.L. Volume viscosity in fluids with multiple dissipative processes. *Phys. Fluids* **2009**, *21*, 12. [CrossRef]
11. Marner, F.; Scholle, M.; Herrmann, D.; Gaskell, P.H. Competing Lagrangians for incompressible and compressible viscous flow. *Roy Soc. Open Sci.* **2019**, *6*, 14. [CrossRef] [PubMed]
12. Scholle, M. A discontinuous variational principle implying a non-equilibrium dispersion relation for damped acoustic waves. *Wave Motion* **2020**, *98*, 11. [CrossRef]
13. Karim, S.M. Second viscosity coefficient of liquid. *J. Acoust. Soc. Am.* **1953**, *25*, 997–1002. [CrossRef]
14. Dukhin, A.S.; Goetz, P.J. Bulk viscosity and compressibility measurement using acoustic spectroscopy. *J. Chem. Phys.* **2009**, *130*, 13. [CrossRef] [PubMed]
15. Holmes, M.J.; Parker, N.G.; Povey, M.J.W. Temperature dependence of bulk viscosity in water using acoustic spectroscopy. *J. Phys. Conf. Ser.* **2011**, *269*, 7. [CrossRef]
16. He, X.; Wei, H.; Shi, J.; Liu, J.; Li, S.; Chen, W.; Mo, X. Experimental measurement of bulk viscosity of water based on stimulated Brillouin scattering. *Opt. Commun.* **2012**, *285*, 4120–4124. [CrossRef]

17. Himr, D.; Habán, V.; Fialová, S. Influence of second viscosity on pressure pulsation. *Appl. Sci.* **2019**, *9*, 5444. [CrossRef]
18. Dörfler, P. Pressure wave propagation and damping in a long penstock. In Proceedings of the 4th International Meeting on Cavitation and Dynamic Problems in Hydraulic Machinery and Systems, Belgrade, Serbia, 26–28 October 2011.
19. Zielke, W. Frequency-dependent friction in transient pipe flow. *J. Basic Eng.* **1968**, *90*, 109–115. [CrossRef]
20. Zielke, W. Frequency Dependent Friction in Transient Pipe Flow. Ph.D. Thesis, The University of Michigan, Ann Arbor, MI, USA, 1966.
21. Záruba, J. *Water Hammer in Pipe-Line Systems*, 2nd ed.; Academia: Prague, Czech Republic, 1993.

Article

DuoTurbo: Implementation of a Counter-Rotating Hydroturbine for Energy Recovery in Drinking Water Networks

Daniel Biner [1,*], Vlad Hasmatuchi [1], Laurent Rapillard [1], Samuel Chevailler [1], François Avellan [2] and Cécile Münch-Alligné [1,*]

1 Institute of Systems Engineering, School of Engineering, HES-SO Valais-Wallis, Rue de l'Industrie 23, 1950 Sion, Switzerland; vlad.hasmatuchi@hevs.ch (V.H.); laurent.rapillard@hevs.ch (L.R.); samuel.chevailler@hevs.ch (S.C.)
2 Technology Platform for Hydraulic Machines PTMH, Ecole Polytechnique Fédérale de Lausanne, Avenue de Cour 33bis, 1007 Lausanne, Switzerland; francois.avellan@epfl.ch
* Correspondence: daniel.biner@hevs.ch (D.B.); cecile.muench@hevs.ch (C.M.-A.)

Citation: Biner, D.; Hasmatuchi, V.; Rapillard, L.; Chevailler, S.; Avellan, F.; Münch-Alligné, C. DuoTurbo: Implementation of a Counter-Rotating Hydroturbine for Energy Recovery in Drinking Water Networks. *Sustainability* **2021**, *13*, 10717. https://doi.org/10.3390/su131910717

Academic Editors: Mosè Rossi, Massimiliano Renzi, David Štefan and Sebastian Muntean

Received: 21 August 2021
Accepted: 23 September 2021
Published: 27 September 2021

Publisher's Note: MDPI stays neutral with regard to jurisdictional claims in published maps and institutional affiliations.

Abstract: To enhance the sustainability of water supply systems, the development of new technologies for micro scale hydropower remains an active field of research. The present paper deals with the implementation of a new micro-hydroelectric system for drinking water facilities, targeting a gross capacity between 5 kW and 25 kW. A counter-rotating microturbine forms the core element of the energy recovery system. The modular in-line technology is supposed to require low capital expenditure, targeting profitability within 10 years. One stage of the DuoTurbo microturbine is composed of two axial counter-rotating runners, each one featured with a wet permanent magnet rim generator with independent speed regulation. This compact mechanical design facilitates the integration into existing drinking water installations. A first DuoTurbo product prototype is developed by means of a Computational Fluid Dynamics based hydraulic design along with laboratory tests to assess system efficiency and characteristics. The agreements between simulated and measured hydraulic characteristics with absolute errors widely below 5% validate the design approach to a large extent. The developed product prototype provides a maximum electrical power of 6.5 kW at a maximum hydraulic head of 75 m, reaching a hydroelectric peak efficiency of 59%. In 2019, a DuoTurbo pilot was commissioned at a drinking water facility to assess its long-term behavior and thus, to validate advanced technology readiness levels. To the best of the authors knowledge, it is the first implementation of a counter-rotating microturbine with independent runner speed regulation and wet rim generators in a real-world drinking water facility. A complete year of operation is monitored without showing significant drifts of efficiency and vibration. The demonstration of the system in operational environment at pre-commercial state is validated that can be attributed to a technology readiness level of 7. The overall results of this study are promising regarding further industrialization steps and potential broad-scale applicability of the DuoTurbo microturbine in the drinking water industry.

Keywords: counter-rotating microturbine; drinking water facilities; system engineering; CFD; performance measurements; prototype endurance tests

1. Introduction

Among today's renewable energy sources, hydropower remains one of the most important suppliers of electricity, meeting over 15% of the global needs. In hydrologically and topographically predestined countries, hydropower constitutes even the bulk of electricity production. In Switzerland, hydropower covers more than 57% of the electricity demand [1]. Thereby, small hydropower (SHP) facilities with a gross capacity below 10 MW constitute about 10% of the Swiss hydroelectric capacity. Globally, energy strategies are evolving towards clean and sustainable technologies that also drives the expansion of SHP [2]. In Switzerland, the installed SHP energy production of about 3.4 TWh/year is planned to be increased by up to 770 GWh/year before 2050. Micro-hydropower (MHP), referring to power units below 500 kW, can contribute to a noticeable extent to the SHP mix. Particularly, the field of water treatment and distribution infrastructures hides a considerable MHP potential [3]. However, different main issues are identified such as variations in flows or turbine efficiency and the necessity of further risk assessments and evaluations of long-term reliability of MHP installations. Therefore, energy recovery in the water industry using MHP remains an active area of research, considering the various technical and economical complexities. In Swiss drinking water supply systems, more than 100 GWh/year are already exploited by hundreds of drinking water turbines, thus, about 50% of the estimated potential [4]. Since 2008, a Swiss electricity supply act is in force providing compensatory feed-in remuneration for renewable energies. The newly introduced feed-in tariffs are encouraging MHP exploitation in lower power ranges. Nevertheless, many drinking water facilities with a power potential of a few kW still remain untapped due to the lack of profitable solutions.

One of the most installed hydraulic machines for energy recovery from water distribution systems is the Pump as Turbine (PAT) [5] which represents an economically interesting solution since small pumps are generally produced in series. Extensive research is dedicated to PATs and their implementation in real-world water distribution systems, for example [6]. Recent studies present methodologies to overcome major obstacles for practical applications of PATs such as the lack of performance data [7,8]. Moreover, several technical solutions have been developed to adapt small pumps into small PATs, and variable speed drives may give them one degree of freedom. A recent case study shows that the adoption of PATs in collective irrigation systems is a viable solution to improve their economic and environmental impacts [9]. The latter research reveals a substantial cost decrease of 74% in the electro-mechanical equipment compared to site-specific designed turbines, whereas the electricity production is decreased by only 19%. To increase the sustainability of water distribution systems, special attention is dedicated to management strategies that minimize the overall energy input, inter alia, by integration of MHP technologies [10]. To point out one result of the proposed management strategies, it is found that a set of independent District Metered Areas locally increase the hydraulic energy (compared to widely open water distribution systems) that favors the installation of energy recovery devices at these locations. To facilitate the function extension of water supply networks to energy production, harmony search algorithms are proposed to optimize the planning of such schemes [11].

In the case of facilities with strong hydraulic variations, the development of adapted low power hydroelectric technologies with additional degrees of freedom may provide more operational flexibility. In comparison, the speed coefficient—discharge coefficient characteristic of a PAT with variable speed is typically clustered on only one S-shaped curve [12] whilst for a counter-rotating micro-turbine with one additional degree of freedom it may cover a large area [13]. Moreover, with specific turbine-oriented hydraulic design, efficiency can be optimized. Low head tubular propeller turbines point out one example of this research field [14]. In the perspective of developing a more flexible micro-hydroelectric system for drinking water facilities, HES-SO Valais-Wallis and EPFL-LMH have initiated the DuoTurbo project, focusing on an output power range between 5 kW and 25 kW. A counter-rotating microturbine forms the core element of the new energy recovery station. One stage of this microturbine consists of two axial counter-rotating runners. One early prototype of this specific turbine design was realized at EPFL-LMH using an elbowed pipe system with external electrical generators. Preliminary CFD simulations of this prototype were carried out [15] showing hydraulic efficiencies over 80% for an output power of 2.65 kW. Numerical and experimental methodologies to evaluate the performance and flow features of such type of turbine, involving Laser Doppler Velocimetry, were presented by [16,17]. The experimental validations including losses from the blade tip radial clearance still showed reasonable hydraulic efficiencies above 60%. The integration of a counter-rotating microturbine into water utility networks, considering analytical hill chart models to maximize energy recovery, was presented by [18]. Further, virtual Energy Recovery Station models were analyzed to simulate the energy recovery on a given installation site [19]. A custom runner design strategy [20] is thereby considered to maximize the electricity production at each potential water network. A second prototype with a more compact "bulb" configuration, comprising in-line electrical generators, was later built at HES-SO Valais-Wallis for ongoing developments. Finally, an even more compact DuoTurbo mechanical concepts with more powerful custom rim generators was developed that facilitates the integration into existing water pipelines. To investigate the new hydro-mechanical and electrical concepts, a DuoTurbo laboratory prototype was built in collaboration with industrial partners and extensively tested on the hydraulic test rig of the HES-SO Valais-Wallis [21,22]. Several mechanical solutions concerning the runner bearings and labyrinth seals were investigated to build a first DuoTurbo product prototype in 2018 [23]. Finally, in May 2019, the first DuoTurbo pilot was commissioned at a drinking water network in Savièse, VS Switzerland. The pilot enables long term monitoring and collection of hydraulic, mechanical and electrical parameters and is therefore essential for advanced system validation.

The aim of this paper is to point out the main technical aspects of the developed DuoTurbo product prototype and to demonstrate its reliability operating in a real drinking water network. First, the hydraulic, mechanical and electrical concepts are presented (see Figure 1). Then, numerical and experimental methods used for the system characterization, including the monitoring of the pilot facility, are explained. This paper only deals with the product prototype, as highlighted in dark blue in Figure 1. The corresponding numerical results, laboratory test results and long-term measurement data from the DuoTurbo pilot facility are then presented and discussed.

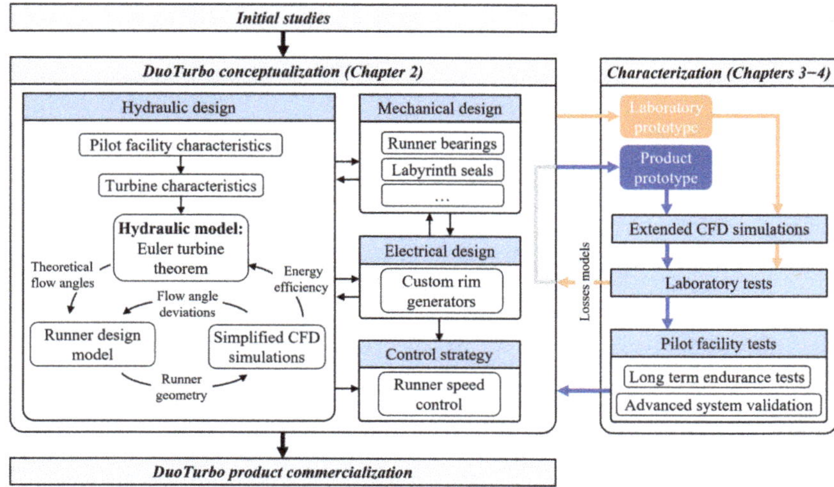

Figure 1. Overview of the DuoTurbo conceptualization and characterization methodology from the initial studies to a potential commercialization phase.

2. DuoTurbo Concept and Specifications

2.1. Concept Overview

In alpine regions, where altitude differences along the water supply chains are significant, a considerable excess of hydraulic energy is dissipated by Pressure Reducing Valves (PRVs). The DuoTurbo turbine is intended to recover a part of this energy surplus for a maximum hydraulic head of currently 75 m per stage. The core element of the developed technology is a counter-rotating axial microturbine with compact dimensions. Operational flexibility is achieved by the individual speed control of each generator to handle fluctuating hydraulic conditions. Commercial converters are currently used for the runner speed control and the power injection into the electrical grid. A controller manages the autonomous operation of the installation and monitors various hydraulic, mechanical and electrical parameters. For drinking water networks with consumption driven discharge, the layout presented in Figure 2 is suggested. Typically two PRV's are necessary to ensure the correct and safe functioning of the energy recovery device without affecting the drinking water supply. These elements would also be required using a PAT as a power generating unit. Two parallel branches are needed to ensure the water supply in case of a turbine failure or high water demand, as well as to protect the turbine from overload. Consequently, each branch must be equipped with a PRV to limit the downstream pressure at part load and while bypassing. The pressure set point of the bypass PRV is set slightly below the one of the turbine branch PRV. This pure hydro-mechanical configuration forces the closure of the bypass branch until reaching the turbine power limit. A solenoid valve is used to switch off the pressure control of the turbine branch's PRV and its deactivation initiates the closure of the turbine branch.

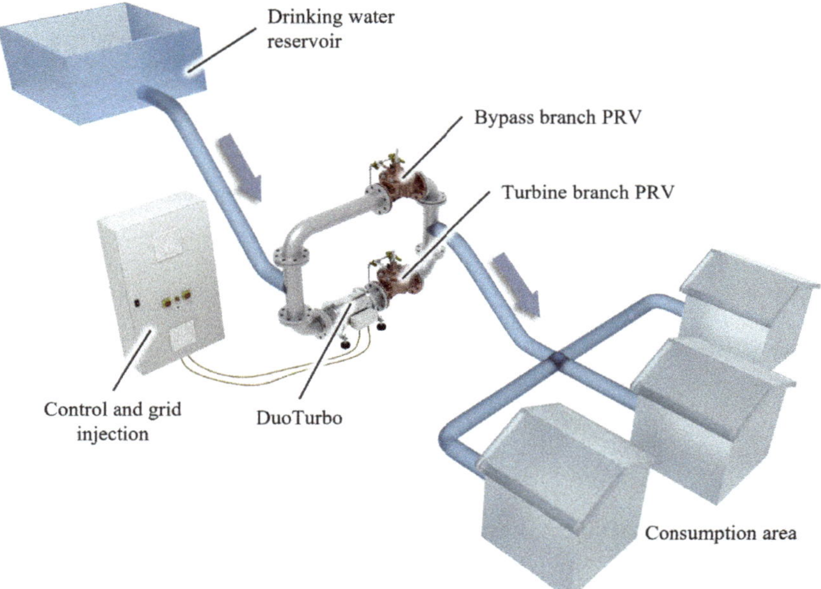

Figure 2. Global schematic with main hydroelectric components of the DuoTurbo energy recovery system in a drinking water network with consumer driven discharge.

2.2. Hydroturbine Concept
2.2.1. Turbine Principle

The DuoTurbo microturbine is an axial flow turbine with two serial counter-rotating runners per stage. In the field of turbomachines, the concept of counter-rotating axial rotors is known since many years and has been considered for potential use in gas turbines or aircraft engines [24]. The elimination of vanes between two counter-rotating runners is beneficial in terms of weight, length and efficiency. The attributes of counter-rotating turbines in such applications are high specific powers, high efficiency and possible balancing of torque and gyroscopic effects. The counter-rotating principle is also applied to aircraft and marine propellers. Various studies on hydraulic counter-rotating axial pico- and microturbines for potential applications in water infrastructures have been published [25–27]. The counter-rotating concept is also known in the field of hydraulic axial flow pumps [28].

From the hydraulic point of view, the main advantages of the counter-rotating architecture of the DuoTurbo microturbine lie in the achievable specific energies, the flexibility of design and the adaptability to variable operating conditions. The speed ratio between the two runners, referred as α, serves as an additional degree of freedom to handle off-design conditions. Further, site specific adaptations of the hydraulic characteristic can be achieved by tuning the nominal speed ratio α_n. To minimize the design complexity, the DuoTurbo microturbine is intended to be vaneless. This implies an axial flow at the inlet of the first runner and the outlet of the second runner at nominal operating conditions.

2.2.2. Energy Transfer

The specific energy E released from the fluid in a horizontal hydraulic turbine is expressed as:

$$E = \frac{\bar{p}_I - \bar{p}_{\bar{I}}}{\rho} + \frac{\bar{C}_I^2 - \bar{C}_{\bar{I}}^2}{2} \tag{1}$$

Thereby, $\bar{p}$ is the average static pressure, $\bar{C}$ is the average absolute flow velocity amplitude and ρ is the density of the fluid. The indices I and $\bar{I}$ refer to the high and the low pressure reference sections of the turbine, respectively, as shown in Figure 3. The indices referring to runner specific pressure reference sections are 1 and $\bar{1}$. The hydraulic power is then defined by the discharge Q passing the turbine:

$$P_h = \rho\,E\,Q \tag{2}$$

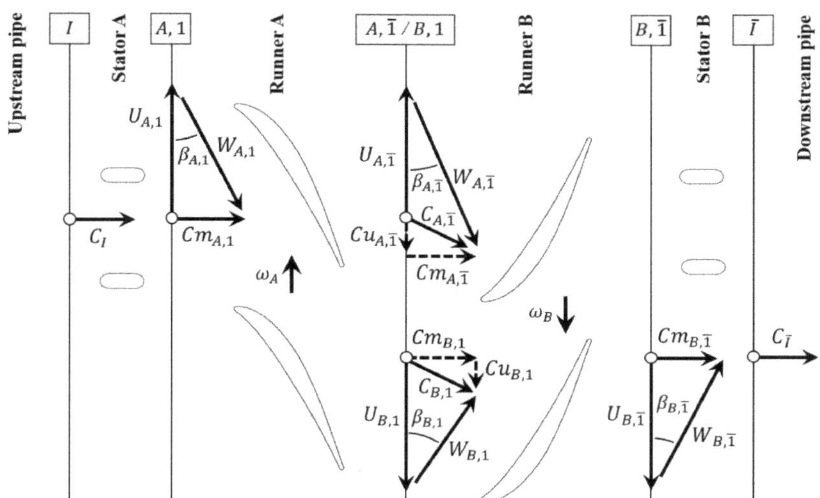

Figure 3. Definition of reference sections and velocity triangles of a counter-rotating turbine stage.

The transferred specific energy E_t transmitted to the turbine runner is then determined by the energy efficiency η_e as a consequence of turbulent dissipation and wall friction:

$$E_t = E\,\eta_e \tag{3}$$

Assuming an inviscid flow passing through the runner on a constant spanwise coordinate, the Euler turbine relation applied to a given streamline yields the correlation between the transferred specific energy E_t and the balance of the flux of angular momentum between the runner's high and low pressure reference sections:

$$E_t = U_1 Cu_1 - U_{\bar{1}} Cu_{\bar{1}} \tag{4}$$

Thereby, U is the tangential runner velocity amplitude and Cu is the peripheral absolute flow velocity amplitude. In Figure 3, these velocity components are used to define the velocity triangles in an ideal situation. W denotes the relative flow velocity amplitude with respect to the rotating reference frame and Cm is the meridional absolute flow velocity amplitude.

2.2.3. Runner Design

On the basis of the velocity triangles defined in Figure 3 and by choosing an appropriate radial distribution of E_t that fulfills the radial equilibrium of the flow, all relative flow angles β can be approximated. The relative flow angles determine to a large extend the runner geometry. The inlet blade angles are generally chosen according to the theoretical relative inlet flow angles, by respecting a certain blade incidence angle. The incidence angles are tuned to optimize the blade design. The specific energy of a runner is very sensitive to its outlet blade angles. To achieve the desired characteristics, flow angle deviations must be taken into account. They can be predicted using empirical data or numerical simulations. They generally depend on the hydrofoil shape, the blade solidity (Equation (5)) and the hydraulic design point. The design process of the DuoTurbo runners is based on iterative CFD simulations, by which the required outlet blade angles, as well as efficient hydrofoil parameters are determined. To generate the blade geometry, hydrofoils are defined only at the hub and shroud diameter using a direct profiling parametrization according to Figure 4. This two section based blade shape definition was chosen since it enables a relatively fast and simple machining process.

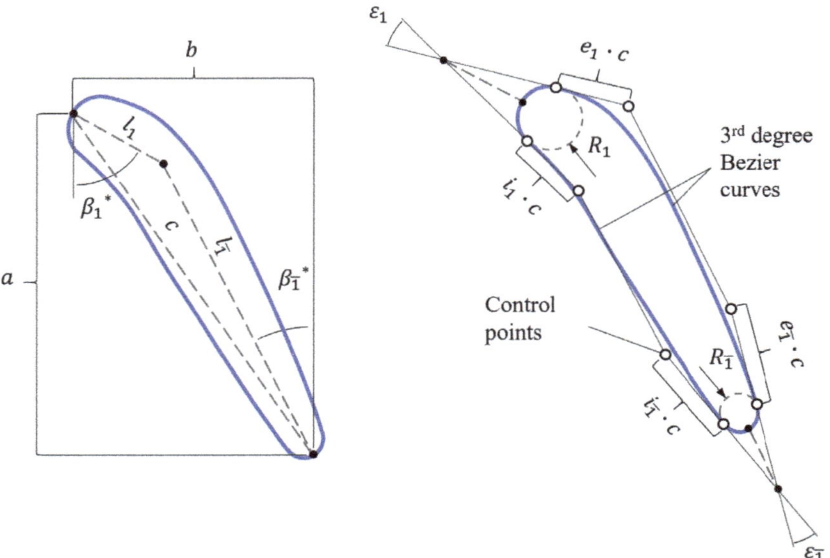

Figure 4. Direct profiling parametrization.

Further, the shroud diameter D_s and the hub diameter D_h of the DuoTurbo runners are constant to minimize the fabrication costs. The main runner design parameters of the product prototype are provided in Table 1. The blade solidity σ is calculated at the shroud diameter according to the number of blades N and the chord length c, see Figure 4.

$$\sigma = \frac{c\,N}{\pi\,D_s} \tag{5}$$

Table 1. Runner specifications of the DuoTurbo product prototype.

Symbole	Description	Value	Unit
D_s	Shroud diameter	100	mm
D_h	Hub diameter	87	mm
b	Axial blade length	30	mm
N_A	Number of blades runner A	5	-
N_B	Number of blades runner B	9	-
σ_A	Blade solidity runner A	1.44	-
σ_B	Blade solidity runner B	1.42	-
$\beta^*_{A,1,s}$	Shroud inlet blade angle A	27.3	deg
$\beta^*_{A,\bar{1},s}$	Shroud outlet blade angle A	10.9	deg
$\beta^*_{B,1,s}$	Shroud inlet blade angle B	67.7	deg
$\beta^*_{B,\bar{1},s}$	Shroud outlet blade angle B	20.7	deg

2.3. Mechanical Concept

In many aspects, the mechanical design is key in regard to the success of a MHP energy recovery device. The most crucial requirement addresses to the product's robustness and lifetime to minimize the occurrence of maintenance operations. Since quite high hydraulic head values are targeted for the DuoTurbo product prototype, the axial loads are significant. An appropriate bearing arrangement with reliable sealing is therefore required, excluding the risk of water contamination by lubricants. Drinking water conform materials are required for all submerged components. Further, the system efficiency is strongly related to the constructive arrangement that is therefore subject to the product optimization process. Accordingly, different mechanical configurations have been tested during the laboratory prototype stage (see Figure 1), along with detailed power loss analyses. Based on these findings, the mechanical components of the product prototype are designed, as presented in Figure 5.

A key feature of the DuoTurbo microturbine are the wet permanent magnet rim generators. Thanks to their design, the mechanical complexity is minimal, reaching a high power density and compact dimensions. A drawback is encountered regarding the additional disc friction losses created by the submerged electrical rotors. Whereas, the cooling effect of the flow increases the power capacity of the permanent magnet generators to a great extent. In the generator gap, a thin polymer tube is placed to separate the electrical stators from the fluid. Labyrinth seals minimize volumetric losses through the generator gap. Actually, the product prototype has an outer diameter of 300 mm and a total axial length of 520 mm.

2.4. Electrical and Electronic Concepts

Due to the particular architecture of the DuoTurbo microturbine, custom permanent magnet synchronous rim generators needed to be designed [29]. The generators are engineered to achieve a maximum electrical power for the given dimensional restrictions and rotational speed values. Specifications of the generators are presented in Table 2. Because of the small axial distance between the runners, partially wounded stators are currently used for the product prototype. The windings are thereby arranged only over $2 \times 90°$ of the stator's circumference. Further, the generator topology is defined by the number of poles, the number of slots, and the number of slots per pole and per phase q. The electromagnetic design was performed and validated using FEM simulations. The simulation showed minimal torque ripple choosing $q = 2$, reaching an average dynamic torque of about 10 Nm at the rated point.

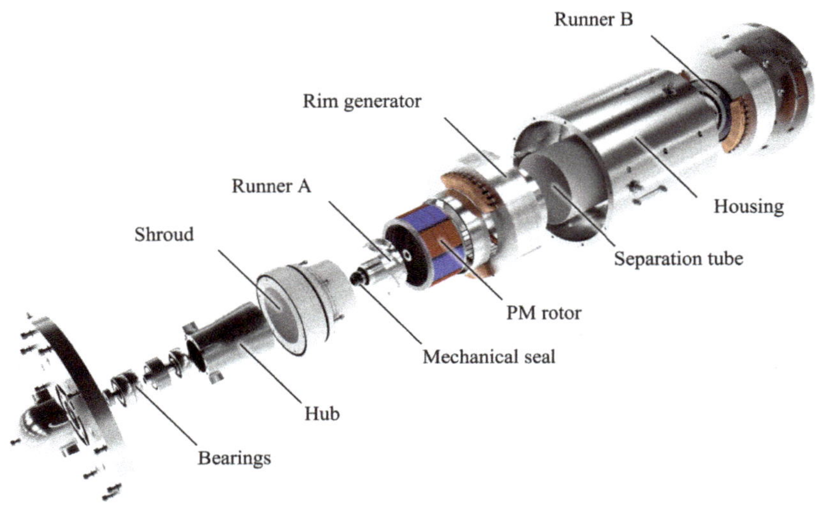

Figure 5. Overview of the mechanical components of the DuoTurbo product prototype.

Table 2. Generator specifications of the DuoTurbo product prototype.

Symbole	Description	Value	Unit
D_{ri}	Inner rotor diameter	100	mm
D_{re}	Outer rotor diameter	126	mm
g	Generator air gap	2.9	mm
l_a	Active length	62	mm
$2p$	Number of poles	8	-
n_n	Nominal rot. speed	3500	1/min
T_n	Nominal torque	10	Nm
U	Line voltage	400	V
$P_{el,n}$	Nominal electrical power	3.3	kW
$\eta_{el,n}$	Nominal efficiency	0.92	-

Power electronics are essential for a micro hydroelectric unit, since the generated electricity is mostly injected to the electrical network and the runner rotational speed may vary. For the DuoTurbo pilot, commercial frequency converters are used for the independent speed control and power regeneration of each generator. A third converter is installed on the grid side for the synchronization and injection of the electrical energy. The two four-quadrant converters on generator side can operate in drive mode for the turbine startup and in regeneration mode during normal operation. The electrical power is transmitted via direct current to the converter on the grid side. The runner speed is controlled in sensorless mode. Because the industrial converters represent an important portion of the total costs of the system, a custom electronic design may be crucial to achieve marketability of the DuoTurbo turbine.

3. Characterization Methodology

3.1. Overview of the Characterization

The product characterization in terms of performance and hydraulic properties is basically assessed by laboratory tests and numerical simulations. Potential characteristic drifts caused by long term degradation are determined by advanced endurance tests carried out at the pilot facility. A diagram of the power flow through the DuoTurbo turbine is presented in Figure 6. Thereby, three types of values are distinguished:

(a)   Experimental data resulting from standard laboratory machine testing such as the hydraulic power, the electrical power and the hydroelectric efficiency η_{h-el};

(b)   Numerical data resulting from CFD simulations as part of the runner design procedure such as the energy efficiency η_e;

(c)   Power loss components estimated or measured by means of particular experimental setups during the prototype investigation phase.

Apart from the efficiency terms, the dimensionless $\varphi - \psi$ hydraulic characteristics according to the IEC standard [30] can also be predicted by the numerical flow simulations and be compared to experimental data. Theses hydraulic characteristics, depending only on the hydraulic head, the discharge and the runner rotational speed, are key to validate the CFD based runner design methodology.

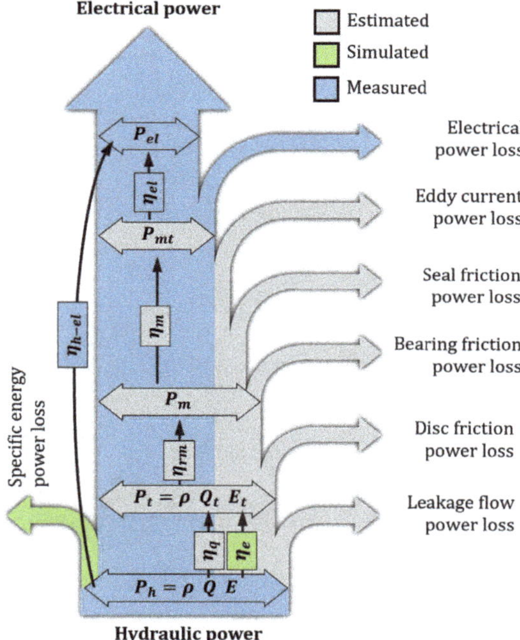

Figure 6. Definition of measured, simulated and indirectly estimated quantities on the power flow of the DuoTurbo turbine.

3.2. CFD Simulations

3.2.1. Numerical Setup

To evaluate the generated hydraulic design of the product prototype, numerical flow simulations have been carried out with the commercial finite volume solver ANSYS CFX 17.0. The software is solving the incompressible steady Reynolds Averaged Navier-Stokes (RANS) equations in their conservative form and the mass conservation equation [31]. Steady state numerical simulations have been performed using the SST turbulence model [32]. The choice of the steady state flow model is justified by the fact that numerous hydraulic design iterations are carried for the development of the product prototype. Steady state simulations for design purposes of hydraulic machinery are widely applied and stated appropriate to capture mean flow features and engineering quantities [33] and are even capable to evaluate complex operating conditions such as for no-load speed determination [34]. A 2nd order spatial scheme with a specified blend factor of 1 is used. A RMS MAX convergence criterion of 10^{-6} is imposed, limiting the the maximum number of iterations to 700 per operating point. The equations are discretized in a full computational domain with a mesh made with the ANSYS ICEM CFD 17.0 commercial software, generating structured grids with hexahedral cells. The meshes of the four subdomains are presented in Figure 7, while specific information is provided in Table 3. The y^+ values, which represent the standardized thickness of the first wall layer, refer to near BEP conditions with $|\omega_A| = |\omega_B| = 2250$ min^{-1} and $Q = 9$ L/s. The grid resolution is chosen according to previous numerical studies of a counter-rotating microturbine presented in [15]. The automatic near wall treatment of the CFX solver accepts relatively large y^+ ranges up to 100 for the SST turbulence model [35]. The presented mesh qualities are therefore considered appropriate for the performed analysis.

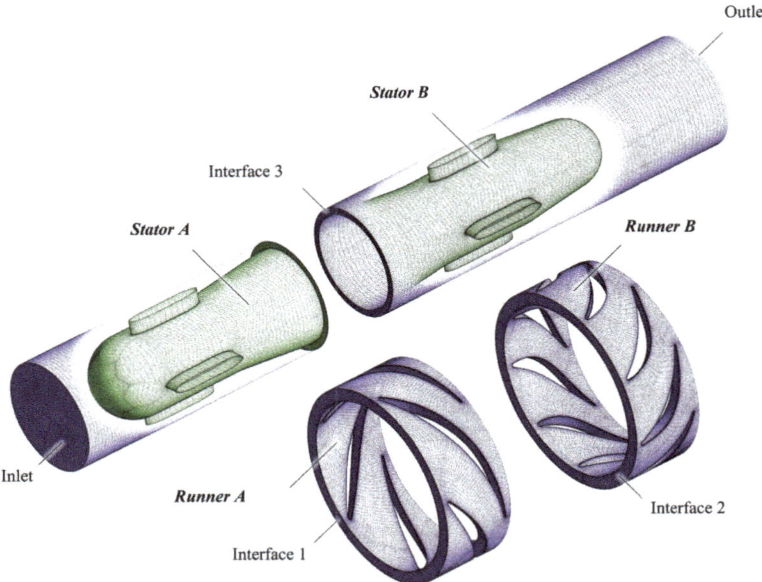

Figure 7. CFD meshes, computational domains and boundary definitions.

Table 3. CFD mesh information.

Domain	Nodes	Elements	y^+ Mean	y^+ Max
Stator A	401'274	378'880	13.64	30.64
Stator B	637'698	608'280	14.30	74.11
Runner A	640'035	601'760	33.16	86.87
Runner B	895'158	839'808	24.32	62.85

The fluxes at the domain interfaces are computed using a General Grid Interface (GGI) algorithm and a frozen rotor formulation is used. Boundary conditions are specified in Table 4. The rotational angular speed values are imposed on the rotating domains.

Table 4. Definition of boundary conditions.

Boundary	Condition
Inlet	Mass flow rate
Outlet	Outlet with average static pressure
Solid walls	Smooth no-slip wall
Interface 1/2/3	Frozen Rotor, GGI

3.2.2. Definitions for the Numerical Simulations Analysis

Since the leakage flow is not considered in the CFD simulations, the relation $Q = Q_t$ applies, see Figure 6. Thus the meridional flow velocity in the runner flow section is:

$$Cm = \frac{4\,Q}{\pi\,(D_s{}^2 - D_h{}^2)} \tag{6}$$

The discharge coefficients are then defined by means of the peripheral runner velocity amplitude at the shroud diameter:

$$\varphi = \frac{Cm}{U_{A,s}} \tag{7}$$

$$\varphi_A = \varphi \qquad \varphi_B = \frac{Cm}{U_{B,s}} \tag{8}$$

The simulated specific energies of the different turbine domains correspond to Equation (1), denoting the domain's high and low pressure reference sections as A_i and $A_{\bar{i}}$, respectively. The position of the reference sections used for the different simulated specific energy values are defined in Table 5 and Figure 7.

Table 5. Reference sections for the specific energy computations.

Variable	Section A_i	Section $A_{\bar{i}}$
E	Inlet	Outlet
E_A	Interface 1	Interface 2
E_B	Interface 2	Interface 3
E_{AB}	Interface 1	Interface 3

Then, the different energy coefficients are computed by means of the corresponding specific energy and the peripheral runner velocity at the shroud diameter:

$$\psi = \frac{2\,E}{U_{A,s}{}^2} \tag{9}$$

$$\psi_A = \frac{2\,E_A}{U_{A,s}{}^2} \qquad \psi_B = \frac{2\,E_B}{U_{B,s}{}^2} \tag{10}$$

The energy efficiency values are then calculated with the simulated torque T acting on the runner walls, composed of the blade, hub and shroud boundaries:

$$\eta_e = \frac{T_A\,\omega_A + T_B\,\omega_B}{\rho\,Q\,E} \tag{11}$$

$$\eta_{e,A} = \frac{T_A\,\omega_A}{\rho\,Q\,E_A} \qquad \eta_{e,B} = \frac{T_B\,\omega_B}{\rho\,Q\,E_B} \tag{12}$$

The design of experiment for the numerical simulations was established with respect to the nominal theoretical discharge coefficient and the speed ratio, as pointed out in Figure 8. To investigate potential Reynolds effects, the same set of points was simulated once close to the lower rotational speed limit and once close to the upper rotational speed limit.

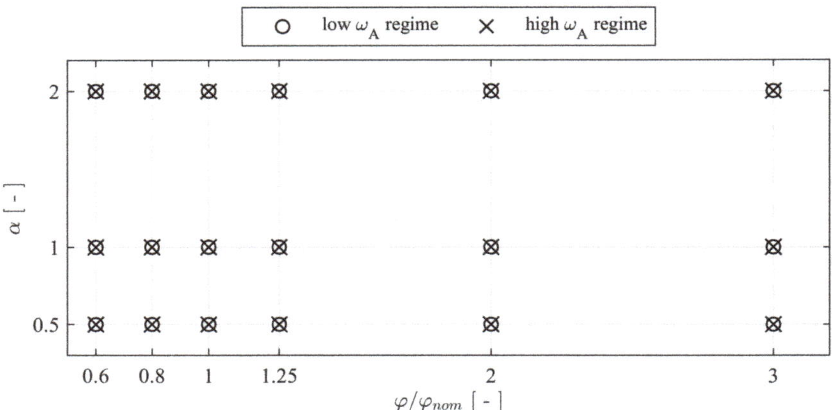

Figure 8. Design of experiment for the numerical simulations.

3.3. *Laboratory Hydraulic Performance Tests*

3.3.1. Experimental Setup

To perform tests of small scale turbines, pumps and other hydraulic components, HES-SO Valais-Wallis employs a hydraulic test rig, detailed in [36]. The DuoTurbo product prototype was installed on the test rig for performance measurements of the complete operating range. The testing section with the experimental setup for the DuoTurbo characterization is shown in Figure 9. The experimental hydraulic layout with two parallel PRV's corresponds to the one implemented at the pilot facility (compare to Figures 2 and 10). This setup also enables analyzing the transient interactions between the PRV's and the turbine. The transient measurements are not part of this publication.

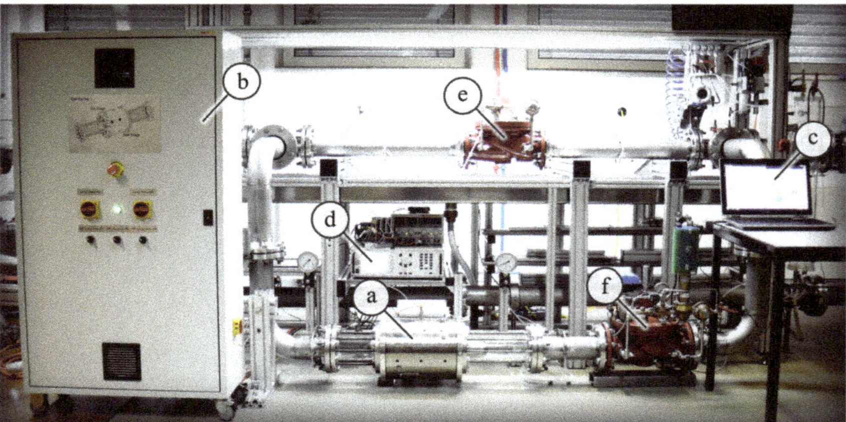

Figure 9. Experimental setup on the hydraulic test rig of HES-SO Valais-Wallis. (**a**) DuoTurbo turbine (**b**) Electrical cabinet (**c**) Control interface (**d**) Precision electrical multimeter (**e**) Bypass PRV (**f**) Turbine branch PRV.

Three parallel recirculating multistage centrifugal pumps with variable speed supply the test circuit with fresh water from a main reservoir. The meanwhile upgraded total pumping power of 42.5 kW delivers a maximum discharge of 130 m^3/h and a maximum pressure of 16 bar. Using a pressurized downstream reservoir, the setting level of the testing model can be adjusted to investigate cavitation phenomena. The test rig is controlled through a customized LabView© interface, whereby either the rotational speed of the pumps, the head on the testing model or the discharge can be regulated. The measurements and the control of the test rig are managed by a wireless data communication structure.

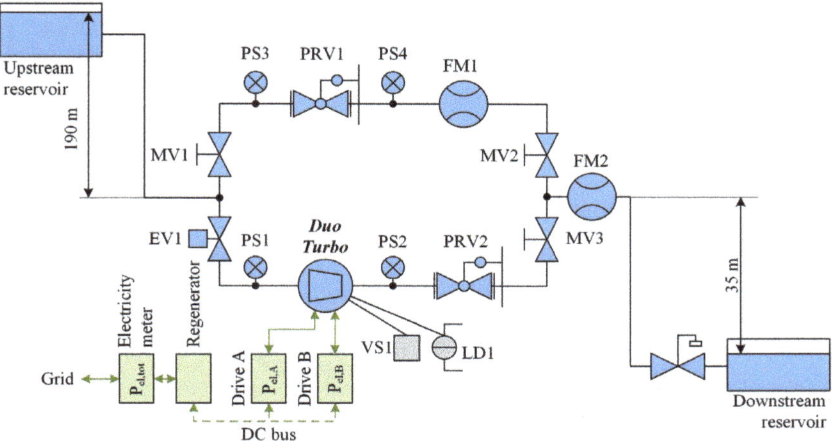

Figure 10. Installation scheme of the DuoTurbo pilot.

3.3.2. Testing Methodology

The instruments used to recover the hydraulic performances of the testing model (detailed in [22]) involve briefly an electromagnetic flowmeter used to recover the discharge and a differential pressure transducer for the testing head. Further, the electrical power and the runner rotational speed values are measured with a precision electrical multimeter. The characteristics of the main measurement instruments are presented in Table 6.

Table 6. Main characteristics of the laboratory measurement equipment.

Sensor Type	Measured Quantity	Range	Accuracy
Electromagnetic flowmeter	Discharge	8–280 m^3/h	±0.2%
Differential pressure sensor	Head	0–16 bar	±0.1%
Precision electrical multimeter	Rotational speed	0–50 kHz	±0.025%
	Electrical power	0–32 Atrms	±0.025%
		0–1000 Vtrms	±0.025%

A static testing methodology is applied, performing measurements at several constant testing heads between 10 m and 80 m. Sets of operating points covering the $\omega_A - \omega_B$ plane were statically measured for each head, implying rotational speeds from 750 to 3500 1/min. The experimentally determined specific energy E_{exp} is based on the static pressure difference between the inlet and outlet pipelines, see Equation (13) at comparable positions to the inlet and outlet boundaries of the CFD model. Since the swirl velocity at the low pressure side is undetermined and therefore not taken into account for the head calculation, a deviation between the measured and simulated specific energy values is possible. However, this deviation only becomes important for significant swirls at severe off-design conditions.

$$E_{exp} = \frac{p_I - p_{\bar{I}}}{\rho} \qquad \psi_{exp} = \frac{2\,E_{exp}}{U_{A,s}^{\;2}} \qquad (13)$$

Consequently, the measured hydroelectric efficiency η_{h-el} is defined as:

$$\eta_{h-el} = \frac{P_{el}}{\rho\, Q\, E_{exp}} \tag{14}$$

Further, the experimentally extracted discharge coefficient φ_{exp} deviates slightly from the simulated term, since the measured discharge includes the leakage discharge Q_l:

$$\varphi_{exp} = \frac{4\,Q}{U_{A,s}{}^2\,\pi\,(D_s{}^2 - D_h{}^2)} = \frac{4\,(Q_t + Q_l)}{U_{A,s}{}^2\,\pi\,(D_s{}^2 - D_h{}^2)} \tag{15}$$

To estimate the volumetric losses, particular tests with sealed runner blade passages were carried out during the laboratory prototype investigations, forcing the discharge to pass trough the labyrinth seals. In this experimental configuration, the leakage discharge as function of the static pressure difference on the labyrinths could be characterized.

3.4. Pilot Endurance and Performance Tests

3.4.1. Hydraulic Scheme and Monitoring Equipment

Endurance tests are crucial to evaluate degradation and reliability of the developed system under real operating conditions. For this purpose, a DuoTurbo pilot is installed at a pressure reducing station situated in the drinking water network of Savièse, VS Switzerland. The pressure reducing station is located between two drinking water reservoirs with an available altitude difference of about 190 m to the upstream free surface. The discharge in the hydraulic network is ruled by the level of the downstream reservoir that varies as function of the water demand in the underlying distribution network. The resulting consumption driven operating mode is typical for drinking water supply systems. The hydraulic scheme of the DuoTurbo pilot, along with the main hydroelectric elements and instrumentation, is shown in Figure 10.

The hydraulic layout with two parallel PRV's corresponds to the one tested in the laboratory framework. To recover the hydraulic variables, two flow meters (FM1 and FM2) and four pressure sensors (PS1–PS4) are used. For security reasons, a solenoid valve (EV1) is placed upstream the hydraulic machine, to fully isolate the turbine branch in case of a severe leakage flow. Thus, a water detector (LD1) needed to be installed to initiate the closure of the solenoid valve. Leakage may occur due to a rupture of the polymer tube in the generator gap or due to advanced wear of the mechanical seals. The mechanical behavior is supervised by a vibration sensor, placed on the housing of the turbine. Electrical parameters are recovered from the variable frequency drives and an electricity meter at the grid connection of the pressure reducing station. A list with specifications of the principal hydro-mechanical sensors is provided in Table 7. An illustration of the DuoTurbo pilot is provided in Figure 11. The installation is controlled and supervised by remote via a GSM antenna.

Table 7. Main hydromechanical instrumentation of the DuoTurbo pilot.

Pos.	Identifier	Sensor Type	Measured Quantity	Range	Accuracy
a	VS1	Vibration	Housing acceleration	-245–245 m/s^2	$\pm 0.2\%$
b	PS2	Abs. pressure	Turbine outlet pressure	0–20 bar	$\pm 0.15\%$
c	PS1	Abs. pressure	Turbine inlet pressure	0–20 bar	$\pm 0.15\%$
d	PS4	Rel. pressure	PRV1 outlet pressure	0–30 bar	$\pm 0.1\%$
e	PS3	Rel. pressure	PRV1 inlet pressure	0–30 bar	$\pm 0.1\%$
f	FM1	Flowmeter	Bypass discharge	3–75 L/s	$\pm 2\%$
g	FM2	Flowmeter	Total discharge	-147–147 L/s	$\pm 0.2\%$

Figure 11. Installed DuoTurbo pilot in a drinking water supply network at Savièse (VS Switzerland). The positions (**a–g**) refer to the instrumentation list given in Table 7.

3.4.2. Pilot Measurement Definitions

The total specific energy E_{tot} theoretically available for electricity generation is defined by the static pressure difference between the inlet and outlet hydraulic junctions of the pressure reducing station and is, inter alia, governed by the setting levels of the PRVs. Neglecting the specific energy losses of the elbows, T-junctions and manual valves, the total specific energy is actually approximated as follows (compare to Figure 10):

$$E_{tot} \approx E_{PRV1} = \frac{p_{PS3} - p_{PS4}}{\rho} \tag{16}$$

Neglecting the residual outlet swirl of the hydraulic machine, according to the laboratory testing methodology, the specific energy extracted by the turbine E_{turb} is calculated as follows:

$$E_{turb} = \frac{p_{PS1} - p_{PS2}}{\rho} \tag{17}$$

The residual head on PRV2 that is significant during part load conditions, is inevitably dissipated. Again, assuming negligible head losses in the piping and valves, the head on PRV2 is approximated as:

$$E_{PRV2} \approx \frac{p_{PS2} - p_{PS4}}{\rho} \tag{18}$$

Finally, the following obvious equilibrium applies for any operating point:

$$E_{tot} \approx E_{turb} + E_{PRV2} \tag{19}$$

The accumulated hydraulic energy released from the hydraulic circuit by the microturbine is specified as:

$$W_h = \rho \int_{t_0}^{t_1} E_{turb}(t)\, Q_{turb}(t)\, dt \tag{20}$$

The accumulated energy recovered by the electrical generators is:

$$W_{el} = \int_{t_0}^{t_1} P_{el}(t)\, dt \tag{21}$$

Considering the auto-consumption of all electronic components like drives, controllers, sensors and solenoid valves yields the net electrical energy $W_{el,net}$ injected to the grid.

4. Results

4.1. CFD Simulation Results

A graphical representation of computed instantaneous relative velocity streamlines at nominal operating conditions is given in Figure 12. The color scale W^* indicates the ratio between the local relative flow velocity amplitude W and the average meridional flow velocity Cm_I at the circular high pressure reference section of the computational domain. Detailed results of the shown nominal operating point, as well as the maximum power point of the product prototype, are provided in Table 8. For both cases, the speed ratio is 1. One may notice that relatively low specific speed values (see Equation (22)) of about 16 are targeted to match the pilot drinking water facility's characteristics. From [37] one may deduce typical specific speeds of 135–320 for axial flow turbines, 50–150 for semi-axial flow turbines and 10–50 for radial flow turbines. Since the DuoTurbo turbine remains a purely axial flow machine, the low specific speed values seem rather unfavorable for the turbine's performance. Nevertheless, very acceptable energy efficiencies above 80% could be found by the CFD simulations. This result may indicate that double rotor axial designs have increased capabilities of handling low specific speed conditions compared to single rotor axial designs. Thus, it can be stated that the design range of a DuoTurbo turbine is considerable.

$$n_q = n[1/min](Q[m^3/s])^{0.5}(H[m])^{-0.75} \tag{22}$$

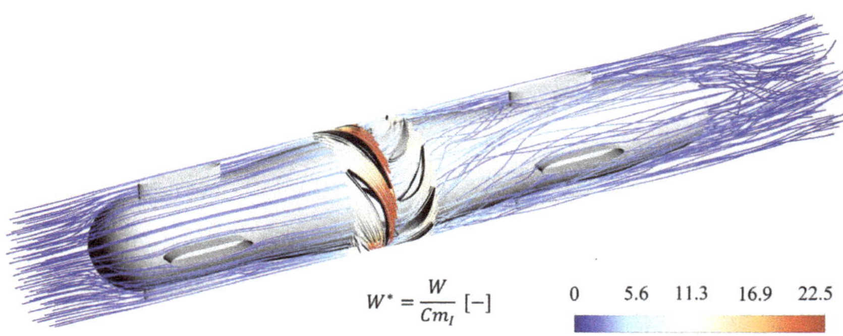

Figure 12. Instantaneous relative velocity streamlines for $Q = 9$ L/s, $n_A = 2250$ min^{-1} and $\alpha = 1$.

Table 8. Simulation results for nominal conditions and maximum power.

Q [L/s]	E [J/kg]	E_A [J/kg]	E_B [J/kg]	P_h [W]	P_t [W]	η_e [-]	n_q [-]
9	302	151	144	2'712	2'239	0.83	16.3
14	727	363	348	10'162	8'509	0.84	16.4

The simulated energy efficiency and energy coefficient of runner A as function of its discharge coefficient are provided in Figure 13a. The three colors represent different values of the speed ratio α and the two symbols indicate the low and high rotational speed regimes close to the lower and upper speed limits of the generators. Since the inlet flow remains always axial for the front runner, the efficiency and energy coefficient curves remain similar for variable speed ratios. Nevertheless, a slight effect of α on the first runner's efficiency close to BEP conditions is observed, thus, an impact of the rear runner's inlet flow condition on the front runner's outlet flow condition can be anticipated. This influence is observed only when the speed ratio is decreased (red curve where $\alpha = 0.5$), causing a decrease of about 2% of the energy efficiency around BEP conditions. In the high velocity regime with increased Reynolds numbers, the energy efficiency of the front runner is increased by about 2% compared to the low velocity regime, regardless the speed ratio.

In Figure 13b, the simulated energy efficiency and hydraulic characteristics of runner B are presented. One can observe a strong effect of the speed ratio on the rear runner's energy efficiency and energy coefficient. Indeed, the second runner's inlet flow is strongly coupled to the first runner's rotational speed. Consequently, clearly distinct characteristics curves are observed for each speed ratio. The maximum simulated energy efficiency of the second runner is about 10% higher compared to the first runner. Moreover, increasing the speed ratio tends to decrease the second runner's peak efficiency. It can further be observed that the efficiency curves decay more rapidly out of BEP conditions compared to the first runner's results, whereas the variation of the speed ratio counteracts the efficiency decrease importantly.

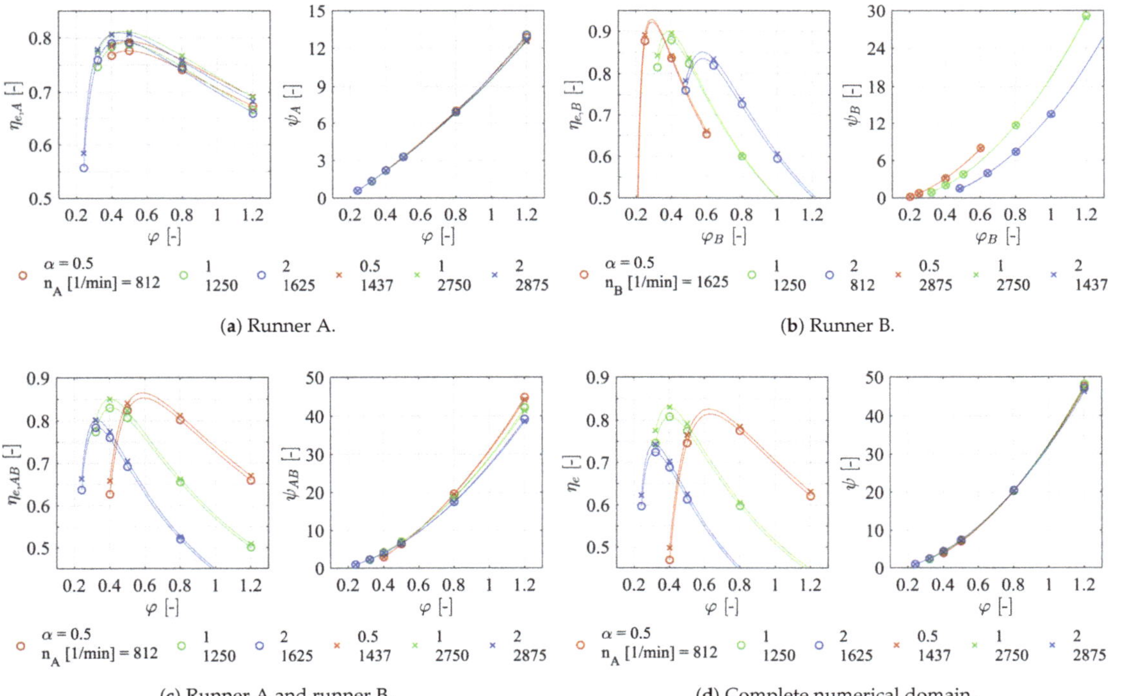

(a) Runner A.

(b) Runner B.

(c) Runner A and runner B.

(d) Complete numerical domain.

Figure 13. Simulated energy efficiency and $\varphi - \psi$ characteristics.

In Figure 13c, the numerical results referring to the two runner's entity are presented. It can be observed that the variation of the speed ratio prevents a significant drop of efficiency for changing flow conditions. Thus, the variation of α significantly increases the operating range and the flexibility of the turbine. The envelope formed by the three efficiency curves resembles to the one of the rear runner presented in Figure 13a. Thereby it must be noticed that the kinetic energy at the outlet of the front runner would not be recovered without the rear runner. It would therefore be misleading to conclude that comparable performances can be achieved with a single vaneless runner regarding the extraction of hydraulic energy. The current hydraulic design shows only a slight change of the hydraulic characteristic curves as function of the speed ratio that indicates, in some way, the limitations of a fixed geometry design in terms of flexibility. For example, the $\varphi - \psi$ representation of a Francis turbine with adjustable guide vanes would typically show distinct characteristics curves for each opening value, comparable to Figure 13b (right).

Finally, in Figure 13d, the results obtained from the total numerical domain between the inlet and outlet reference sections are presented. In contrast to Figure 13c, the two stator domains are included, each one featured with 4 axially aligned stay vanes to support the turbine hub. Obviously, the outlet stator is expected to interact with the flow when a residual swirl is exiting the rear runner. Indeed, the axial realignment of the flow by the outlet stator slightly modifies the energy coefficient curves at high discharge coefficients. The three curves are almost perfectly matching regardless the speed ratio. Additionally, a slight decrease of efficiency occurs due to the additional head losses in the stator domains. For all numerical results, a slight Reynolds effect is observable, leading to an increase of the energy efficiencies in the order of 2–3% at the high rotational speed regime, thus, at the high power regime.

4.2. Laboratory Test Results

According to the testing strategy, operating points covering the $\omega_A - \omega_B$ plane are captured for each constant testing head. Correspondingly, for each measured plane, the electrical power and the discharge are interpolated. Using these interpolation surfaces, the maximum electrical power as function of the discharge can be deduced for each head value. The resulting curves are then used to construct a global hill chart of maximum electrical power, provided in Figure 14. The corresponding efficiency values on the maximum power hill chart are provided in Figure 15. Operating limits are imposed by the admissible rotational speed and generator torque. It can be expected that the low specific speed of the current hydraulic design (see Table 8) is causing the rather narrowly shaped operating band. On both hill charts, the Maximum Power Line (MPL) is indicated that represents the most beneficial operating conditions in terms of energy output. The MPL is used for the programming of the rotational speed control algorithm implemented at the pilot facility. The speed ratio varies from 0.4 to 1.6 inside the presented hill charts. The maximum specific energy for long term operation is actually limited to 750 J/kg to reduce the risk of premature wear of the runner bearings. A hydroelectric peak efficiency of 59% is detected that lets anticipate non-negligible amounts labyrinth leakage, disc friction and mechanical power losses.

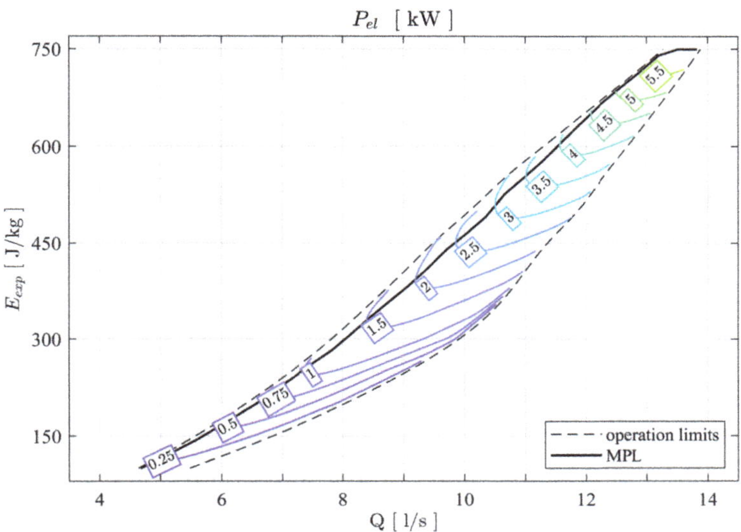

Figure 14. Measured maximum electrical power hill chart along with the Maximum Power Line (MPL).

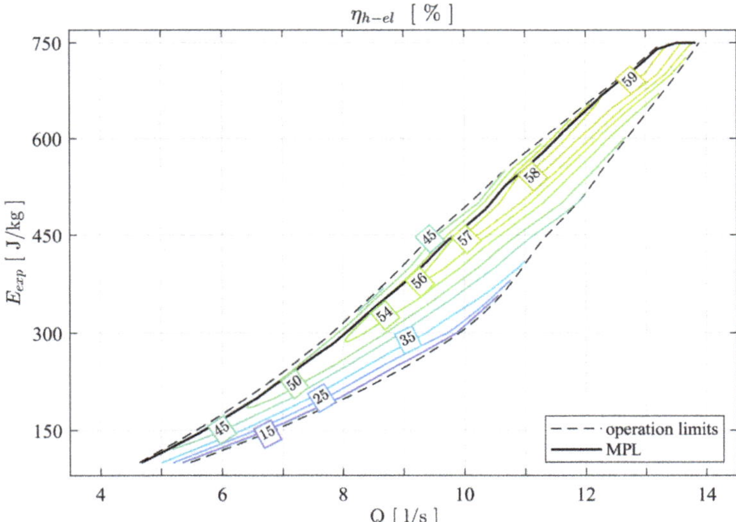

Figure 15. Measured hydroelectric efficiency hill chart at maximum electrical power along with the Maximum Power Line (MPL).

4.3. Comparison between Laboratory Tests and Numerical Simulations

To validate the CFD based hydraulic design methodology, a comparison between the numerical and experimental hydraulic characteristics is performed. In contrast, a comparison regarding the energy efficiency remains challenging, since a direct measurement of the transmitted mechanical power (see variable P_t defined in Figure 6) is not possible with the actual mechanical configuration. Keeping in mind that additional optimization steps may be justified for a final commercialization, the main objective of the product prototype development concentrates on the proof of advanced readiness levels of the technology. A detailed analysis of the various power losses is therefor not part of the present paper. The comparison between the simulated and measured correlations between the energy coefficient and the discharge coefficient for different ranges of the speed ratio are presented in Figure 16. The absolute errors are calculated according to Equation (23) with respect to the maximum energy coefficient ψ_{exp} = 50.6 contained in the analyzed data set.

$$\epsilon_{abs} = \frac{\psi_{sim} - \psi_{exp}}{max(\psi_{exp})} \tag{23}$$

It should be mentioned that discrepancies in the compared data occur due to the distinctions between the measured and simulated energy and discharge coefficients. If the volumetric losses would be captured in the laboratory tests, the experimental characteristic curves would shift towards the left, resulting in a positive offset of the energy coefficients. Actually all presented energy coefficients are already greater than the simulated ones, thus the integration of the volumetric losses would further increase the discrepancies. On the contrary, the experimental energy coefficient is overestimated if an outlet swirl is present, which would lead to a better agreement of the compared values. Moreover, the machining tolerance of the runners as well as surface roughness effects are additional sources of deviations. The maximum absolute error in the presented data sets is about 8.1% for the maximum measured energy coefficient value. The error tends to increase with increasing discharge coefficients that could be related to the limited capability of the steady state flow model far from BEP conditions. Around the BEP conditions, the observed errors indicate a quite fair agreement between simulation and laboratory tests, that lets conclude that the hydraulic design procedure is reasonably reliable.

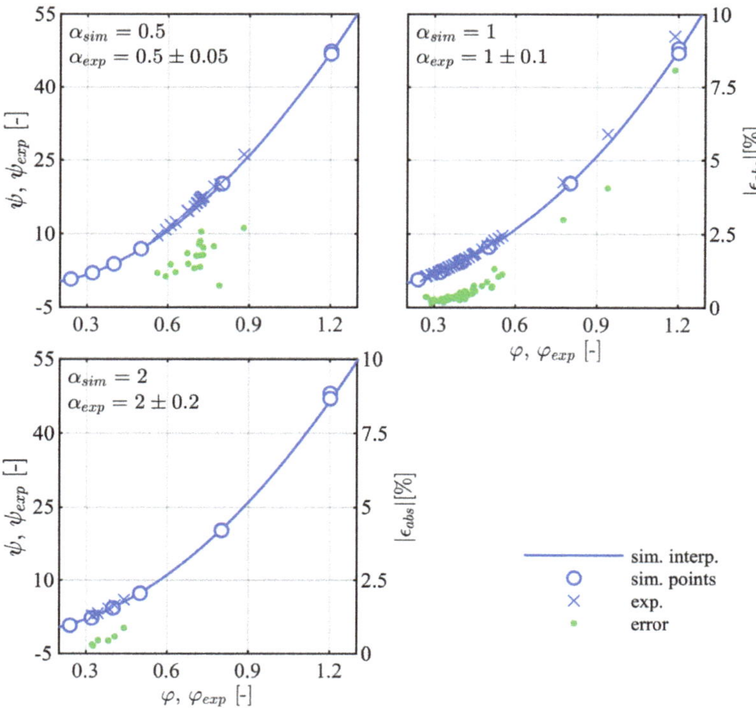

Figure 16. Comparison between simulated and measured hydraulic characteristics.

4.4. Pilot Monitoring Results

The following monitoring results were registered between 15 May 2019 and 20 May 2020. During this period of 52 weeks, the DuoTurbo pilot was almost permanently in operation, without requiring any significant human interventions. The reliability and functionality of the system under real conditions is therefore demonstrated in operational environment at pre-commercial state that can be attributed to a technology readiness level of 7 [38]. At steady flow conditions, the rotational speed of the turbine runners is controlled through a Maximum Power Point Tracking (MPPT) method [19] to constantly maximize the generated power. In case of transient flow, a sensorless control loop brings the machine rapidly back on the MPL illustrated in Figures 14 and 15. This approach is comparable to the idea of modified affinity laws in hydraulic machines towards the Best Efficiency Line (BEL) proposed by [39], whereas the BEL and MPL may not coincide depending on the hydraulic scheme. Indeed, if head reserves are available for a given flow (if a PRV is operating downstream the turbine), the operating point may be adapted to increase the head recovered by the turbine. This may lead to an increased power output, not necessarily respecting the BEL.

The evolution of the specific energies of the turbine, PRV2 and the complete pressure reducing station as function of the total discharge, along with the relative cumulative discharge frequency, is pointed out in Figure 17. A time frame of 4 weeks is used for the head statistics, whereas the discharge data of the whole year is respected to derive the frequency of the flow rates. The variability of the data within intervals of 1 L/s is used for the evaluation of the sample standard deviation s. Exceeding a threshold of 5.5 L/s, the turbine start-up is initiated and the turbine branch is opened, see point (a). Falling below 5 L/s, the turbine branch is closed and the turbine is put to standby mode. At point (b), that corresponds to the maximum admissible discharge of the machine, the turbine head and the total net head tend to cross. In this situation the turbine branch PRV is no longer able to maintain the adjusted pressure level at the outlet of the station. Therefore, the bypass PRV will gradually open if the discharge increases further. It can be observed that the residual head loss of PRV2 reduces the potential output power of the turbine in the discharge range above point (b). From Figure 17 one can further deduce that about 10% of the flow rates are below point (a) which do not contribute to the energy production. About 83% of the flow rates are situated between point (a) and point (b) and about 7% of the flow rates exceed the maximum discharge capacity of the machine at point (b).

The maximum and average vibration velocity amplitudes $\hat{v}_v$ for weekly measurement periods as function of the elapsed operating time t_{op} are presented in Figure 18 (left). The intensity of vibration is directly related to the rotational speed, as shown in Figure 18 (right). An eigenmode of the structure is excited around $n_a = 2000$ min^{-1}. The alarm threshold for the temporary vibration amplitude is fixed at 2 mm/s. After about 3400 h of operation, a decrease of the vibration intensity can be observed. This is due to the decrease of discharge during the winter period resulting in a shift of the rotational speed range.

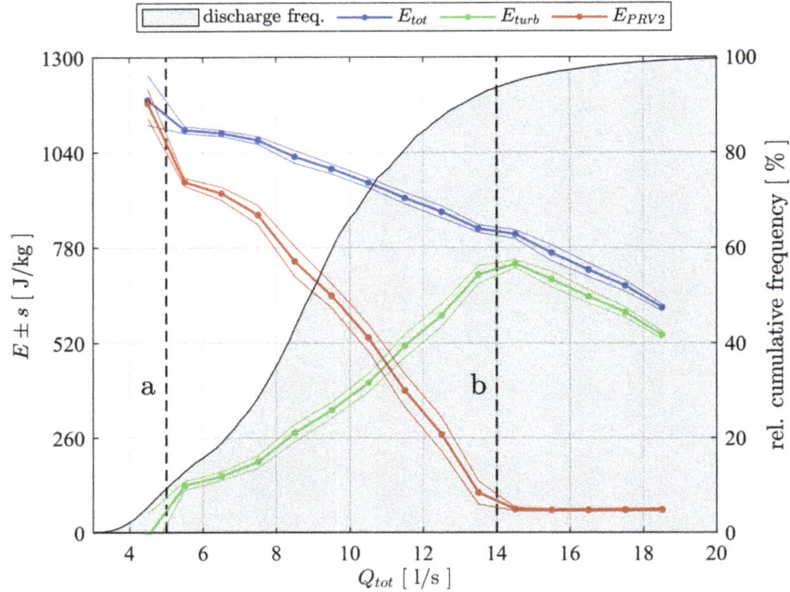

Figure 17. Monitored behavior of the specific energy of the pressure reducing station, the Duo-Turbo turbine and PRV2 as function of the total discharge, along with the relative cumulative discharge frequency.

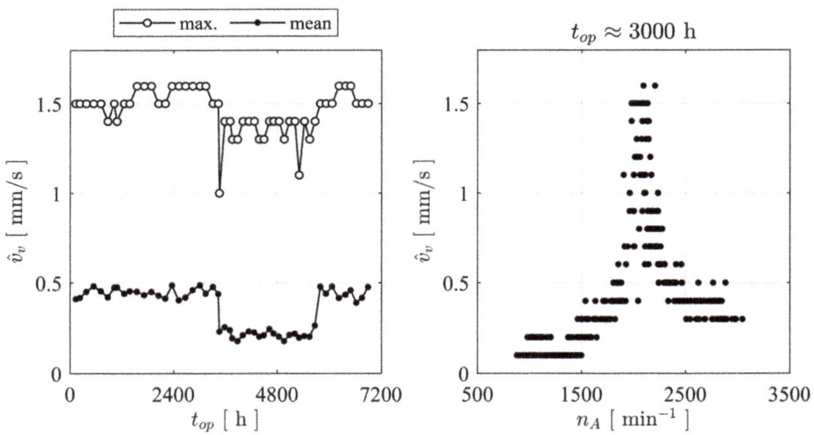

Figure 18. Pilot vibration monitoring. Weekly maximum and average vibration velocity amplitudes (**left**) and example of the vibration variation during one week (**right**).

In contrast to the laboratory tests, the monitored hydroelectric efficiency seems to decrease towards the maximum power levels, as pointed out in Figure 19 (right). This is probably due to a misjudgment of the turbine discharge close to the opening point of the bypass, since the bypass flow meter is not reliable for small discharge values. In Figure 19 (left), the trend of the weekly hydroelectric efficiency statistics inside the hydraulic power range between 7 kW and 8 kW is shown. Mean values fluctuate within a range of about

7%, whereas, no systematic drift can be observed. It is therefore assumed that no important degradation of the runner blades or labyrinth seals has developed so far.

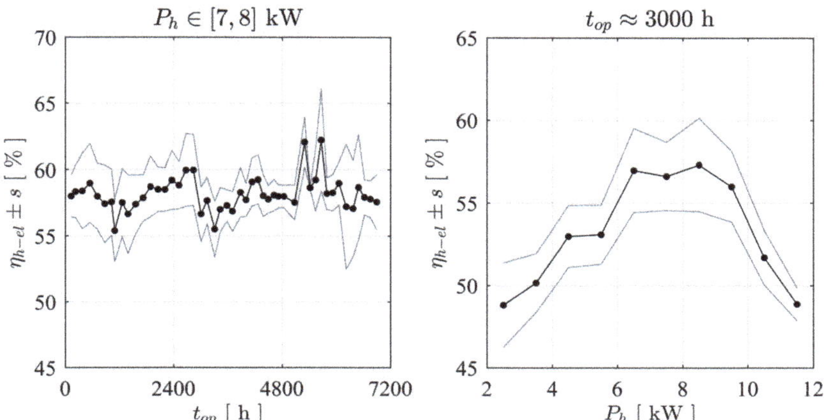

Figure 19. Efficiency monitoring of the DuoTurbo pilot. Weekly statistics for a fixed power interval (**left**) and example of a weekly efficiency variation as function of the hydraulic power (**right**).

The evolution of the different accumulated energies of the DuoTurbo pilot are presented in Figure 20. The meteorological seasons are indicated by colors. The drop of discharge at the end of autumn is significant and consequently the production rate is quite low during winter. About 13 MWh of injected electrical energy are counted during the pilot's first year of operation. The registered hydraulic energy exploited by the turbine corresponds to 27 MWh. The average injected electrical power is 1.88 kW with an average overall system efficiency of 48% (water to wire).

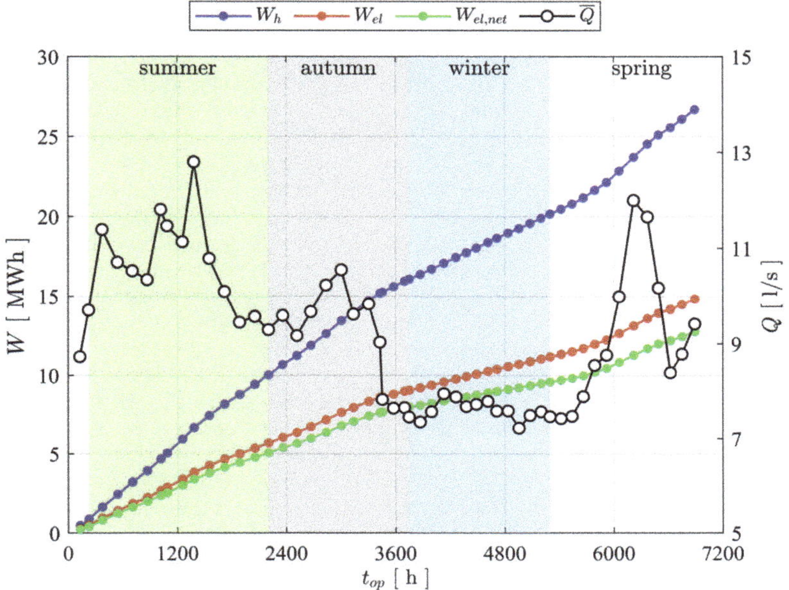

Figure 20. Monitored accumulated energies and weekly discharge average of the DuoTurbo pilot.

5. Conclusions

The drinking water industry hides a considerable hydraulic energy potential that remains mostly unexploited for gross capacities of a few kW. In this context, the DuoTurbo research project was initiated to provide a complementary solution for energetic exploitation of drinking water networks in the power range between 5 kW and 25 kW. A new hydroelectric system with a counter-rotating microturbine is developed and installed for long term tests at a drinking water facility. The counter-rotating architecture with particular variable speed rim generators enables a compact design, reaching relatively high specific energy values with extended operational flexibility compared to single degree of freedom solutions. The present study provides an overview of the different technical concepts and methodologies applied for the development of a product prototype with advanced technology readiness level. The particular double rotor hydraulic design is developed and characterized by steady state CFD RANS simulations. The numerical investigations point out the significance of the speed ratio of the system that allows for an extension of the high efficiency range facing important variations of the discharge coefficient. No significant offsets between the characteristic lines of each simulated speed ratio in the $\varphi - \psi$ plane are observed considering the complete turbine domain. This result indicates the limitations on hydraulic flexibility of a fixed geometry hydraulic concept. Despite the currently very low specific speed value of about 16, an energy efficiency greater than 83% could numerically be revealed. This indicates an increased capability of double rotor axial designs handling low specific speed regimes compared to single rotor solutions. The specific speed range of the DuoTurbo turbine could be increased to more than 135 according to typical axial flow designs. Thus, the design range of the DuoTurbo turbine is considerable. In a complementary development step, advanced laboratory tests are performed on a hydraulic test rig to validate the design methodology and to reveal the final characteristics of the product prototype. The tests are conducted at several constant heads between 10 m and 80 m while varying the rotational speeds of each runner to capture the entire characteristics and hydroelectric efficiency of the system. A peak hydroelectric efficiency of 59% is measured that lets estimate a still relatively high amount of power losses from labyrinth leakage flow, disc friction and mechanical friction. The measured energy coefficients and discharge coefficients are then compared to the steady state RANS simulation results, showing a maximum absolute error of almost 8% in the high energy coefficient range, but displaying fair agreements close to BEP conditions with absolute errors widely far below 5%. The hydraulic design procedure is therefore considered reasonably accurate. For the final step of the system validation, a DuoTurbo pilot was commissioned in May 2019 at a pressure reducing station of a drinking water supply network. A complete year of operation counting more than 6'890 operating hours is monitored that confirms the functioning, stability and reliability of the system to a large extend. The vibration level on the turbine housing as well as the hydroelectric efficiency do not show significant deviation trends during the long-term endurance testing. The demonstration of the system in operational environment at pre-commercial state is validated that can be attributed to a technology readiness level of 7. This state includes briefly the in-field operation of the demonstrator adaptable to different operational environments and the verification of the expected efficiency and system stability under long-term real-time operational conditions. The present work demonstrates the technological capability of a counter-rotating micro turbine to recover hydraulic energy from existing water facilities in the power range of a few kW. Subsequent to the endurance tests, research on the required commercialization steps to transform the DuoTurbo concept into a viable MHP solution would typically be addressed. The further steps would also require the broadening and fine tuning of the hydraulic, mechanical and electrical designs along with system scaling considerations in order to target a wide range of potential drinking water networks.

Author Contributions: Conceptualization, D.B., V.H., L.R., S.C, F.A. and C.M.-A.; methodology, D.B., V.H., L.R., S.C., F.A. and C.M.-A.; software, D.B.; validation, D.B., V.H., L.R. and S.C.; formal analysis, D.B. and V.H.; investigation, D.B. and V.H.; resources, V.H.; data curation, D.B. and V.H.; writing—original draft preparation, D.B.; writing—review and editing, D.B., V.H., F.A. and C.M.-A.; visualization, D.B.; supervision, S.C., F.A. and C.M.-A.; project administration, F.A. and C.M.-A.; funding acquisition, F.A. and C.M.-A. All authors have read and agreed to the published version of the manuscript.

Funding: This research is part of the KTI DuoTurbo project number 17197.1 PFEN IW and received funding from the Swiss Commission for Technology and Innovation.

Institutional Review Board Statement: Not applicable.

Informed Consent Statement: Not applicable.

Data Availability Statement: Not applicable.

Acknowledgments: This project was performed in the framework of SCCER-SoE. The authors would like to thank the industrial partners Telsa SA, Valelectric Farner SA, Jacquier-Luisier SA and LAMI SA for their collaboration. Finally, the onsite measurements were performed in the framework of the Savièse project, financially supported by Cimark and by the program The Ark Energy of the Ark, Valais, Switzerland. ClaVal Switzerland is also thanked for all their support concerning the pressure reducing valves.

Conflicts of Interest: The authors declare no conflict of interest.

References

1. SFOE. Swiss Federal Office of Energy, Hydropower. 2021. Available online: https://www.bfe.admin.ch/bfe/en/home/supply/renewable-energy/hydropower.html (accessed on 23 September 2021).
2. United Nations Industrial Development Organization, World Small Hydropower Development Report 2019. 2021. Available online: https://www.unido.org/our-focus-safeguarding-environment-clean-energy-access-productive-use-renewable-energy-focus-areas-small-hydro-power/world-small-hydropower-development-report (accessed on 23 September 2021).
3. McNabola, A.; Coughlan, P.; Corcoran, L.; Power, C.; Williams, A.P.; Harris, I.; Gallagher, J.; Styles, D. Energy recovery in the water industry using micro-hydropower: An opportunity to improve sustainability. *Water Policy* **2014**, *16*, 168–183. [CrossRef]
4. Vincent Denis, A.C.; Punys, P. *Integration of Small Hydro Turbines into Existing Water Infrastructures*; Hydropower, IntechOpen: Rijeka, Croatia, 2012; [CrossRef]
5. Carravetta, A.; Derakhshan Houreh, S.; Ramos, H.M. *Pumps as Turbines—Fundamentals and Applications*; Springer: Cham, Switzerland, 2018.
6. Postacchini, M.; Darvini, G.; Finizio, F.; Pelagalli, L.; Soldini, L.; Di Giuseppe, E. Hydropower Generation Through Pump as Turbine: Experimental Study and Potential Application to Small-Scale WDN. *Water* **2020**, *12*, 958. [CrossRef]
7. Fontanella, S.; Fecarotta, O.; Molino, B.; Cozzolino, L.; Della Morte, R. A Performance Prediction Model for Pumps as Turbines (PATs). *Water* **2020**, *12*, 1175. [CrossRef]
8. Perez-Sanchez, M.; Sánchez-Romero, F.J.; Ramos, H.M.; López-Jiménez, P.A. Improved planning of energy recovery in water systems using a new analytic approach to PAT performance curves. *Water* **2020**, *12*, 468. [CrossRef]
9. Algieri, A.; Zema, D.A.; Nicotra, A.; Zimbone, S.M. Potential energy exploitation in collective irrigation systems using pumps as turbines: A case study in Calabria (Southern Italy). *J. Clean. Prod.* **2020**, *257*, 120538. [CrossRef]
10. Giudicianni, C.; Herrera, M.; di Nardo, A.; Carravetta, A.; Ramos, H.M.; Adeyeye, K. Zero-net energy management for the monitoring and control of dynamically-partitioned smart water systems. *J. Clean. Prod.* **2020**, *252*, 119745. [CrossRef]
11. Kougias, I.; Patsialis, T.; Zafirakou, A.; Theodossiou, N. Exploring the potential of energy recovery using micro hydropower systems in water supply systems. *Water Util. J.* **2014**, *7*, 25–33.
12. Hasmatuchi, V.; Bosioc, A.I.; Luisier, S.; Münch-Alligné, C. Dynamic Approach for Faster Performance Measurements on Hydraulic Turbomachinery Model Testing. *Appl. Sci.* **2018**, *8*, 1426. [CrossRef]
13. Hasmatuchi, V.; Münch-Alligné, C.; Gabathuler, S.; Chevailler, S.; Avellan, F. New Counter-Rotating Micro-Hydro Turbine for Drinking Water Systems. In Proceedings of the Hidroenergia 2014, Instanbul, Turkey, 21–23 May 2014.
14. Samora, I.; Hasmatuchi, V.; Münch, C.; Franca, M.J.; Schleiss, A.J.; Ramos, H.M. Experimental characterization of a five blade tubular propeller turbine for pipe inline installation. *Renew. Energy* **2016**, *95*, 356–366. [CrossRef]
15. Münch-Alligné, C.; Richard, S.; Meier, B.; Hasmatuchi, V.; Avellan, F. Numerical simulations of a counter rotating micro turbine. *Adv. Hydroinform.* **2013**, *32*, 363–373.
16. Vagnoni, E.; Andolfatto, L.; Delgado, J.; Münch-Alligné, C.; Avellan, F. Application of Laser Doppler Velocimetry to the development of a counter rotating micro-turbine. In Proceedings of the 36th IAHR World Congress, Hague, The Netherlands, 28 June–2 July 2015.
17. Vagnoni, E.; Andolfatto, L.; Richard, S.; Münch-Alligné, C.; Avellan, F. Hydraulic performance evaluation of a micro-turbine with counter rotating runners by experimental investigation and numerical simulation. *Renew. Energy* **2018**, *126*, 943–953. [CrossRef]

18. Andolfatto, L.; Delgado, J.; Vagnoni, E.; Münch-Alligné, C.; Avellan, F. Analytical hill chart towards the maximisation of energy recovery in water utility networks with counter-rotating runners micro-turbine. In Proceedings of the 36th IAHR World Congress, Hague, The Netherlands, 28 June–2 July 2015.
19. Andolfatto, L.; Delgado, J.; Vagnoni, E.; Münch-Alligné, C.; Avellan, F. Simulation of energy recovery on water utilitynetworks by a micro-turbine with counter-rotating runners. *IOP Conf. Ser. Earth Environ. Sci.* **2016**, *49*, 102012. [CrossRef]
20. Andolfatto, L.; Euzena, C.; Vagnoni, E.; Münch-Alligné, C.; Avellan, F. A mixed standard/custom design strategy to minimize cost and maximize efficiency for Picohydro power potential harvesting. In Proceedings of the 2015 5th International Youth Conference on Energy (IYCE), Pisa, Italy, 27–30 May 2015.
21. Biner, D.; Hasmatuchi, V.; Violante, D.; Richard, S.; Chevailler, S.; Andolfatto, L.; Avellan, F.; Münch-Alligné, C. Engineering and performance of duoturbo: Microturbine with counter-rotating runners. *IOP Conf. Ser. Earth Environ. Sci.* **2016**, *49*, 102013. [CrossRef]
22. Hasmatuchi, V.; Biner, D.; Avellan, F.; Münch-Alligné, C. Performance measurements on the duoturbo microturbine for drinking water systems. In Proceedings of Hydro 2016 Conference, Montreux, Switzerland, 10–12 October 2016.
23. Biner, D.; Andolfatto, L.; Hasmatuchi, V.; Rapillard, L.; Chevailler, S.; Avellan, F.; Münch-Alligné, C. DuoTurbo: A New Counter-Rotating Microturbine for Drinking Water Facilities. In Proceedings of the International Conference on Innovative Applied Energy, Oxford, UK, 14–15 March 2019.
24. Louis, J.F. Axial Flow Contra-Rotating Turbines. In Proceedings of ASME 1985 International Gas Turbine Conference and Exhibit, Houston, TX, USA, 18–21 March 1985.
25. Lee, N.J.; Choi, J.W.; Hwang, Y.H.; Kim, Y.T.; Lee, Y.H. Performance analysis of a counter-rotating tubular type micro-turbine by experiment and CFD. *IOP Conf. Ser. Earth Environ. Sci.* **2012**, *15*, 042025. [CrossRef]
26. Nan, D.; Shigemitsu, T.; Zhao, S. Investigation and analysis of attack angle and rear flow condition of contrarotating small hydro-turbine. *Energies* **2018**, *11*, 1806. [CrossRef]
27. Sonohata, R.; Fukutomi, J.; Shigemitsu, T. Study on the contra-rotating small-sized axial flow hydro turbine. *Open J. Fluid Dyn.* **2012**, *11*, 318–323. [CrossRef]
28. Linlin, C. High Performance Design of a Contra-Rotating Axial Flow Pump with Different Rotor-Speed Combination. Ph.D. Thesis, Kyushu University Institutional Repository, Fukuoka, Japan, 2014.
29. Violante, D.; Farner, L.; Chevailler, S. Design of a pm-generator for a straight flow counter-rotating microhydro turbine. In Proceedings of the 19th European Conference on Power Electronics and Applications (EPE'17 ECCE Europe), Warsaw, Poland, 11–14 September 2017.
30. IEC60193:1999. *Hydraulic Turbines, Storage Pumps and Pump-Turbines—Model Acceptance Tests*; International Electrotechnical Commission: Geneva, Switzerland, 1999.
31. Launder, B.E.; Spalding, D.B. The numerical computation of turbulent flow. *Comput. Methods Appl. Mech. Eng.* **1974**, *3*, 269–289. [CrossRef]
32. Menter, F.R. Two equation eddy viscosity turbulence models for engineering application. *AIAA J.* **1994**, *32*, 1598–1650. [CrossRef]
33. Stoessel, L.; Nilsson, H. Steady and unsteady numerical simulations of theflow in the Tokke Francis turbine model, at three operating conditions. *J. Phys. Conf. Ser.* **2015**, *579*, 012011. [CrossRef]
34. Hosseinimanesh, H.; Devals, C.; Nennemann, B.; Guibault, F. Comparison of steady and unsteady simulation methodologies for predicting no-load speed in Francis turbines. *Int. J. Fluid Mach. Syst.* **2015**, *8*, 155–168. [CrossRef]
35. Menter, F.R.; Kuntz, M.; Langtry, R. Ten Years of Industrial Experience with the SST Turbulence Model. In *Proceedings of the 4th International Symposium on Turbulence, Heat and Mass Transfer*; Begell House Inc.: Redding, CT, USA, 2003; pp. 625–632.
36. Hasmatuchi, V.; Gabathuler, S.; Botero, F.; Münch-Alligné, C. Design and control of a new hydraulic test rig for small hydro turbines. *Hydropower Dams* **2015**, *22*, 54–60.
37. Chapallaz, J.M.; Eichenberger, P.; Fischer, G. *Manual on Pumps Used as Turbines*; MHGP Series Harnessing Water Power on a Small Scale; Informatica International, Incorporated: Singapore, 1992.
38. De Rose, A.; Buna, M.; Strazza, C.; Olivieri, N.; Stevens, T.; Peeters, L.; Tawil-Jamault, D. *Technology Readiness Level: Guidance Principles for Renewable Energy Technologies*; Technical report; EU Publications Office, European Commission: Luxembourg, 2017.
39. Pérez-Sánchez, M.; López-Jiménez, P.A.; Ramos, H.M. Modified Affinity Laws in Hydraulic Machines towards the Best Efficiency Line. *Water Resour. Manag.* **2018**, *32*, 829–844. [CrossRef]

sustainability

Article

A Numerical Investigation into the PAT Hydrodynamic Response to Impeller Rotational Speed Variation

Maxime Binama [1], Kan Kan [2,*], Hui-Xiang Chen [3], Yuan Zheng [1,2], Da-Qing Zhou [1,2], Wen-Tao Su [4], Xin-Feng Ge [2] and Janvier Ndayizigiye [1]

1 College of Water Conservancy and Hydropower Engineering, Hohai University, Nanjing 210098, China; binamamaxime@hhu.edu.cn (M.B.); zhengyuan@hhu.edu.cn (Y.Z.); zhoudaqing@hhu.edu.cn (D.-Q.Z.); janvier.ndayizigiye@outlook.com (J.N.)
2 College of Energy and Electrical Engineering, Hohai University, Nanjing 210098, China; gexinfeng@hhu.edu.cn
3 College of Agricultural Science and Engineering, Hohai University, Nanjing 210098, China; chenhuixiang@hhu.edu.cn
4 College of Petroleum Engineering, Liaoning Shihua University, Fushun 113001, China; wentaosu@lnpu.edu.cn
* Correspondence: kankan@hhu.edu.cn

Abstract: The utilization of pump as turbines (PATs) within water distribution systems for energy regulation and hydroelectricity generation purposes has increasingly attracted the energy field players' attention. However, its power production efficiency still faces difficulties due to PAT's lack of flow control ability in such dynamic systems. This has eventually led to the introduction of the so-called "variable operating strategy" or VOS, where the impeller rotational speed may be controlled to satisfy the system-required flow conditions. Taking from these grounds, this study numerically investigates PAT eventual flow structures formation mechanism, especially when subjected to varying impeller rotational speed. CFD-backed numerical simulations were conducted on PAT flow under four operating conditions (1.00 Q_{BEP}, 0.82 Q_{BEP}, 0.74 Q_{BEP}, and 0.55 Q_{BEP}), considering five impeller rotational speeds (110 rpm, 130 rpm, 150 rpm, 170 rpm, and 190 rpm). Study results have shown that both PAT's flow and pressure fields deteriorate with the machine influx decrease, where the impeller rotational speed increase is found to alleviate PAT pressure pulsation levels under high-flow operating conditions, while it worsens them under part-load conditions. This study's results add value to a thorough understanding of PAT flow dynamics, which, in a long run, contributes to the solution of the so-far existent technical issues.

Keywords: flow dynamics; numerical simulation; pressure pulsation; pump as turbine; rotational speed

Citation: Binama, M.; Kan, K.; Chen, H.-X.; Zheng, Y.; Zhou, D.-Q.; Su, W.-T.; Ge, X.-F.; Ndayizigiye, J. A Numerical Investigation into the PAT Hydrodynamic Response to Impeller Rotational Speed Variation. *Sustainability* **2021**, *13*, 7998. https://doi.org/10.3390/su13147998

Academic Editor: Alessandro Franco

Received: 21 June 2021
Accepted: 12 July 2021
Published: 17 July 2021

1. Introduction

The use of fossil fuels as main energy sources goes with a number of shortfalls, environmental disruption, and possible reserves for depletion, among others [1]. Fossil fuels are linked to the widely condemned greenhouse gas emission, i.e., gases which are believed to be at the source of climate change and other detrimental effects [2]. Of course, the non-renewable nature of fossil fuels itself poses a threat of resources exhaustion in the near or far future [3]. With the international will on reducing the CO_2 emissions, renewable energy sources (REN) have been advocated, and countries have taken action to not only involve RENs in their power systems, but to continue increasing RENs annual productions [4,5]. Among the so far available RENs, hydropower is one of the most mature REN technologies that dominates the market with a multitude of hydrosites already expanding the whole world over [6,7]. Hydropower technology is linked to different advantages when compared to other REN technologies, including their huge energy storage ability and operational flexibility, among others. Depending on different aspects, the targeted production and site-available potential among others, hydropower technology exploits natural waters from all sorts, such as seas, lakes, rivers, and estuaries. This also somehow leads to the classification of hydrosites as large- or small-scale hydrosites [8]. While large hydropower plants are generally known to feed large-size power grids such as the national or even international grids, small hydropower is mostly used under off-grid operational mode, contributing to the electrification of rural areas which not only provides electric lighting for households but also contributes to other economic activities such as the agricultural irrigation [9–11].

With the continued research on hydropower development, water supply and irrigation systems have been found to be another alternative for hydropower production on small scales [12]. In these systems, water flows continuously, thus presenting a stable hydropower potential. Besides, system excess energy in terms of pressure is usually wasted through pressure-reducing valves (PRVs) to avoid water leakages [13]. The installation of hydraulic turbines in these networks not only replaces the formerly utilized valves in regulating the system's energy levels but also uses that same energy to produce usable electric power [14–17].

Now, as also mentioned by different investigators, for small-scale hydropower (SHP) projects, the initial installation budget is the most expensive as the electromechanical equipment takes up a considerably high percentage of the total investments, in the range from 35% to 40% or even beyond [18,19]. In line with this, the utilization of normal hydraulic pumps in reverse mode, as an economic alternative to SHP installation budget, has recently attracted the attention of many REN field players [20,21]. In addition to its cost-effectiveness, pump as turbines (PATs) present different advantages over comparable mini-scale hydroturbines, including the easy availability of the machine itself and spare parts, structural simplicity, easy maneuverability, short delivery time, and a long life span [22,23].

Nevertheless, PAT technology has also shown different shortfalls. Among others, the so far controversial conversion of pump-mode performance characteristics to turbine mode is the most researched, where the so far available analytical conversion methods present errors in the range of $\pm20\%$ [24]. Scientific research on this matter still goes on as of today. For more about PAT turbine mode performance predictions, review articles by Jain et al. [24], Elbatran et al. [25], Adu et al. [26], and Ntiri et al. [27] are recommended.

Another problem about PAT technology resides in its structural design that does not necessarily meet the energy production requirements. This aspect itself plays a big role in the above-mentioned difficulty in PAT's turbine mode performance prediction. To be more specific, the pump impeller blades are curved in the opposite direction of its rotation (backward). This blade design becomes forward-curved when the flow direction changes in turbine mode, thus serving the source to large flow separations and associated hydraulic losses as a result of flow impingement on the blade's pressure side in the vicinities of its leading edge. In addition, hydraulic pumps are not necessarily provided with flow control mechanism. Therefore, the machine is unable to regulate the flow rate or the water flow incidence angle to satisfy the demanded operating conditions. This also explains its inability to handle off-design operating conditions. Therefore, PATs have been found to optimally function in a tiny range of flows around the best efficiency point (BEP), while they operate poorly under off-design conditions [28]. This constitutes one of the biggest shortfalls of PAT technology and has attracted the attention of many researchers, mainly targeting the understanding of PAT flow dynamics and subsequently attempting to widen its optimum operating range. Among the recently published findings, one would cite a study conducted by Štefan et al. [29] on PAT internal flow field under both pump and turbine modes. In total agreement with the literature, the turbine mode was found to attain its BEP under a higher head and flow rate, as compared to the pumping mode (41% and 27%, respectively), while the efficiency was slightly lower ($\eta_T = 0.96\,\eta_P$). Moreover, a steep drop on the PAT efficiency curve has been noticed for deep part-load operating conditions, supposedly owing to the intensification of swirling flows within the PAT's outlet pipe. Based on the fact that, for fluid machinery, adverse flow phenomena mostly occurring under unstable conditions led to increased energy loss rates, Lin et al. [30] has used the entropy dissipation theory to study PAT operating characteristics under off-design operating conditions. As expected, unsteady flow phenomena such as the backflows, shock flows, and vortical flows, have inflicted higher entropy and energy dissipation, where flow zones with the highest energy losses have been pointed out as the vicinities of the blade leading edge, the cavity zone (impeller's back and front chambers), and volute throat. In the same respect, Guan et al. [31] and Morabito et al. [32] found that the position and geometry of the volute throat may influence the machine energy loss characteristics and its efficiency.

More studies have also been conducted to try and decrease PAT hydraulic losses through different technics, geometry modification, and optimization, among others. To say the least, Ghorani et al. [33,34], after finding that irreversible hydraulic losses associated with flow shock at the blade leading edge and other secondary flow phenomena such as backflow and vortices take place within PAT flow zones, have opted to optimizing the PAT impeller geometry, leading to substantial reduction in the above-stated unsteady flow phenomena and a corresponding increase in the machine hydraulic efficiency for different operating conditions (3.8% and 5.72% for Q_{BEP} and 1.3 Q_{BEP} flow conditions, respectively). In the same respect, Wang et al. [35] mentioned that one of most cost-effective ways to improve PAT's turbine mode performance would be by redesigning its blade geometry without touching other components. Therefore, they designed a special impeller with forward-swept blades and compared its performance with the original back-swept bladed one. The special impeller, owing to the associated considerable reduction in impeller hydraulic and shock losses, exhibited a flatter efficiency-flow rate curve (η-Q curve), thus promising a larger range of optimum operating conditions around the BEP (1.5% efficiency variation for flow conditions within the range from 0.9 Q_{BEP} to 1.2 Q_{BEP}). Moreover, a 7.9% increase in the maximum attainable efficiency was achieved, thus confirming the special impeller's superiority over the original one. In the same respect, Sengpanich et al. [36] proposed "an impulse pump-as-turbine" using an injector as a flow control mechanism. This technique also assured a continuously high efficiency for a wide range of flows.

Coming back to the application of PAT technology in water distribution networks (WDNs), different studies have so far been carried out [25,37–40], which concern the above-discussed PAT technology-associated problems. Consequently, the current technology readiness level (TRL) for PAT technology is estimated at TRL-4 [41,42]. The utilization of PATs in WDNs must consider the variability of pressure and flow rate as imposed by the system, due to the instantaneity in water demands [43–45]. In this respect, Carraveta et al. [46–48] have developed a so-called "variable operating strategy" or VOS, with the aim of efficiently using PAT technology to regulate the WDN energy. This strategy consists of two distinct regulation methods, namely hydraulic regulation (HR) and electric regulation (ER), though they can also be combined to give new regulation features. The HR method consists of the usage of PAT in a series-parallel circuit with PRVs where the system energy regulation is mainly achieved by just acting on the utilized PRVs. The valve in series with PAT is used to dissipate the surplus energy leading to a lower system flow as compared to PAT's BEP flow ($Q_S < Q_T$), while the valve in parallel with PAT bypasses a quantity of flow thus shifting the system operating point to higher flows ($Q_S > Q_T$). As for the ER method, an electric inverter is used to regulate the impeller rotational speed (IRS) where, for instance, one would increase the IRS to shift the system operating point to lower flows, or decrease it to obtain the system operating point to higher flow conditions. In this method, however, owing to a comparatively narrow range of system operating conditions between the runaway (T = 0 Nm) and resistance (RRS = 0 rpm) characteristic curves, high IRS values may lead to a system operating point where the runner rotational speed is close to the runaway value, possibly causing PAT's poor performance. To avoid this situation, a series-PRV can be used to shift the IRS to lower values, while respecting the system-required operating point. VOS functionality discussions and applications have been presented through a number of studies by Fecarrota et al. [49] and Stefanizzi et al. [41], among others.

Specifically, the utilization of variable-speed PATs in WDNs as a means to control the flow while keeping a good system operational efficiency has so far been researched by different authors, including Delgado et al. [50], Lima et al. [51], and Tahani et al. [52], among others. In most of these studies, the main focus has been the optimization of selection, position, and utilization of variable-speed PATs, targeting the improved WDN energy regulation and efficient energy recovery. On the other hand, a few more studies have been conducted on the effect of PAT speed variation, but considering its application to normal hydropower production within small-scale hydroelectric plants, where the machine external performance characteristics have been the focus [53,54]. What is also worth mentioning is a study on the effect of PAT rotational speed variation on tip leakage vortex (TLV) dynamics under pump mode, as conducted by Han et al. [55], where the TLV was found to worsen with an increase in PAT impeller rotational speed. To the authors' best knowledge; literature is so far quite limited on the impact of impeller rotational speed variation on PAT flow dynamics under turbine operating mode.

Therefore, by targeting a thorough understanding of PAT flow dynamics under these circumstances, especially considering PAT's off-design operating conditions, the present article presents a numerical investigation on the formation mechanism of PAT flow structures, and the effect of impeller rotational speed variation on the same. CFD-backed numerical simulations have been conducted on PAT's 3D turbulent flow under optimal condition and three part-load conditions (1.00 Q_{BEP}, 0.82 Q_{BEP}, 0.74 Q_{BEP}, and 0.55 Q_{BEP}), where five impeller rotating speeds, namely 110 rpm, 130 rpm, 150 rpm, 170 rpm, and 190 rpm, were considered.

The investigated model is a mixed flow PAT with a three-bladed impeller and a flow control system of seven guide vanes. This article is composed of three main sections. Section 1 provides an extensive background on PAT technology and the associated techno-scientific issues, as well as the so far attained level of knowledge. Section 2 describes both the investigated case and utilized methods. Section 3 provides a thorough discussion on the present study's findings, which finally led to a number of concluding remarks.

2. Materials and Methods

In order to achieve this study's objective, a roadmap was designated and followed, as shown by the flowchart in Figure 1. The PAT prototype was first experimentally tested under various operating conditions to finally obtain the pump as turbine's external performance characteristics in terms of Head (H) and Efficiency (η).

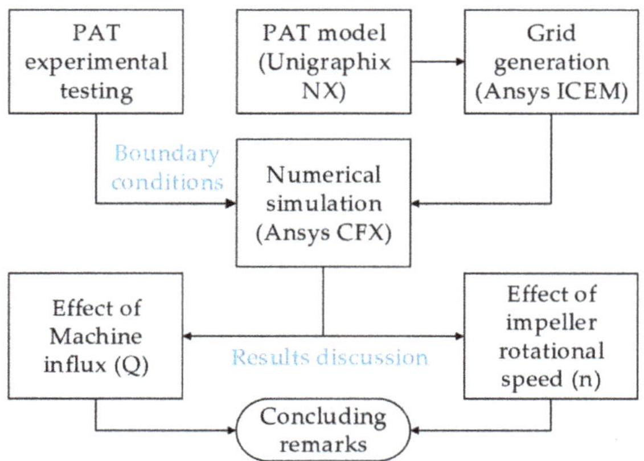

Figure 1. Study roadmap.

Then, a PAT geometrical model was built, from which a predominantly hexahedral computational grid was later generated. The resultant grid was taken to ANSYS CFX-Solver (developed by Ansys, Inc., an American company based in Canonsburg, Pennsylvania) for PAT flow simulation under various operating conditions, as also mentioned in the last sections of the article.

During simulation, the experimental test-extracted data were used as system boundary conditions, and the utilized numerical scheme validation was later performed by comparing experimental and numerical simulation results. Finally, farther study results in terms of machine flow and pressure field characteristics with gradually changing machine influx (Q) and impeller rational speed (n) were extracted, and deeply analyzed, leading to a number of concluding remarks.

2.1. PAT Geometry

The research object of this study is a mixed-flow pump as turbine (PAT) composed of four main components, namely the inlet pipe (IP), the impeller, the guide vanes (GV) zone, and the outlet pipe (OP). The impeller had 3 blades (Z_R = 3) while the guide vanes zone was composed of 7 guide vanes (Z_G = 7). Both the inlet and outlet pipes were split type tubes. The model's specific speed (N_Q) was 163.9 min^{-1}. The impeller diameter was 320 mm, while its outlet diameter was 300 mm. The guide vanes inlet diameter was 350 mm, and there was no existent blade tip clearance. A corresponding computational domain was built and simulated using Unigraphix NX8.0 (owned by Siemens PLM, an American company based in Plano, TX, USA) and Ansys CFX softwares, respectively. Figure 2 shows the utilized PAT computational domain and its components.

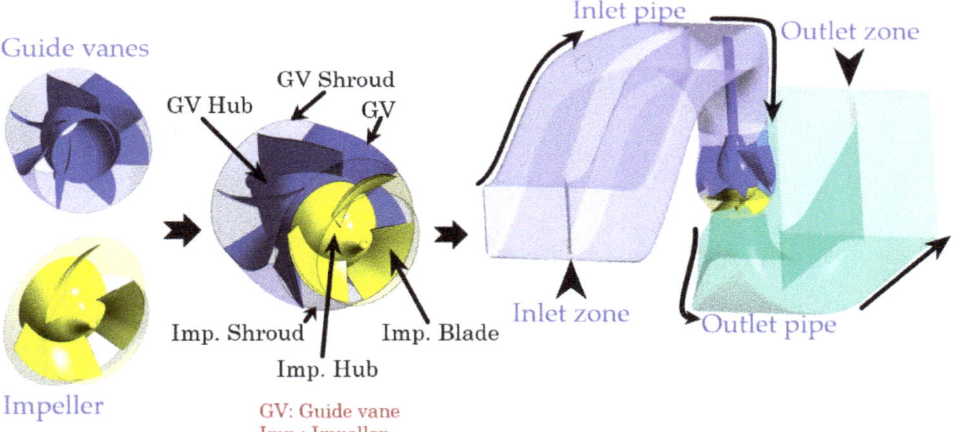

Figure 2. PAT computational domain and components.

2.2. Grid Generation

Grid generation constitutes one of very important phases of the numerical simulation process. The generated grid's quality is a very crucial parameter, as it plays a fundamental role on different aspects, numerical solution convergence, and accuracy within numerical simulation results, among others. Therefore, great attention is always paid to this phase. Depending on different factors such as the geometry complexity or the nature of simulated system's operations, grid characteristics can be varied at different flow zones.

For the case of an axial-flow PAT, which is this study's case, flow separations are likely to take place in rotating parts and the adjacent flow zones. Therefore, for the sake of accurately capturing the flow dynamics at those zones, a finer grid was created around the impeller blades and guide vanes, leading to a global y+ value less than 50. Ansys

ICEM software was used to generate this grid, where the hexahedral mesh type was utilized for all of the components of the computational domain. In order to avoid the possibility of numerical simulation results being influenced by the utilized grid number, a grid independence test was carried out where six different grid numbers ranging from 4 to 8 million were tested. Having been simulated under the same boundary conditions, different grid numbers have led to different values of PAT's hydraulic head. Table 1 displays the head variation with the increasing grid number from 4 to 8 million elements. It can be seen that the head increased with the grid number until 7.2 million, after which the head value approximately stabilized. Therefore, in line with the available computational resources, a grid number of 7.2 million was chosen for further transient simulations. Details of the selected grid are presented in Table 2.

Table 1. Hydraulic head variation with different grid numbers.

Testing Parameter	Grid No. 1 (3.9 M)	Grid No. 2 (4.63 M)	Grid No. 3 (5.84 M)	Grid No. 4 (7.19 M)	Grid No.5 (7.8 M)	Grid No.6 (8.1 M)
Head (m)	6.3	7.53	8.26	9.2	9.21	9.19

Table 2. Selected grid details for each PAT component.

Grid Details	Inlet Pipe	Impeller	Guide Vanes	Outlet Pipe
Grid type	Hexahedral	Tetrahedral	Hexahedral	Hexahedral
Grid number (Million)	1.2	2.51	2.17	1.31
Orthogonal Quality (0–1)	0.45	0.32	0.4	0.4
Skewness (0–1)	0.45	0.3	0.4	0.35

2.3. Numerical Simulation Scheme

The water flow that passes through a PAT's full computational domain is three dimensional. This study considers the working fluid as the water at 25 °C. It is considered incompressible, isothermal, and is governed by the Reynolds-averaged Navier-Stokes equations, otherwise known as RANS. The RANS-associated continuity and momentum equations are shown in Equations (1) and (2).

Continuity equation:

$$\frac{\partial}{\partial x_i}(\rho u_i) = 0 \tag{1}$$

Momentum equation:

$$\frac{\partial}{\partial t}(\rho u_j) + \frac{\partial}{\partial x_j}(\rho u_j u_i) = -\frac{\partial p}{\partial x_j} + \frac{\partial \tau_{ij}}{\partial x_j} + S_{ij} \tag{2}$$

In these equations, u_i and u_j stand for velocity components, τ_{ij} stand for shear stress, S_{ij} for the source term, p for pressure, and ρ for density. In this study, the shear stress transport (SST) computational model was selected for the closure of the RANS equations [56]. This model combines the advantages of the renowned k-ε and k-ω models [57,58], thus giving it superiority over both of them. This is due to its ability to deliver high-accuracy predictions for flow dynamics in the wall vicinities and flow zones away from it, especially when solving for study cases involving adverse pressure gradients and flow separations [59]. Equations (3)–(5) are the mathematical expressions of SST's turbulent kinetic energy, its specific dissipation rate, and the kinematic eddy viscosity, respectively.

$$\frac{\partial k}{\partial t} + u_j \frac{\partial k}{\partial x_j} = P_k - \beta^* k\omega + \frac{\partial}{\partial x_j}\left[(\nu + \sigma_k \nu_T)\frac{\partial k}{\partial x_j}\right] \tag{3}$$

$$\frac{\partial \omega}{\partial t} + u_j \frac{\partial \omega}{\partial x_j} = \alpha S^2 - \beta \omega^2 + \frac{\partial}{\partial x_j}\left[(v + \sigma_\omega v_T)\frac{\partial \omega}{\partial x_j}\right] + 2(1 - F_1)\sigma_{\omega,2}\frac{1}{\omega}\frac{\partial k}{\partial x_i}\frac{\partial \omega}{\partial x_i} \tag{4}$$

$$v_T = \frac{a_1 k}{\max(a_1 \omega, S F_2)} \tag{5}$$

The 3D numerical simulation of PAT flow under different operating conditions is conducted using the CFD's commercial code Ansys CFX 18.0. Simulations were set in such a way that the water flow entered the computational domain at the inlet zone of the inlet pipe (see Figure 1), then passed through the guide vanes inter-spaces before whirling through the impeller inter-blade channels, and flowed down to the outlet pipe to finally exit the computational domain through the outlet pipe's outlet zone.

For the sake of a quick solution convergence, steady state numerical simulations were initially run, where their results files served the initial states for the followed transient simulations. A constant mass flow (kg s^{-1}) was imposed at the inlet of the computational domain as the inlet boundary condition, while a gauge pressure (Pa) was selected as the outlet boundary conditions. Different interfaces were utilized between components of the computational domain, depending on the location and simulation type. The frozen rotor interface was used at both ends of the impeller under steady state numerical simulations, while it was changed to transient rotor-stator interface for transient numerical simulations. The general grid interface (GGI) was used between the stationary components for both simulation types. All the remaining solid walls were given a non-slip boundary condition.

The length of a single simulation session was set to ten impeller revolutions, where the selected timestep (Δt) was equivalent to the impeller's one degree rotation ($1°$), which therefore means that a single impeller revolution was equivalent to $360 \times \Delta t$. Five internal loops were selected for each timestep. The solution convergence criteria was chosen as the root mean square of residuals inferior to 1×10^{-4}. For the sake of checking the trustworthiness of generated simulation results by the above-presented numerical scheme, experimental and numerical simulation results in terms of machine external performance characteristics were compared.

Experiments were conducted on a PAT test rig, as shown in Figure 3. In this figure, the test rig components are shown on the right side, and their respective positions on a fully assembled PAT system are shown by letters A, B, and C. The test rig had the highest head (H) and flow rate (Q) of 100 m and 1000 L/s, respectively. The adoptable impeller speed range was 0 to 1500 rpm, which could accommodate impeller diameters in the range from 250 to 460 mm.

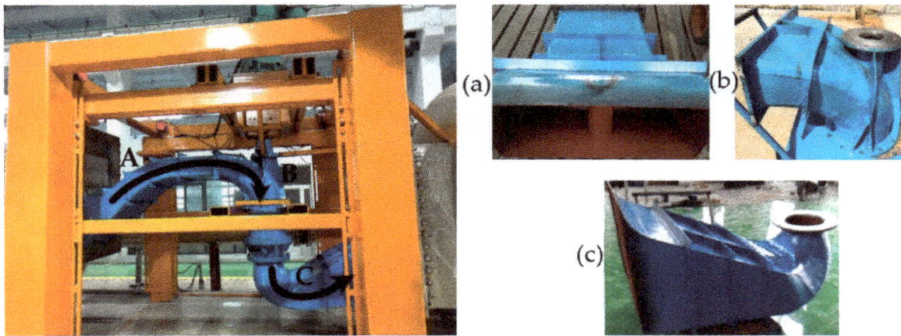

Figure 3. PAT Experimental test rig and components (**a**) Inlet pipe (**b**) Shaft casing (**c**) Outlet pipe.

As shown in Figure 4, a global error of less than 8% and 6.5% for head and efficiency predictions was recorded, respectively. The overestimation of experimental results by the utilized numerical scheme is believed to take source for different reasons, including the computational geometry simplification, among others. Nevertheless, the utilized numerical scheme showed the ability to generate reasonably accurate results, thus proving its trustworthiness.

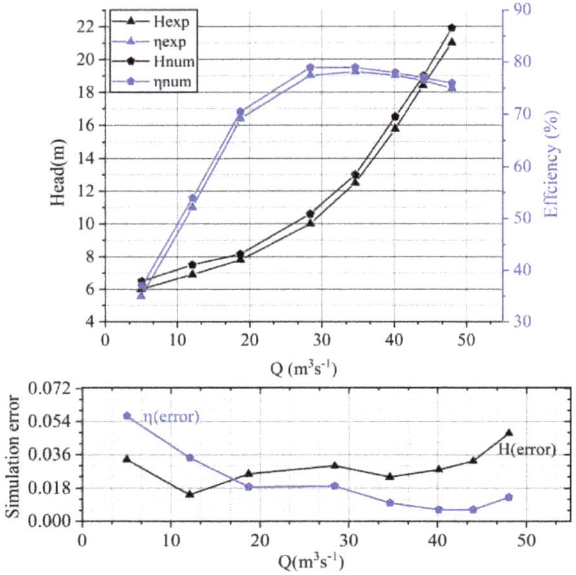

Figure 4. Comparison of test and numerical results.

3. Results

3.1. Flow Field Characteristics Evolution within the Impeller and Guide Vanes

Though the experimental tests were conducted on a wide range of flows, this study has only considered four operating conditions, namely 1 Q_{BEP}, 0.82 Q_{BEP}, 0.74 Q_{BEP}, and 0.55 Q_{BEP}, considering a constant impeller rotational speed of 150 rpm. This was carried out with a target of investigating the eventual changes in machine flow dynamics for a gradually decreasing machine influx. To do so, three consecutive and equidistant turbo-surfaces were selected in the streamwise direction from the impeller inlet to the outlet.

Considering the impeller inlet and outlet positions as 0 and 1, respectively, the 1st plane (S1) was at 0.2, the 2nd (S2) at 0.5, and the 3rd (S3) at 0.8. Each of the three planes expanded from the impeller hub to the shroud. During the investigation of local flow parameter, changes took place as the machine flow rate decreased, and the above-mentioned surfaces were taken as the reference flow zones within the impeller. In this respect, an attempt was completed to showcase some of these changes through both Figures 5 and 6.

Figure 5 shows the pressure distribution mode and associated flow pattern within the impeller flow zone under 0.55 Q_{BEP}. In this figure, from left to right, are plane S3 (impeller outlet vicinal zone), plane S2 (impeller most central zone), and plane S1 (impeller inlet vicinal zone).

It can be seen that the flow region in the vicinities of the impeller inlet (Plane S1) was marked by considerably low-pressure zones on the blade pressure side and high-pressure zones on the blade's suction side. This, as confirmed by Figure 5b, is linked to the occurred flow separations on the blade pressure side around the blade leading edge under this condition. However, low-pressure zones and corresponding flow separations weakened in the downstream direction. The evolution of local pressure distribution mode, and associated flow structures as the machine influx decreased, are presented in Figure 6. In this figure, the selected plane is S2 which is, as already mentioned, located in the impeller's most central zone (blade mid-length zone). As presented in Figure 6 left side, high-pressure zones were located on the blade pressure surface (BPS) in the vicinal flow zones of the impeller shroud. On the other hand, low-pressure zones were located on each blade's suction side (BSS), in the shroud's vicinal flow zones.

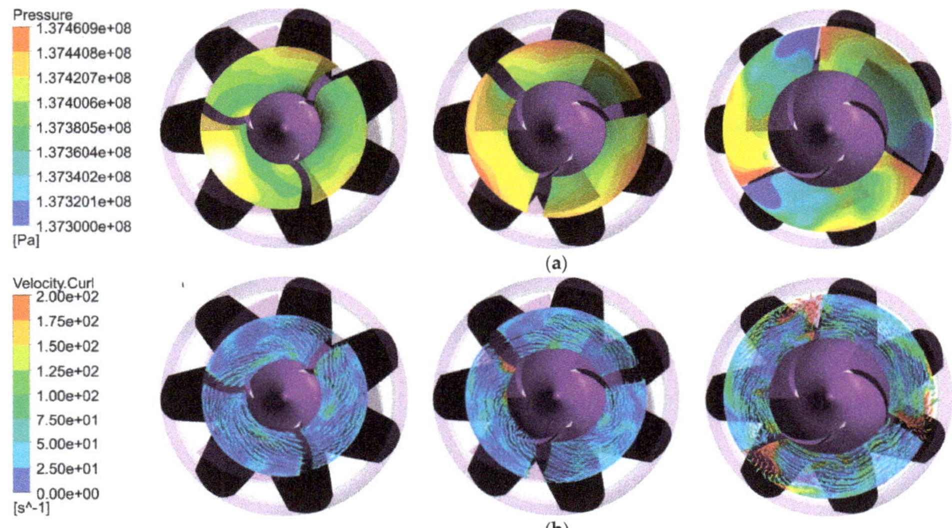

Figure 5. Impeller flow field characteristics at different planes in the streamwise direction for 0.55 Q_{BEP} condition (**a**) pressure distribution contours (**b**) velocity curl-colored flow streamlines.

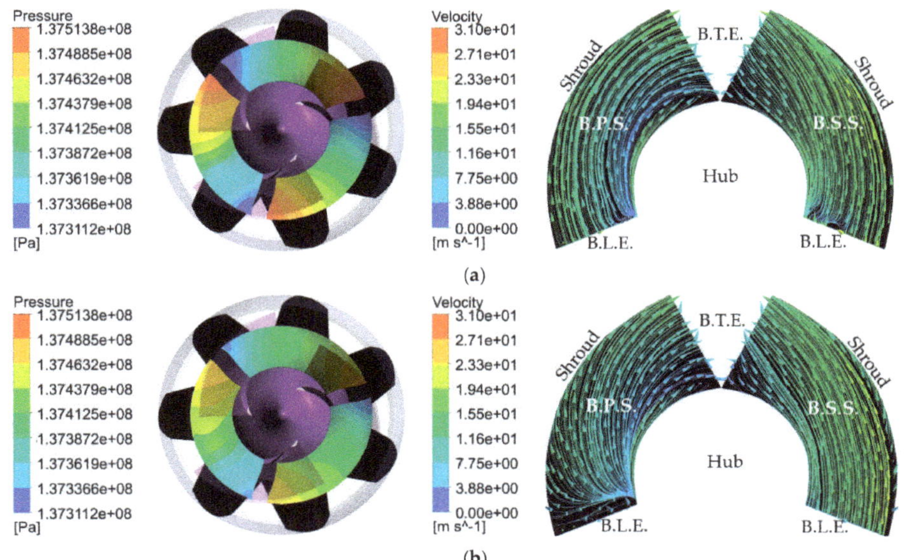

Figure 6. *Cont.*

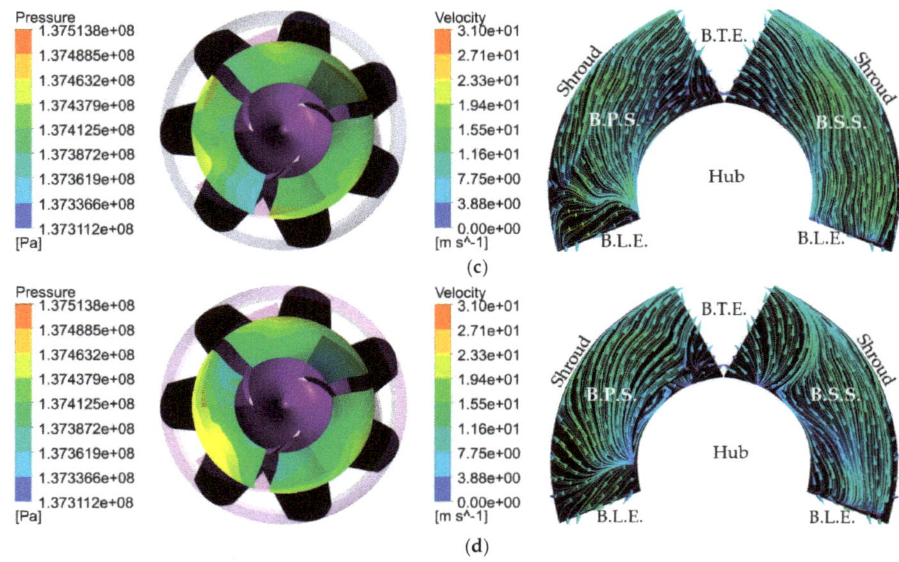

Figure 6. Evolution of Impeller flow field characteristics (**a**) 1.00 Q_{BEP} (**b**) 0.82 Q_{BEP} (**c**) 0.74 Q_{BEP} and (**d**) 0.55 Q_{BEP}.

With the machine influx decrease from 1.00 Q_{BEP} to 0.55 Q_{BEP}, local pressure on the BSS was found to gradually increase, while the BPS pressure correspondingly decreased. The right side of Figure 6 presents the evolution of velocity-colored flow streams on the blade's two surfaces (BPS and BSS). Keeping in mind that the plane in the left hand of Figure 6 was located in the impeller's most central zone, the velocity-colored streams at the same zone showed high and comparatively low speed flows on the BSS and BPS, respectively. This explains the presence of low- and high-pressure zones on respective surfaces, as presented in Figure 6 left side. In addition, the level of flow unsteadiness was found to gradually worsen with the machine influx decrease. Under 0.82 Q_{BEP} conditions, there was an emergence of flow separations within the vicinal flow zones of the blade leading edge (BLE), especially on the BPS.

Under 0.74 Q_{BEP} conditions, the area occupied by BLE flow separations widened up, especially in the shroud vicinities, and extended towards downstream, eventually reaching the blade's mid-length zone. In addition, another vortex flow zone appeared at the blade trailing edge (BTE) within the hub region and vicinities. Under this condition, the flow on the BSS also became slightly disturbed, especially in the hug region towards the BTE. Under 0.55 Q_{BEP} conditions, the BLE flow separation became even wider, while the BTE one almost did not grow. However, flow vortices on the BSS grew far larger, both at the BLE's hub region and the BTE's whole zone from hub to shroud, probably causing the blockage of the flow passage in the blade's vicinal zone. It is therefore obvious that the impeller flow field becomes more and more disturbed with the decreasing machine influx, where flow separations in form of shock and vortical flows emerge and eventually grow. It was found that these secondary flow structures are more likely to take place at the impeller blade's leading edge and flow zones in the vicinities of the impeller shroud, respectively.

As for the flow characteristics in the guide vanes region, Figure 7 shows the eventual changes in local flow structures as the machine influx decreased. The left side shows the velocity-colored streams at different flows. It can be seen that, as is the case with the impeller flow zone, the guide vanes flow pattern becomes messy when the flow rate decreases, and where vortical structures are emerge and eventually grow at certain flow zones, especially the vaneless space zone and the vane pressure side's shroud vicinities. For high-flow conditions (1.00 Q_{BEP} and 0.82 Q_{BEP}), flow vortices in shroud vicinities occupy a tiny area (D), while flow wakes (E) formed at the vane's trailing edge expand to the vaneless space.

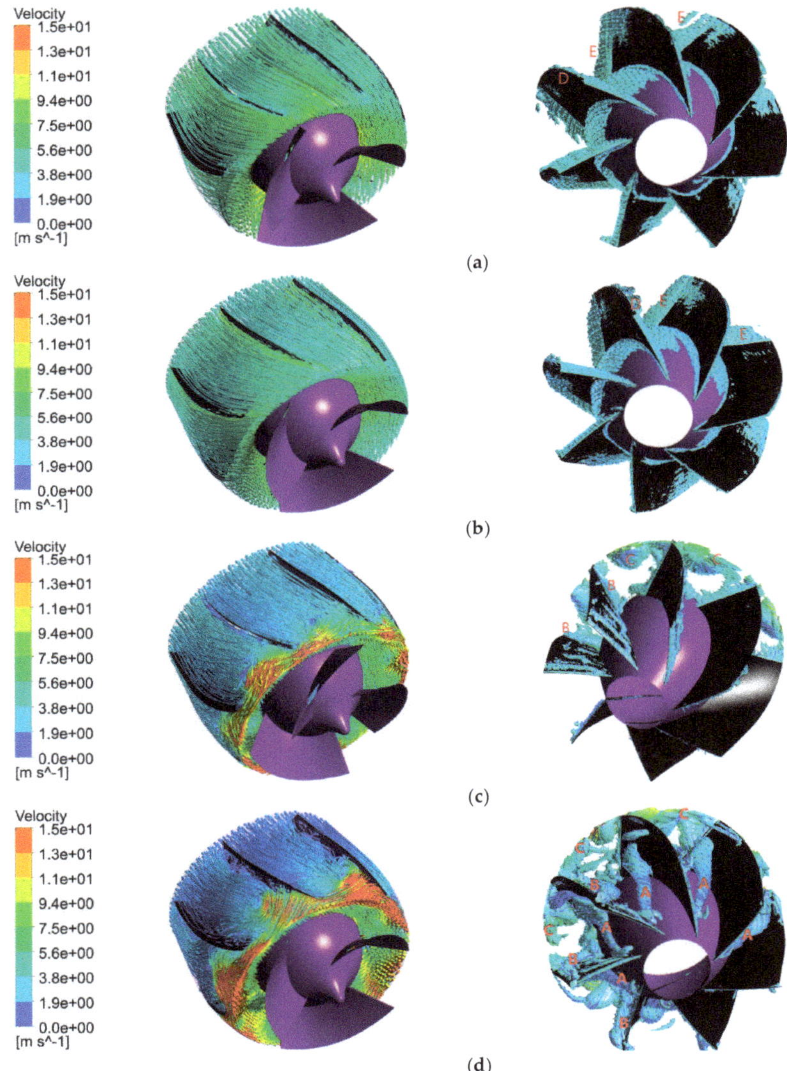

Figure 7. Guide vane flow structures evolution for different operating conditions (**a**) 1 Q_{BEP}, (**b**) 0.82 Q_{BEP}, (**c**) 0.74 Q_{BEP}, and (**d**) 0.55 Q_{BEP}.

Under 0.74 Q_{BEP}, there was a new tubular vortex that emerged in the vicinity of the guide vane trailing edge, expanding spanwise from hub to the shroud side (B). In addition, another vortex structure appeared near the guide vane's outlet zone, within the guide vane shroud region (C). Under deep part-load condition (0.55 Q_{BEP}), type C vortex grew even larger, occupying the flow zones in the layer below, while type B grew bigger and slightly shifted its location towards upstream zones. In addition, a new and large vortical structure (A) developed from the guide vane's leading edge on its suction side, and travelled downstream obliquely, finally touching the next vane's pressure side in the vicinities of its trailing edge. The Type A vortex occurred mostly within the hub region.

It can therefore be briefly concluded that the guide vane flow pattern deteriorates with the decreasing flow where most of vortical structures form in the close proximity of the vaneless space.

3.2. Effect of Impeller Rotational Speed Variation

In line with this study's main objectives, an effort was also made to investigate the effect of impeller rotational speed variation on the above discussed PAT flow dynamics under part-load conditions. In this respect, 0.74 Q_{BEP} was selected as the tested part-load flow condition where five impeller rotational speed ranging from 110 to 190 rpm were considered. For the sake of investigating the occurred changes in impeller flow field characteristics as its rotational speed increased, a mid-span plane (Span = 0.5) was taken as a reference flow zone.

Figure 8 displays the evolution of local pressure distribution within the impeller, as the impeller rotational speed gradually increased. It is generally found that low-pressure zones are located on BLE's pressure side and BSS towards the trailing edge. With a continuously increasing impeller rotational speed, low-pressure zones grew increasingly larger at both mentioned zones, leading to a 190 rpm situation where they can then obviously extend to the draft tube flow zone (adjacent flow zone). In addition, high-pressure zones are found to be on the BLE's suction side and continuously enlarge within the increase in impeller rotational speed.

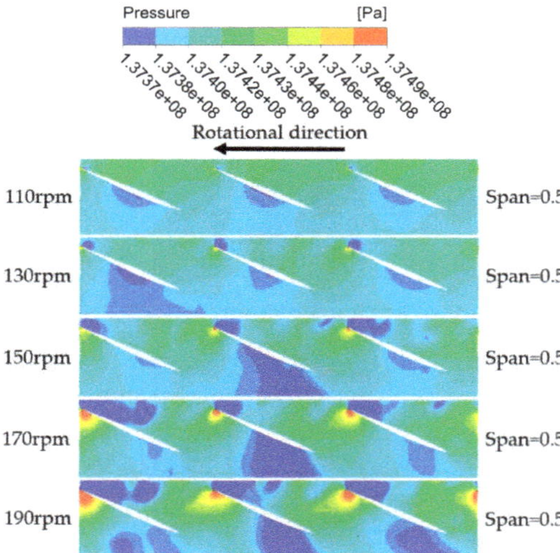

Figure 8. Local pressure variation with the impeller rotating speed under 0.74 Q_{BEP} flow condition.

In Figure 9, velocity-colored flow streamlines and correspondent turbulence kinetic energy distribution on the BPS are presented. Having considered a part-load condition (0.74 Q_{BEP}), the above-discussed flow separations were already present for the lowest of five tested speed values (110 rpm). Those were the BLE's pressure side shock flow, its extension downstream in shroud vicinities, and the BTE's hub vortex. With the increasing impeller rotational speed, the above-cited vortical structures strengthened and occupied even wide areas. This aspect is confirmed through Figure 9 right side where the evolution of BPS turbulence kinetic energy with the increasing speed is presented. It is shown that a high-BPS turbulence kinetic energy zone was located at the BLE close to the shroud under 110 rpm condition.

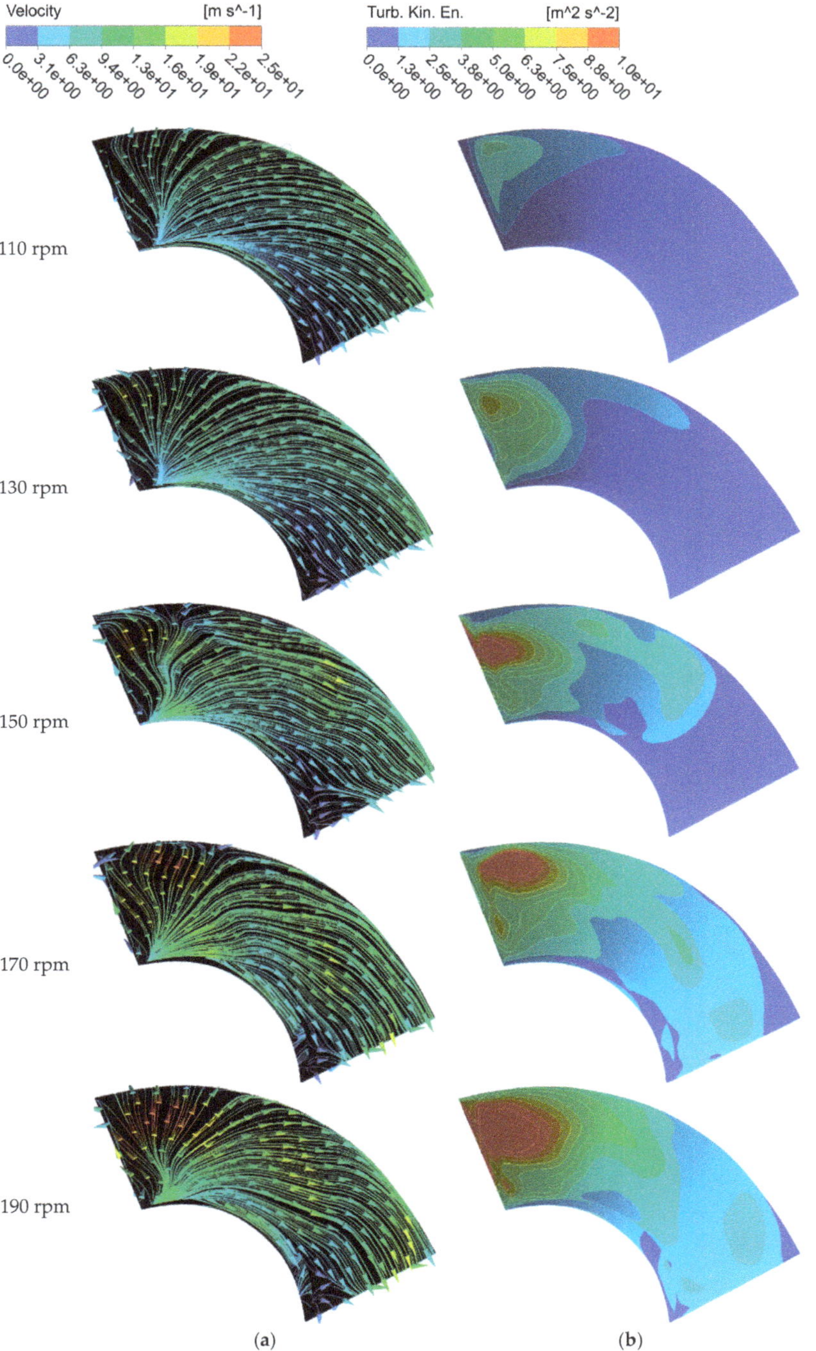

Figure 9. Variation in flow patterns under 0.74 Q_{BEP} operating condition (**a**) BPS velocity-colored flow streamlines (**b**) BPS turbulence kinetic energy contours.

With the increasing rotational speed, this zone gradually widened up in both the spanwise and streamwise directions, reaching the blade mid-length zone under 190 rpm condition. It is therefore obvious that, for a PAT already operating under part-load conditions, the increase in impeller rotational speed worsens the level of flow instability.

3.3. Pressure Pulsation Characteristics

In order to study the evolution of pressure pulsation characteristics and its distribution mode for the above-discussed operating conditions, different flow zones were selected, at which pressure-monitoring points were positioned. To be more specific, nine pressure monitors were positioned on the mid-span plane of the inter-blade flow zone, as shown in Figure 10a. In addition, considering three consecutive span-wise planes from hub to the shroud, 18 monitors were positioned on the BPS, where each of the planes held six equidistant monitors expanding from the blade's leading edge to the trailing edge as shown in Figure 10b.

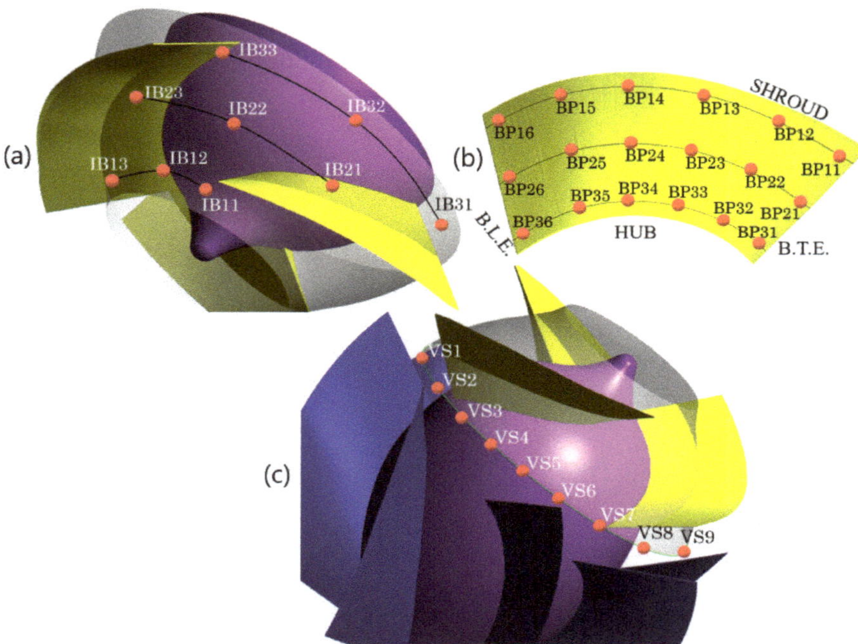

Figure 10. Pressure monitoring point locations. (**a**) Inter-blade channel (**b**) Blade pressure side and (**c**) Vaneless space.

Six monitors, BP16 to BP11, were located within the shroud's vicinal zone. The other six, BP26 to BP21, were positioned on the mid-span plane, and the last six were located in the hub's vicinal zone. Finally, 24 equidistant monitors were positioned around the whole vaneless space between the impeller and the guide vanes as shown in Figure 10c.

Figure 11 shows the pressure pulsation characteristics in the vaneless space zone for the four investigated flow conditions. The VS pressure pulsation spectra were found to be dominated by 7 *fn* component for both BEP and 0.82 Q_{BEP} conditions (*fn*: impeller rotational frequency).

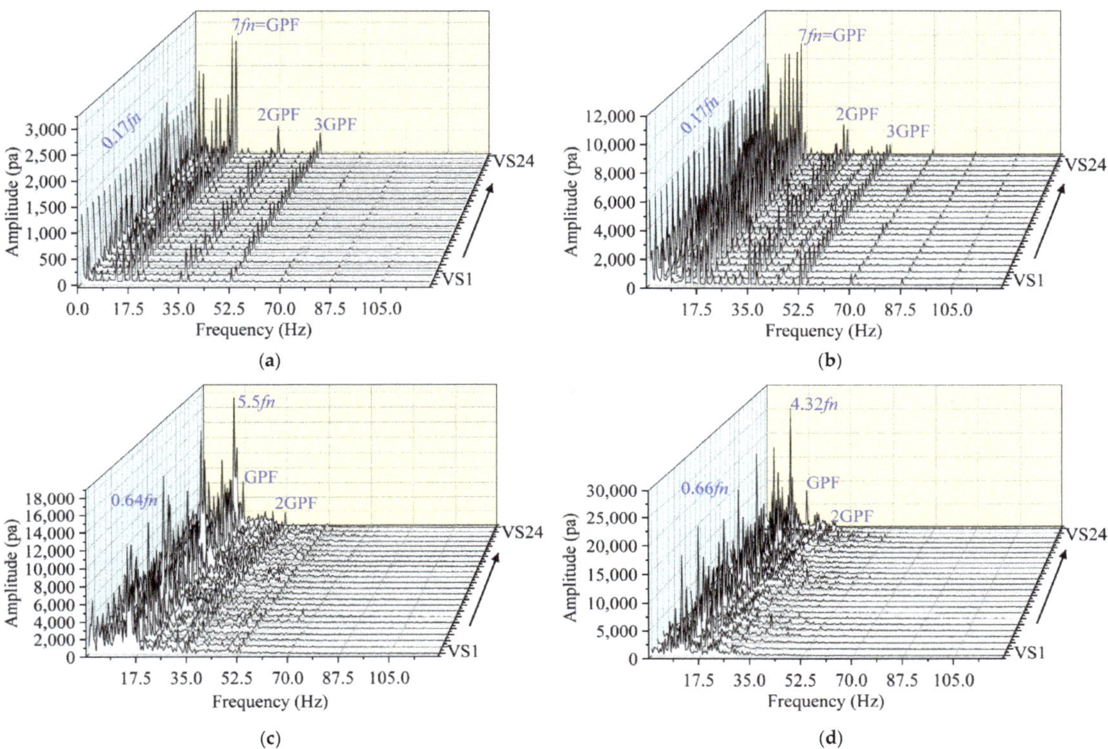

Figure 11. Vaneless space pressure pulsation spectra under different flow conditions (**a**) 1.00 Q$_{BEP}$ (**b**) 0.82 Q$_{BEP}$ (**c**) 0.74 Q$_{BEP}$ and (**d**) 0.55 Q$_{BEP}$.

Considering the impeller rotational speed of 150 rpm, this component represented the effect of rotor-stator interaction (RSI), known as the guide vane passing frequency (GPF = $Z_G \times fn$). Under BEP conditions, the VS pressure pulsation spectrum was marked by GPF as the 1st dominant frequency, followed by a low-frequency component (LFC) 0.17 *fn*. Moreover, more LFCs, which are multiples of *fn*, emerged between these two components (0.17 *fn* and GPF), adding to GPF harmonics which are considered high-frequency components. When the machine influx decreased from 1.00 Q$_{BEP}$ to 0.82 Q$_{BEP}$, the above-described components persisted, but their amplitudes considerably increased, reaching to pulsation amplitude values four times higher.

Considerably remarkable changes in VS pressure pulsation characteristics were again recorded when the machine flow conditions reached 0.74 Q$_{BEP}$. Under this condition, unlike the first two conditions, the dominant frequency component shifted to LFC class, where 5.5 *fn* serves as the 1st dominant frequency component, followed by 0.64 *fn* as the 2nd dominant frequency. The RSI-born components are still present but with comparatively small amplitudes.

It is also worth mentioning that, compared to previous flow conditions, there was a global increase in pressure pulsation amplitudes, and a multitude of more LFCs emerged. This corresponds to increased local flow unsteadiness (LFU), as also discussed in the above sections. Under 0.55 Q$_{BEP}$ condition, pulsation characteristics from the precedent flow condition persisted where, however, pulsation amplitudes continued to grow, and the 1st dominant frequency shifted to 4.32 *fn*, which is still classified as the LFC, while 0.66 *fn* served the 2nd dominant frequency.

In brief, the vaneless space pressure pulsation characteristics were found to be consistent with PAT flow dynamics evolution as the machine influx decreased. Under the first two high flow conditions, PAT flow experienced low-level of unsteadiness. Therefore, correspondent vaneless space pressure pulsation spectra were dominated by RSI-born frequency components.

Moreover, the number of LFCs was still low under these two conditions, which means a low level of local flow instability. On the other hand, the last two low flow conditions were marked by the dominance of LFCs and the increase in their numbers.

Though the RSI-born components were still present, their amplitudes were comparatively very low. This coincides with the increase in local flow unsteadiness under these conditions, where the impact of RSI on VS pressure pulsation characteristics became negligible, as compared to that inflicted by local flow instabilities.

In the same respect, an attempt was made to investigate the evolution of pressure pulsation levels in the vicinities of the impeller blade pressure surface for different flow conditions, as shown in Figure 12. This was performed using the relative pressure pulsation coefficient *CP*, for which the mathematical expression is presented in Equation (6).

$$CP = \frac{(P_{imax} - P_{imin})}{\rho g H} \tag{6}$$

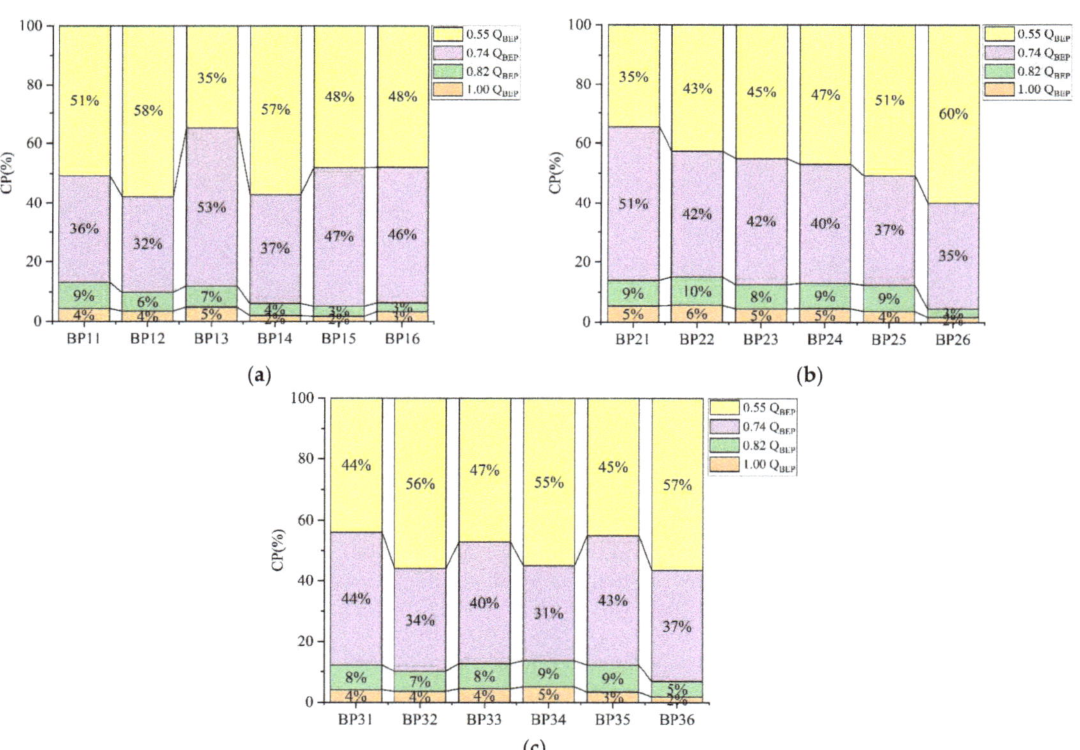

Figure 12. Pressure pulsation extent in BPS vicinities at three span-wise planes under different flow conditions (**a**) Shroud vicinal zone (**b**) Mid-span zone and (**c**) Hub vicinal zone.

In this formula, P_{imax}, and P_{imin}, ρ, g, and H stand for the individual point's maximum and minimum pressure pulsation amplitudes, fluid density, gravitational acceleration, and head, respectively. Figure 12 considers BPS monitoring points as shown in Figure 10b. In this figure, pressure pulsation amplitudes are found to decrease in the streamwise direction from the blade's leading edge (vaneless space vicinities) towards its trailing edge (impeller outlet zone vicinities).

As also discussed in the above sections, it is also shown that the first two high flow conditions (1.00 Q_{BEP} and 0.82 Q_{BEP}) exhibited the lowest of pressure pulsation levels, while the last two (0.74 Q_{BEP} and 0.55 Q_{BEP}) were marked by high-pressure pulsation levels. Generally speaking, BPS pressure pulsation levels worsened with the decrease in machine flow conditions.

In the next section, an attempt was made to investigate the evolution in pressure pulsation distribution modes at different PAT flow zones, with the increasing impeller rotational speed, under two operating conditions, namely 1.00 Q_{BEP} and 0.74 Q_{BEP}. Figure 13 shows the pressure pulsation distribution mode within the vaneless space zone for different speeds, considering the above-mentioned flow conditions.

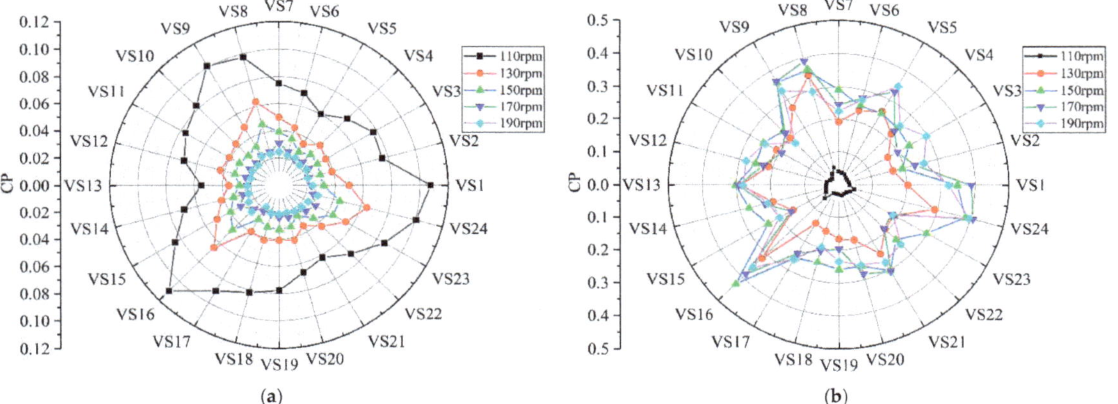

Figure 13. Vaneless space pressure pulsation distribution mode for different impeller rotational speeds (**a**) 1.00 Q_{BEP} (**b**) 0.74 Q_{BEP}.

In this figure, as also explained in the above sections, the VS pressure pulsation level under optimum operating conditions were far lower compared to the 0.74 Q_{BEP} case. In addition, Figure 13 shows that the impeller rotational speed inflicts opposite reactions from both conditions' VS pressure pulsation characteristics. It is shown that, under optimum operating conditions, the increase in impeller rotational speed leads to correspondingly decreasing vaneless pressure pulsation levels.

On the other hand, under 0.74 Q_{BEP} conditions, speed increases from 110 to 130 rpm led to the increase in pulsation levels, while the situation was complex for the rest of speeds. The pressure pulsation distribution mode for the rest of speeds (150, 170, and 190 rpm) was quite similar and their amplitudes quite randomly varied, but within short ranges.

Nevertheless, one would globally assume the worsening of pressure pulsation levels with the increase in impeller rotational speed under this condition, which would also be supported by correspondent flow field characteristics evolution, as presented in both Figures 8 and 9.

In Figure 14, the effect of impeller rotational speed on pressure pulsation distribution mode along PAT's full flow passage is displayed. In the direction of PAT water flow from the computational domain's inlet zone towards the outlet (see Figure 1), different monitoring points were successively positioned along the full flow passage as follows: inlet pipe (OP4, OP33, OP32, and OP31), inter-guide vane channel (GV24, GV23, GV22, and GV21), vaneless space zone (VS), impeller inter-blade channel (IB32, IB22, and IB12), and outlet pipe (IP12, IP22, IB32, and IP4).

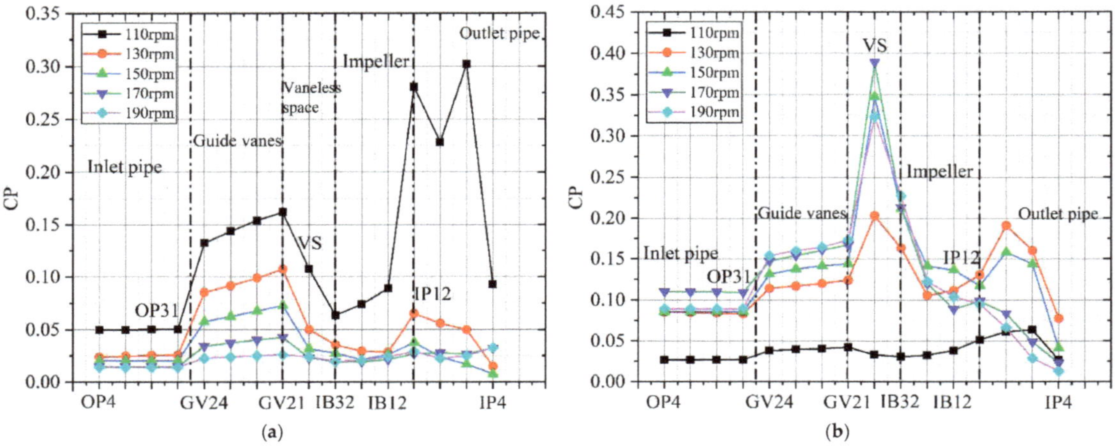

Figure 14. Pressure pulsation distribution mode along PAT flow passage for different impeller rotational speeds (**a**) 1.00 Q_{BEP} (**b**) 0.74 Q_{BEP}.

Under optimum operating conditions, pressure pulsation amplitudes within the guide vanes flow zone were globally found to be the highest, with the exception of 130 rpm case where the outlet pipe presented the highest of pressure pulsations. Under this condition, it is again confirmed that the increase in impeller rotational speed led to a global weakening of pressure pulsations along PAT's full flow passage.

Under the 0.74 Q_{BEP} condition, however, the lowest of pressure pulsations were recorded for 110 rpm cases, followed by 130 rpm for the majority of flow zones. The situation is complex for the rest of speed cases, where, however, unlike the case optimum flow condition case, the highest pressure pulsation level was recorded within the vaneless space.

Therefore, one would globally conclude that, for a PAT under optimum operating conditions, the impeller rotational speed increase would improve the pressure pulsation levels along the machine at full flow passage, while causing an opposite effect for the case of part-load operating conditions.

4. Conclusions

This study numerically investigated the evolution of flow dynamics in a mixed-flow PAT under turbine operating mode, and the effect of impeller rotational speed on the same. Four different operating conditions ranging from optimum to part-load flow conditions (1.00 Q_{BEP}, 0.82 Q_{BEP}, 0.74 Q_{BEP}, and 0.55 Q_{BEP}) were selected, and five impeller rotational speeds ranging from 110 to 190 rpm were tested. After a thorough discussion and deep analysis of study results, a number of concluding remarks were drawn as follows:

With the machine influx decrease from optimum to 0.82 Q_{BEP} condition, PAT flow pattern slightly deteriorated. A further decrease to lower flow conditions inflicted the emergence of very large flow instabilities where, of the four investigated conditions, 0.55 Q_{BEP} presented the least stable flow structures. Moreover, the vaneless space zone between the impeller and guide vanes was found to experience the highest flow unsteadiness under part-load conditions.

Vaneless space pressure pulsation characteristics depended on the machine operating conditions, where pulsation amplitudes continuously increased as the machine influx decreased. For high flow conditions, vaneless space pressure pulsation spectrum was dominated by RSI-born frequency components, while the LFU-born frequency components dominated for low flow conditions.

The impeller rotational speed was found to differently impact the machine flow and pressure fields characteristics, depending on the considered operating condition. Under optimum operating conditions, the impeller rotational increase led to a global decrease in pressure pulsations at a majority of flow zones. On the other hand, the increase in impeller rotational speed worsened the pressure pulsation level at different flow zones, under part-load conditions.

It is therefore obvious that the impeller rotational speed exercises a considerable impact of the machine flow and pressure field characteristics, which under certain conditions may end up affecting the overall machine operational efficiency. This, for instance, sends an alert to electric regulation (ER) users within WDNs, where careful impeller rotational speed change is recommended. Further, CFD-supported parametric studies are still needed to reveal the flow dynamics of PATs under turbine mode, and to compare and establish a relationship with the pump mode. This would lead to a complete understanding of PAT flow dynamics, and probably contribute to an improvement in terms of PAT performance prediction accuracy, for the so far existent pump-to-turbine performance conversion expressions.

Author Contributions: Conceptualization, M.B., K.K. and H.-X.C.; Methodology, M.B., H.-X.C., W.-T.S. and X.-F.G.; Software, M.B., K.K., X.-F.G. and J.N.; Formal analysis, M.B., K.K., H.-X.C., Y.Z. and D.-Q.Z.; Writing-original draft, M.B., X.-F.G. and J.N.; Writing—review & editing, Y.Z., D.-Q.Z., H.-X.C. and K.K.; Resources, K.K., Y.Z. and D.-Q.Z.; Supervision, Y.Z., D.-Q.Z., and K.K.; Funding acquisition, M.B., K.K. and H.-X.C. All authors have read and agreed to the published version of the manuscript.

Funding: This research was funded by the Postdoctoral Research Fund of Jiangsu Province (2020Z152), National Natural Science Foundation of China (52009033; 52006053), Natural Science Foundation of Jiangsu Province (BK20200509; BK20200508), and the Fundamental Research Funds for Central Universities (B210202066).

Institutional Review Board Statement: Not applicable.

Informed Consent Statement: Not applicable.

Data Availability Statement: Not applicable.

Acknowledgments: The support from the College of Water Conservancy and Hydropower Engineering at Hohai University, China is gratefully acknowledged.

Conflicts of Interest: The authors declare no conflict of interest.

Nomenclature

PAT	Pump as Turbine
RANS	Reynolds Averaged Nevier-Stokes equations
BEP	Best Efficiency Point
SST	Shear Stress Transport turbulence model
IRS	Impeller Rotational Speed
GGI	General Grid Interface
VS	Vaneless Space
LFC	Low Frequency Component
CFD	Computational Fluid Dynamics
TRL	Technology Readiness Level
HR	Hydraulic Regulation
LFU	Local Flow Unsteadiness
GPF	Guide vane Passing Frequency
RSI	Rotor-Stator Interactions

ER	Electric Regulation
BLE	Blade Leading Edge
BPS	Blade Pressure Surface
BTE	Blade Trailing Edge
BSS	Blade Suction Surface
GV	Guide Vane
WDN	Water Distribution Networks
GSS	Guide vane Suction Surface
IEC	International Electrotechnical Commission
GPS	Guide vane Pressure Surface
IP	Inlet Pipe
OP	Outlet Pipe
VOS	Variable Operating Strategy
PRV	Pressure Relieve Valve
$y+$	Dimensional height from the wall
fn	Runner rotational frequency
Q	Discharge
n	Runner rotational speed
H	Head
g	Gravitational acceleration
η	Efficiency
P	Static pressure
ρ	Density
Z_R	Number of runner blades
Z_G	Number of guide vanes

References

1. Algieri, A.; Zema, D.A.; Nicotra, A.; Zimbone, S.M. Potential energy exploitation in collective irrigation systems using pumps as turbines: A case study in Calabria (Southern Italy). *J. Clean. Prod.* **2020**, *257*, 120538. [CrossRef]
2. Kandi, A.; Moghimi, M.; Tahani, M.; Derakhshan, S. Optimization of pump selection for running as turbine and performance analysis within the regulation schemes. *Energy* **2021**, *217*, 119402. [CrossRef]
3. Al-Shetwi, A.Q.; Hannan, M.A.; Jern, K.P.; Mansur, M.; Mahlia, T.M.I. Grid-connected renewable energy sources: Review of the recent integration requirements and control methods. *J. Clean. Prod.* **2020**, *253*, 119831. [CrossRef]
4. Namahoro, J.P.; Wu, Q.; Xiao, H.; Zhou, N. The asymmetric nexus of renewable energy consumption and economic growth: New evidence from Rwanda. *Renew. Energy* **2021**, *174*, 336–346. [CrossRef]
5. Namahoro, J.P.; Wu, Q.; Xiao, H.; Zhou, N. The Impact of Renewable Energy, Economic and Population Growth on CO_2 Emissions in the East African Region: Evidence from Common Correlated Effect Means Group and Asymmetric Analysis. *Energies* **2021**, *14*, 312. [CrossRef]
6. Hamududu, B.; Killingtveit, A. Assessing Climate Change Impacts on Global Hydropower. *Energies* **2012**, *5*, 305–322. [CrossRef]
7. Hamududu, B.H.; Killingtveit, Å. Hydropower Production in Future Climate Scenarios: The Case for Kwanza River, Angola. *Energies* **2016**, *9*, 363. [CrossRef]
8. Haidar, A.M.A.; Senan, M.F.M.; Noman, A.; Radman, T. Utilization of pico hydro generation in domestic and commercial loads. *Renew. Sustain. Energy Rev.* **2012**, *16*, 518–524. [CrossRef]
9. Ismail, M.A.; Othman, A.K.; Islam, S.; Zen, H. End Suction Centrifugal Pump Operating in Turbine Mode for Microhydro Applications. *Adv. Mech. Eng.* **2014**, *6*, 139868. [CrossRef]
10. Motwani, K.H.; Jain, S.V.; Patel, R.N. Cost Analysis of Pump as Turbine for Pico Hydropower Plants–A Case Study. *Procedia Eng.* **2013**, *51*, 721–726. [CrossRef]
11. Ranjitkar, G.; Huang, J.; Tung, T. In Proceedings of the Application of Micro-Hydropower Technology for Remote Regions, EIC Climate Change Technology. Ottawa, ON, Canada, 10–12 May 2006; pp. 1–10.
12. Renzi, M.; Rudolf, P.; Štefan, D.; Nigro, A.; Rossi, M. Installation of an axial Pump-as-Turbine (PaT) in a wastewater sewer of an oil refinery: A case study. *Appl. Energy* **2019**, *250*, 665–676. [CrossRef]
13. Moazeni, F.; Khazaei, J. Optimal energy management of water-energy networks via optimal placement of pumps-as-turbines and demand response through water storage tanks. *Appl. Energy* **2021**, *283*, 116335. [CrossRef]
14. Du, J.; Shen, Z.; Yang, H. Study on the effects of runner geometries on the performance of inline cross-flow turbine used in water pipelines. *Sustain. Energy Technol. Assess.* **2020**, *40*, 100762. [CrossRef]
15. Du, J.; Shen, Z.; Yang, H. Effects of different block designs on the performance of inline cross-flow turbines in urban water mains. *Appl. Energy* **2018**, *228*, 97–107. [CrossRef]
16. Jiyun, D.; Zhicheng, S.; Hongxing, Y. Numerical study on the impact of runner inlet arc angle on the performance of inline cross-flow turbine used in urban water mains. *Energy* **2018**, *158*, 228–237. [CrossRef]

17. Jiyun, D.; Zhicheng, S.; Hongxing, Y. Study on the effects of blades outer angle on the performance of inline cross-flow turbines. *Energy Procedia* **2019**, *158*, 1039–1045. [CrossRef]
18. Hosseini, S.M.H.; Forouzbakhsh, F.; Rahimpoor, M. Determination of the optimal installation capacity of small hydro-power plants through the use of technical, economic and reliability indices. *Energy Policy* **2005**, *33*, 1948–1956. [CrossRef]
19. Mishra, S.; Singal, S.K.; Khatod, D.K. Optimal installation of small hydropower plant—A review. *Renew. Sustain. Energy Rev.* **2011**, *15*, 3862–3869. [CrossRef]
20. Williams, A.A. Pumps as turbines for low cost micro hydro power. *Renew. Energy* **1996**, *9*, 1227–1234. [CrossRef]
21. Ramos, H.; Borga, A. Pumps as turbines: An unconventional solution to energy production. *Urban Water* **1999**, *1*, 261–263. [CrossRef]
22. Mercier, T.; Hardy, C.; Van Tichelen, P.; Olivier, M.; De Jaeger, E. Control of variable-speed pumps used as turbines for flexible grid-connected power generation. *Electr. Power Syst. Res.* **2019**, *176*, 105962. [CrossRef]
23. Kusakana, K. A survey of innovative technologies increasing the viability of micro-hydropower as a cost effective rural electrification option in South Africa. *Renew. Sustain. Energy Rev.* **2014**, *37*, 370–379. [CrossRef]
24. Jain, S.V.; Patel, R.N. Investigations on pump running in turbine mode: A review of the state-of-the-art. *Renew. Sustain. Energy Rev.* **2014**, *30*, 841–868. [CrossRef]
25. Patelis, M.; Kanakoudis, V.; Gonelas, K. Pressure Management and Energy Recovery Capabilities Using PATs. *Procedia Eng.* **2016**, *162*, 503–510. [CrossRef]
26. Adu, D.; Jianguo, D.; Darko, R.O.; Boamah, K.B.; Emmanuel, A. Investigating the state of renewable energy and concept of pump as turbine for energy generation development. *Energy Rep.* **2020**, *6*, 60–66. [CrossRef]
27. Asomani, S.N.; Yuan, J.; Wang, L.; Appiah, D.; Zhang, F. Geometrical effects on performance and inner flow characteristics of a pump-as-turbine: A review. *Adv. Mech. Eng.* **2014**, *12*. [CrossRef]
28. Kaunda, C.S.; Kimambo, C.Z.; Nielsen, T.K. A technical discussion on microhydropower technology and its turbines. *Renew. Sustain. Energy Rev.* **2014**, *35*, 445–459. [CrossRef]
29. Štefan, D.; Rossi, M.; Hudec, M.; Rudolf, P.; Nigro, A.; Renzi, M. Study of the internal flow field in a pump-as-turbine (PaT): Numerical investigation, overall performance prediction model and velocity vector analysis. *Renew. Energy* **2020**, *156*, 158–172. [CrossRef]
30. Lin, T.; Li, X.; Zhu, Z.; Xie, J.; Li, Y.; Yang, H. Application of enstrophy dissipation to analyze energy loss in a centrifugal pump as turbine. *Renew. Energy* **2021**, *163*, 41–55. [CrossRef]
31. Hongyu, G.; Wei, J.; Yuchuan, W.; Hui, T.; Ting, L.; Diyi, C. Numerical simulation and experimental investigation on the influence of the clocking effect on the hydraulic performance of the centrifugal pump as turbine. *Renew. Energy* **2021**, *168*, 21–30. [CrossRef]
32. Morabito, A.; Vagnoni, E.; Di Matteo, M.; Hendrick, P. Numerical investigation on the volute cutwater for pumps running in turbine mode. *Renew. Energy* **2021**, *175*, 807–824. [CrossRef]
33. Ghorani, M.M.; Sotoude Haghighi, M.H.; Maleki, A.; Riasi, A. A numerical study on mechanisms of energy dissipation in a pump as turbine (PAT) using entropy generation theory. *Renew. Energy* **2020**, *162*, 1036–1053. [CrossRef]
34. Ghorani, M.M.; Sotoude Haghighi, M.H.; Riasi, A. Entropy generation minimization of a pump running in reverse mode based on surrogate models and NSGA-II. *Int. Commun. Heat Mass Transf.* **2020**, *118*, 104898. [CrossRef]
35. Wang, T.; Wang, C.; Kong, F.; Gou, Q.; Yang, S. Theoretical, experimental, and numerical study of special impeller used in turbine mode of centrifugal pump as turbine. *Energy* **2017**, *130*, 473–485. [CrossRef]
36. Sengpanich, K.; Bohez, E.L.J.; Thongkruer, P.; Sakulphan, K. New mode to operate centrifugal pump as impulse turbine. *Renew. Energy* **2019**, *140*, 983–993. [CrossRef]
37. Patelis, M.; Kanakoudis, V.; Gonelas, K. Combining pressure management and energy recovery benefits in a water distribution system installing PATs. *J. Water Supply Res. Technol. Aqua* **2017**, *66*, 520–527. [CrossRef]
38. Cimorelli, L.; D'Aniello, A.; Cozzolino, L.; Pianese, D. Leakage reduction in WDNs through optimal setting of PATs with a derivative-free optimizer. *J. Hydroinform.* **2020**, *22*, 713–724. [CrossRef]
39. Postacchini, M.; Darvini, G.; Finizio, F.; Pelagalli, L.; Soldini, L.; Di Giuseppe, E. Hydropower Generation Through Pump as Turbine: Experimental Study and Potential Application to Small-Scale WDN. *Water* **2020**, *12*, 958. [CrossRef]
40. Pérez-Sánchez, M.; Sánchez-Romero, F.J.; Ramos, H.M.; López-Jiménez, P.A. Improved Planning of Energy Recovery in Water Systems Using a New Analytic Approach to PAT Performance Curves. *Water* **2020**, *12*, 468. [CrossRef]
41. Stefanizzi, M.; Capurso, T.; Balacco, G.; Binetti, M.; Camporeale, S.M.; Torresi, M. Selection, control and techno-economic feasibility of Pumps as Turbines in Water Distribution Networks. *Renew. Energy* **2020**, *162*, 1292–1306. [CrossRef]
42. Kougias, I.; Aggidis, G.; Avellan, F.; Deniz, S.; Lundin, U.; Moro, A.; Muntean, S.; Novara, D.; Pérez-Díaz, J.I.; Quaranta, E.; et al. Analysis of emerging technologies in the hydropower sector. *Renew. Sustain. Energy Rev.* **2019**, *113*, 109257. [CrossRef]
43. Lima, G.M.; Luvizotto, E.; Brentan, B.M. Selection and location of Pumps as Turbines substituting pressure reducing valves. *Renew. Energy* **2017**, *109*, 392–405. [CrossRef]
44. Polák, M. The Influence of Changing Hydropower Potential on Performance Parameters of Pumps in Turbine Mode. *Energies* **2019**, *12*, 2103. [CrossRef]
45. Spedaletti, S.; Rossi, M.; Comodi, G.; Salvi, D.; Renzi, M. Energy recovery in gravity adduction pipelines of a water supply system (WSS) for urban areas using Pumps-as-Turbines (PaTs). *Sustain. Energy Technol. Assess.* **2021**, *45*, 101040.

46. Carravetta, A.; Del Giudice, G.; Fecarotta, O.; Ramos, H.M. Pump as Turbine (PAT) Design in Water Distribution Network by System Effectiveness. *Water* **2013**, *5*, 1211–1225. [CrossRef]
47. Carravetta, A.; Del Giudice, G.; Fecarotta, O.; Ramos, H.M. Energy Production in Water Distribution Networks: A PAT Design Strategy. *Water Resour. Manag.* **2012**, *26*, 3947–3959. [CrossRef]
48. Carravetta, A.; del Giudice, G.; Fecarotta, O.; Ramos, H. PAT Design Strategy for Energy Recovery in Water Distribution Networks by Electrical Regulation. *Energies* **2013**, *6*, 411–424. [CrossRef]
49. Fecarotta, O.; Carravetta, A.; Ramos, H.M.; Martino, R. An improved affinity model to enhance variable operating strategy for pumps used as turbines. *J. Hydraul. Res.* **2016**, *54*, 332–341. [CrossRef]
50. Delgado, J.; Ferreira, J.P.; Covas, D.I.C.; Avellan, F. Variable speed operation of centrifugal pumps running as turbines. Experimental investigation. *Renew. Energy* **2019**, *142*, 437–450. [CrossRef]
51. Lima Gustavo, M.; Luvizotto, E.; Brentan Bruno, M.; Ramos Helena, M. Leakage Control and Energy Recovery Using Variable Speed Pumps as Turbines. *J. Water Resour. Plan. Manag.* **2018**, *144*, 04017077. [CrossRef]
52. Tahani, M.; Kandi, A.; Moghimi, M.; Houreh, S.D. Rotational speed variation assessment of centrifugal pump-as-turbine as an energy utilization device under water distribution network condition. *Energy* **2020**, *213*, 118502. [CrossRef]
53. Jain, S.V.; Swarnkar, A.; Motwani, K.H.; Patel, R.N. Effects of impeller diameter and rotational speed on performance of pump running in turbine mode. *Energy Convers. Manag.* **2015**, *89*, 808–824. [CrossRef]
54. Fernández, J.; Blanco, E.; Parrondo, J.; Stickland, M.T.; Scanlon, T.J. Performance of a centrifugal pump running in inverse mode. *Proc. Inst. Mech. Eng. Part A J. Power Energy* **2004**, *218*, 265–271. [CrossRef]
55. Han, Y.; Tan, L. Influence of rotating speed on tip leakage vortex in a mixed flow pump as turbine at pump mode. *Renew. Energy* **2020**, *162*, 144–150. [CrossRef]
56. Menter, F.R. Two-Equation Eddy-Viscosity Turbulence Models for Engineering Applications. *AIAA J.* **1994**, *32*, 1598–1605. [CrossRef]
57. Wilcox, D.C. Reassessment of the scale-determining equation for advanced turbulence models. *AIAA J.* **1988**, *26*, 1299–1310. [CrossRef]
58. Jones, W.P.; Launder, B.E. The prediction of laminarization with a two-equation model of turbulence. *Int. J. Heat Mass Transf.* **1972**, *15*, 301–314. [CrossRef]
59. Menter, F. Zonal Two Equation k-w Turbulence Models For Aerodynamic Flows. In Proceedings of the 23rd Fluid Dynamics, Plasmadynamics, and Lasers Conference, American Institute of Aeronautics and Astronautics, Orlando, FL, USA, 6–9 July 1993.

 *sustainability*

Article

Investigation into Influence of Wall Roughness on the Hydraulic Characteristics of an Axial Flow Pump as Turbine

Kan Kan [1,2,*], **Qingying Zhang** [1], **Yuan Zheng** [1,2], **Hui Xu** [2,3], **Zhe Xu** [2], **Jianwei Zhai** [1] and **Alexis Muhirwa** [4]

[1] College of Energy and Electrical Engineering, Hohai University, Nanjing 211100, China; 211306020020@hhu.edu.cn (Q.Z.); zhengyuan@hhu.edu.cn (Y.Z.); 18936163392@163.com (J.Z.)
[2] College of Water Conservancy and Hydropower Engineering, Hohai University, Nanjing 210098, China; hxu@hhu.edu.cn (H.X.); xuzhe@hhu.edu.cn (Z.X.)
[3] College of Agricultural Science and Engineering, Hohai University, Nanjing 211100, China
[4] Department of Renewable Energy, Rwanda Polytechnic—IPRC Tumba, Kigali 6579, Rwanda; amuhirwa@iprctumba.rp.ac.rw
* Correspondence: kankan@hhu.edu.cn; Tel.: +86-15151862390

Abstract: Pump as turbine (PAT) is a factual alternative for electricity generation in rural and remote areas where insufficient or inconsistent water flows pose a threat to local energy demand satisfaction. Recent studies on PAT hydrodynamics have shown that its continuous operations lead to a progressive deterioration of inner surface smoothness, serving the source of near-wall turbulence build-up, which itself depends on the level of roughness. The associated boundary layer flow incites significant friction losses that eventually deteriorate the performance. In order to study the influence of wall roughness on PAT hydraulic performance under different working conditions, CFD simulation of the water flow through an axial-flow PAT has been performed with a RNG k-ε turbulence model. Study results have shown that wall roughness gradually decreases PAT's head, efficiency, and shaft power. Nevertheless, the least wall roughness effect on PAT hydraulic performance was experienced under best efficiency point conditions. Wall roughness increase resulted in the decrease of axial velocity distribution uniformity and the increase of velocity-weighted average swirl angle. This led to a disorderly distribution of streamlines and backflow zones formation at the conduit outlet. Furthermore, the wall roughness impact on energy losses is due to the static pressure drop on the blade pressure surface and the increase of turbulent kinetic energy near the blade. Further studies on the roughness influence over wider range of PAT operating conditions are recommended, as they will lead to quicker equipment refurbishment.

Keywords: wall roughness; axial flow pump; pump as turbine; hydraulic characteristics

Citation: Kan, K.; Zhang, Q.; Zheng, Y.; Xu, H.; Xu, Z.; Zhai, J.; Muhirwa, A. Investigation into Influence of Wall Roughness on the Hydraulic Characteristics of an Axial Flow Pump as Turbine. *Sustainability* **2022**, *14*, 8459. https://doi.org/10.3390/su14148459

Academic Editors: Mosè Rossi, Massimiliano Renzi, David Štefan and Sebastian Muntean

Received: 30 May 2022
Accepted: 7 July 2022
Published: 11 July 2022

Publisher's Note: MDPI stays neutral with regard to jurisdictional claims in published maps and institutional affiliations.

1. Introduction

Under low-head operating conditions, axial flow pumps are more likely to exhibit good hydraulic performance characteristics. Hence, these pumps are suitable for large- and medium-sized pumping stations in plain areas, where they play a vital role in urban water supply and agricultural irrigation [1,2]. According to the installation position of the pump shaft, axial flow pumps can be classified into three types: vertical, horizontal, and slanted pumps. Conversely to the vertical type, slanted axial flow pump's inlet conduit does not need 90° elbow, which relatively explains its low hydraulic loss. Compared with the horizontal type, the slanted axial flow pump has better performance, since the connected electric motor does not have to be installed underwater [3,4]. Axial flow pump has been widely applied in the eastern route of China's South-to-North Water Diversion Project. Due to the landform and climate reasons, wet seasons are dominated by a lot of floods and stagnant waters. To achieve considerable economic benefits, pumping stations can make full use of such water energy resources by means of pump as turbine (PAT) technology [5,6]. Thanks to the control valve utilization, the flow direction can be reversed for power generation. The pump technology has grown mature to reach a convenient maintenance routine and a fair affordability [7]. Many recent projects have endeavored to replace conventional small hydropower units with pumps in order to meet the energy demand in remote regions. However, the efficiency of PATs is still relatively low [8]. In addition, due to the influence of manufacturing processes, the inner wall surface cannot be perfectly smooth; i.e., the wall roughness of the machine is not uniform. After a certain operation period, PAT performance ends up deteriorating as sedimentation, oxidation, and corrosion accumulate over time. Such accumulation jointly increases the surface irregularities of the conveying structure and leads to declining hydraulic performance [9]. With the improvement of water supply projects and drainage irrigation, the safety and stable operation of axial flow pumps have been a research priority [10]. Therefore, the study of the influence of wall roughness on hydraulic performance of the axial flow PAT can be very useful for the daily maintenance of pump stations.

Lin et al. [11] used numerical methods to predict PAT performance curves under both the pump and turbine operating modes, and verified them experimentally. Yang et al. [12] used CFD to simulate the influence of splitter blades on PATs. The addition of splitter blades came up with a great performance improvement in operation efficiency. Comparison between experimental and numerical results showed that CFD can be used in the performance prediction and optimization of PAT. A novel impeller design has been proposed to retrofit conventional double suction pump. With the open source CFD code OpenFOAM, the novel impeller has been investigated in turbine mode using 3D URANS simulations and the characteristic curves of the two machines have been calculated and compared [13]. Derakhshan et al. [14] redesigned the shape of impeller blades leading its efficiency improvement as compared to experimental data. It was noted that PAT efficiency can be improved just by impeller modification. On the other hand, PAT volute casing's cutwater design can also influence its performance characteristics, as noted by Morabito et al. [15].

As shown in the above literature, CFD has been an important tool in studies on PAT flow dynamics. In the same respect, it has crucially contributed to the attainment of an in-depth understanding about wall roughness effect and the underlying flow instability development mechanism. Gu et al. [16] found that the influence of surface roughness on pump hydraulic performance depends on the actual roughness of the structure, velocity distribution, and near-wall turbulence along the flow path. A study by Bai Tao et al. [17] revealed a close relationship between the surface roughness in the boundary layer and the operational Reynolds number, for a wind turbine case. At low Reynolds numbers, surface roughness could weaken separation bubbles, lowering the aerodynamic losses. On the other hand, operations at high Reynolds numbers suffered significant losses due to transient processes triggered by surface roughness. For the case of hydraulic pumps, Bellary et al. [18] found that wall roughness increases with the head and shaft power, while hydraulic efficiency correspondingly decreased. In the same respect, Deshmukh et al. [19] studied the influence of wall roughness on centrifugal pumps hydraulic performance under optimal and off-design conditions. For completely rough surfaces, the hydraulic performance of centrifugal pumps has displayed a progressive deterioration. When the roughness critical value was reached, the hydraulic performance stopped declining, where further decrease yielded slight head increase. Within the same context, Juckelandt et al. [20] comparatively analyzed the influence of surface roughness on the pump performance by means of experiments and simulations. The influence of roughness on flow parameters such as the average velocity and turbulence pulsations was investigated. He et al. [21] conducted a series of numerical calculations of different turbulence models and surface roughness for a multistage centrifugal pump based on CFD. The results came up with the same conclusion that the surface roughness strongly affects the head and efficiency. However, the impact gradually slows down, and the effect of surface roughness on efficiency is greater than that of head. Limbach et al. [22] carried out experimental measurements and numerical simulation on cavitation flow within a low specific speed centrifugal pump under different working conditions and different roughness surface. The results showed that, in the flow without cavitation, the head measured and simulated adamantly stays consistent, and the experimental measurements drew an inference that with the worsening of surface roughness, the net positive suction head (NPSH) increases significantly. Nonetheless, this finding has not been reproduced by simulations of the wall function.

To cut short this overview of the literature relevant to the present study, Lim et al. [23] have carried out the study on the similar investigation on a dual-suction centrifugal pump. They came up with a conclusion that the roughness has the greatest influence on the impeller, while the inlet conduit flow does not reflect that high ascendancy. In addition, the pump performance has been found to largely depend on the surface roughness on the impeller shroud. So far, the existent literature about the influence of wall roughness on hydraulic machinery performance is predominantly focused on centrifugal pumps, where comparatively few studies have been conducted on axial-flow pumps, and the generating mode of PATs is rarely investigated. Moreover, most of studies have investigated the effect of surface roughness on the machine's external performance characteristics (EPCs), using experimental and numerical simulation methods without digging deeper to the root cause of recorded changes in machine EPCs. Therefore, the influence of surface roughness on PAT's internal flow dynamics is still inadequately studied.

In this respect, the present article evaluates the hydraulic performance of a slanted axial flow PAT subjected to different wall roughness scenarios. The influence of wall roughness on the internal flow state of an axial flow PAT under different flow conditions for a wide range of roughnesses is analyzed in order to improve the equipment integrity rate and operational life span. This study's results provide a useful reference for pump station optimization. The article structure comprises four sections organized as follows: Section 2 details the applied methodology concept, where the concerned PAT's geometric model, as well as the utilized numerical method, are well presented. For an analogic understanding of the subject matter, the theory of equivalent sand-grain roughness is briefly recapitulated, as well. The results of this study, including the external performance characteristics of the slanted axial flow PAT and the influence of wall roughness on the same, are presented and discussed within Section 3. Finally, to draw this work to a close, relevant conclusions and further research recommendations have been pointed out through Section 4.

2. Numerical Model and Methods

2.1. Governing Equations

The continuity and momentum equations that are extensively used in flow diagnostics are stated through Equations (1) and (2) [24,25]:

$$\frac{\partial \rho}{\partial t} + \frac{\partial(\rho u_i)}{\partial x_i} = 0 \tag{1}$$

$$\frac{\partial(\rho u_i)}{\partial t} + \frac{\partial(\rho u_i u_j)}{\partial x_j} = -\frac{\partial p}{\partial x_i} + \frac{\partial}{\partial x_j}(\mu \frac{\partial u_i}{\partial x_j}) + \frac{\partial \tau_{ij}}{\partial x_j} \tag{2}$$

where the following designation applies ρ: the density, t: the physical time, p: the local pressure, μ: the dynamic viscosity, τ_{ij}: the Reynolds stress; whereas x_i and x_j represent the Cartesian coordinate components in the i and j directions, respectively, u_i and u_j represent the corresponding components of the time-averaged velocity.

2.2. Equivalent Sand-Grain Roughness

The wall roughness is indicated through the consideration of peaks and valleys under different shapes and sizes. It is convenient to simulate that surface texture with an equivalent analogy of sand-grain roughness. Figure 1a depicts that similarity by using a wall with a layer of closely packed spheres [26]. The average roughness height of the spheres is K_s [µm], also known as equivalent sand height [27]. The friction effect only occurs in the upper part of balls, and the equivalent sand-grain roughness only affects the flow in its vicinity. Therefore, the actual flow surface may be rounded as shown in Figure 1b, where the x-axis represents the physical wall surface.

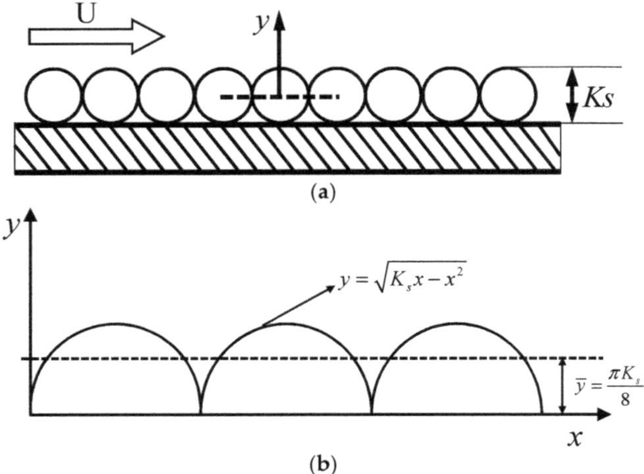

Figure 1. Equivalent sand-grain roughness (**a**) and the surface profile diagram (**b**).

Roughness is the measure of the micro spacing and peak-valley unevenness on the surface texture. The arithmetic average of absolute value (R_a) that is used to quantify the surface roughness is couched in Equation (3).

$$R_a = \frac{1}{n}\sum_{i=1}^{n}|y_i|$$ (3)

where y_i is the distance from the average height of a profile (the mean line) for measurement i, and n is the number of measurements [28]. As the number of measurements approaches infinity, Equation (3) tends to look like Equation (4).

$$R_a = \frac{1}{K_s}\int_{x=0}^{K_s}|y - \overline{y}|dx$$ (4)

Considering parameters in Equation (4), the equation of wall roughness profile can be formulated as to Equation (5) and the mean line as to Equation (6).

$$y = \sqrt{K_s x - x^2}$$ (5)

$$\overline{y} = \frac{\pi K_s}{8}$$ (6)

Substitute Equations (7) and (8) into Equation (6) and integrate:

$$R_a = \frac{K_s}{2}\left(\frac{\pi}{2} - \cos^{-1}\left(1 - \frac{\pi^2}{16}\right)^{1/2} - \frac{\pi}{4}\left(1 - \frac{\pi^2}{16}\right)^{1/2}\right)$$ (7)

Simplify the Equation (7), the average roughness height of the spheres (K_s) can be expressed as to Equation (8).

$$K_s = 11.0293 R_a$$ (8)

2.3. The Geometric Model Description

The wall roughness involves the entire hydraulic circuit of a slanted axial flow pump. Therefore, the computational domain consists of four main parts of inlet conduit, impeller, guide vane and outlet conduit, as shown in Figure 2. Design and operation parameters of the investigated hydraulic machine are specified in Table 1.

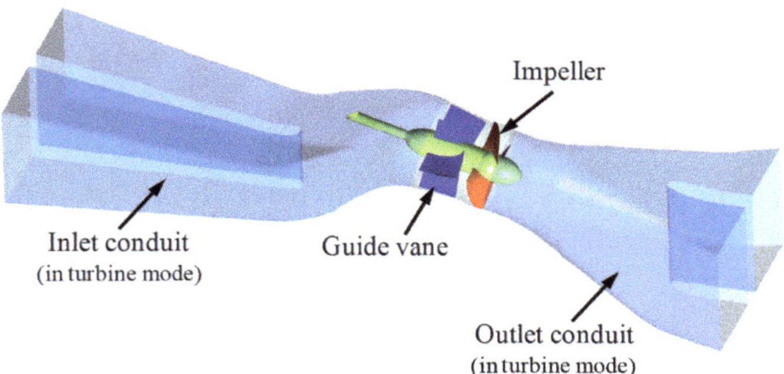

Figure 2. Computational domain of the axial flow PAT.

Table 1. Parameters of the slanted axial flow pump.

Parameter	Value
Impeller nominal diameter D [m]	3.25
Design head H [m]	2.6
Design flow rate Q [m^3/s]	45.5
Design power P [kW]	1319
Design efficiency η [%]	88
Rotational speed n [r/min]	122

2.4. The Grid Generation Methodology

To generate the mesh of PAT's computational domain, the unstructured type of computational grid (Figure 3) has been adopted to deeply resolve the hectic flow dynamicity in the rough wall's vicinal flow zones. To capture the effect of roughness on near-wall flow more precisely, local grid refinement was carried out, and wall function method was adopted within that critical zone of activity. In order to determine the appropriate grid number for further simulations; a grid independence test has been conducted. To do this, six grid size sets have been generated and simulated under similar boundary conditions, leading to six different PAT performances in terms of a single common parameter, namely the PAT hydraulic efficiency (η). The curve in Figure 4 is the delineation of how PAT hydraulic efficiency varies with the change in grid number under pump-mode operation for the similar flow rate.

From Figure 4, it is obvious that when the grid number reaches 7.48 million, the efficiency tends to display a relatively stable pattern. To that end, the grid size of 7.48 million has been selected for further numerical simulations, as it features the desired advantages of calculation accuracy and the computation cost in terms of simulation time [29]. The grid number in each part of the computational model is shown in Table 2.

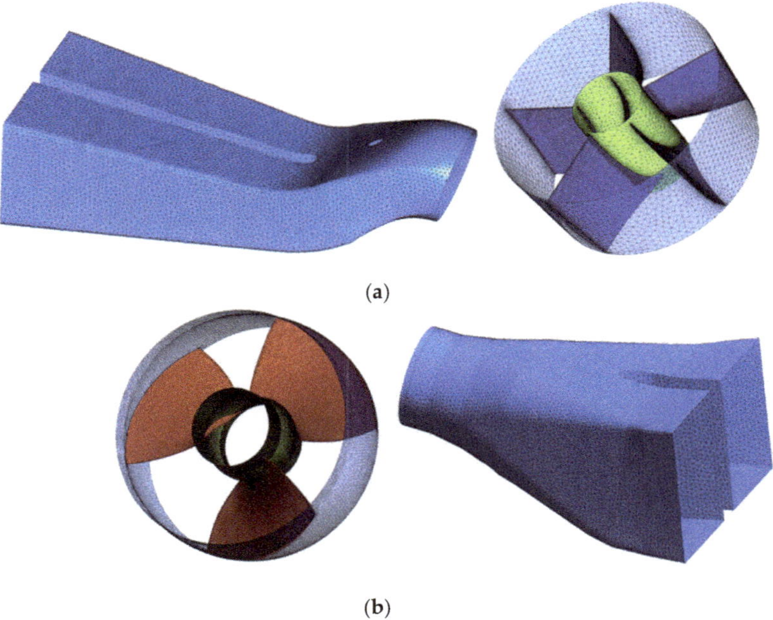

(**a**)

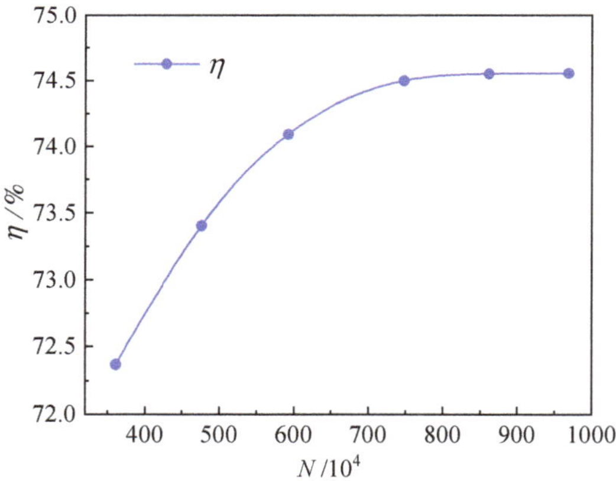

(**b**)

Figure 3. Unstructured mesh of the slanted axial flow pump components. (**a**)inlet conduit, guide vane (**b**) impeller, outlet conduit.

Figure 4. The grid independence assessment against the efficiency.

Table 2. Grid repartition per each component.

Hydraulic Circuit Components	Grid Size [10^4]
Inlet conduit	79
Guide vane	241
Impeller	334
Outlet conduit	94
An all-embracing grid size	748

2.5. Boundary Conditions

ANSYS Fluent is applied to simulate the steady incompressible flow of the slanted axial-flow PAT. The finite volume method is used for space discretization, and the pressure velocity coupling is realized by SIMPLEC algorithm. The inlet boundary condition of the computational domain has been set to be the mass flow inlet, and the outlet boundary condition was set as pressure outlet. The impeller domain was based on the rotating reference frame, while the other domains were based on the stationary reference frame. All wall boundaries adopted non-slip walls, and the convergence criterion is set to 1×10^{-5}.

3. Results and Discussion

3.1. Experimental Validation of the Numerical Scheme

Existing turbulence models featured with ANSYS Fluent are Standard k-ε, RNG k-ε, SST k-ω, etc. Different turbulence models are suitable for different flow situations and need to be selected according to the actual situation. On that account, the performance of different turbulence models has been numerically assessed and compared with experimental results under flow rates ranging from 0.7 Q–1.2 Q. As illustrated by Figure 5, the numerical results of the RNG k-ε turbulence model mostly fall in good agreement with the experimental results. Therefore, the RNG k-ε turbulence model is the best option for this specific numerical simulation. In the range of 40–70 m^3/s; the maximum relative errors of pump head and efficiency between experiment and numerical results are 1.60% and 2.89%, respectively. Hence, both the selected mesh arrangement and numerical methods can be considered reliable enough to guarantee the accuracy and reliability of numerical outcomes.

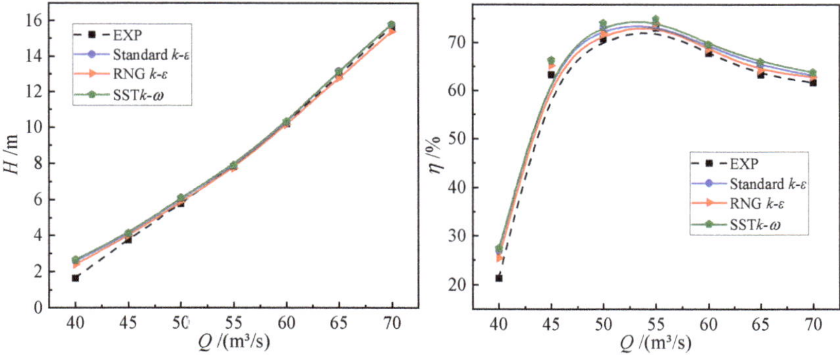

Figure 5. Experimental and numerical results with different turbulent models.

3.2. Effect of Wall Roughness on PAT Performance

For analytical purpose, the actual situation of the slanted axial flow PAT is simplified by modeling the wall roughness as a uniform distribution throughout the hydraulic components. To study the effect of the wall roughness on the slanted axial flow pump performance, different values of the wall roughness ($R_a = 0$ μm, 6 μm, 60 μm, 120 μm, 240 μm, 480 μm, 960 μm) are selected for the numerical calculation. For each wall roughness, Figure 6a,b show the corresponding performance characteristics (head and efficiency) variation with respect to the activated flow rate. The general remark for both figures is that the order of roughness influence on the head and efficiency for each flow rate is directly connected to the level of roughness; i.e., the deeper the roughness, the higher the performance reduction.

As to normal flow behavior, Figure 6a shows the proportionality between the flow rate and the head. That is, the increase of flow rate is concomitant to the head rise. The optimal condition of the investigated axial flow PAT is Q = 55 m^3/s. When the flow rate is in the range of 0.7 Q–1.2 Q, the head at the same working point decreases gradually with the increase in wall roughness. When wall roughness increases from 0–960 μm, the head in each flow rate condition decreases by 21.0%, 14.6%, 9.9%, 7.9%, 9%, 6.9%, and 5% respectively. This demonstrates wall roughness has the maximum effect on pump head at the small flow rate condition, and the effect gradually decreases with increasing flow rate.

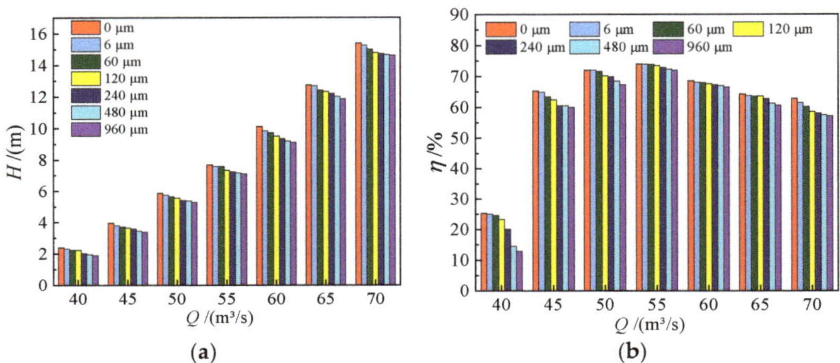

Figure 6. The slanted axial flow PAT performance varies with flow rate at different wall roughness. (**a**) Head variation with the flow rate (**b**) efficiency variation with the flow rate.

For all roughness scenarios in Figure 6b, it is clear that with the increase in flow rate, the efficiency displays an overall trend of increasing and decreasing before and after the optimum condition, respectively. In the range of 0.7 Q–1.2 Q, the efficiency gradually decreases with the increase in wall roughness at the same flow rate. When wall roughness increases from 0–960 μm, the efficiency in each of flow conditions decreases by 12.53%, 5.31%, 4.76%, 2.07%, and 5.72%, respectively. This shows that the wall roughness has the least effect on PAT performance at the optimal operating condition, while it inflicts a greater impact on PAT performance under off-design operating conditions, especially under low flow conditions. Therefore, the influence of wall roughness on the performance of an axial-flow PAT should be considered when the machine has been operating under off-design conditions for a long time.

3.3. Effect of Wall Roughness on Shaft Power

Having demonstrated the direct influence of the wall roughness on the head and efficiency, it is reasonable to anticipate its impact on the shaft power, as well. Figure 7 shows the effect of different wall roughness depths on PAT's shaft power (P_t) under different flow conditions. As noticed previously (Figure 6), the increase in flow rate corresponds to the same increase in the power of the water flow doing work on the impeller, which actuates the shaft power continuously. Once more, the extent of the wall roughness tends to have the same influence on the shaft power. This is evidenced while considering each flow rate alone; the shaft power displays a gradually decreasing pattern with the increasing wall roughness depth.

When the wall roughness increases from 0–960 μm, the shaft power for each of the investigated flow conditions correspondingly decreases by 50.5%, 15.6%, 9.3%, 6.5%, 4.2%, 3.7%, and 1%. It is therefore obvious that the effect of wall roughness on the shaft power is most significant under low flow conditions (Q = 40 m^3/s), whereas it gradually weakens with further flow rate increase. Reflecting back to Figure 6, the roughness influence on the head is relatively small at the small flow rate, and wall roughness leads to greater shaft power decline. This corresponds to the main reason for the efficiency reduction at small flow rates.

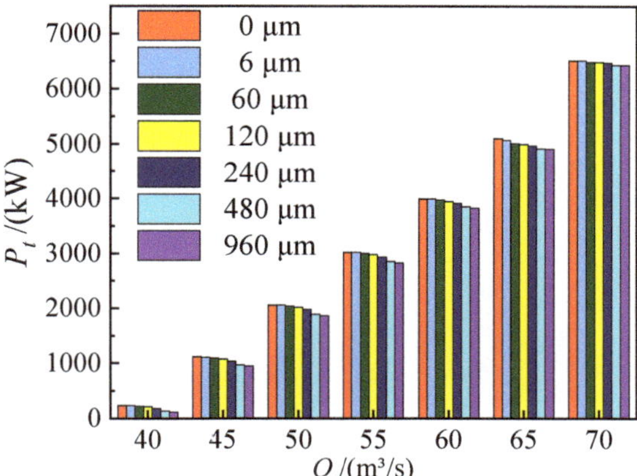

Figure 7. Shaft power varies with flow rate at different wall roughness.

3.4. Effect of Wall Roughness on Internal Flow

3.4.1. Wall Roughness Effect on Relative Flow Velocity over the Blade Pressure Surface

While investigating the wall roughness effect on the performance of a slanted axial flow PAT, the impact on the internal flow must be regarded as a key focus. For that purpose, monitoring points have been placed over the blade pressure surface along 3 lines at a span of 0.9, 0.5, and 0.1, respectively. As illustrated by Figure 8, equidistant 5 monitoring points (designated by "S") are evenly arranged along the outermost chord direction of the blade and along the mid-span (designated by "M"), whereas 4 monitoring points (designated by "H") have been evenly arranged over the innermost span.

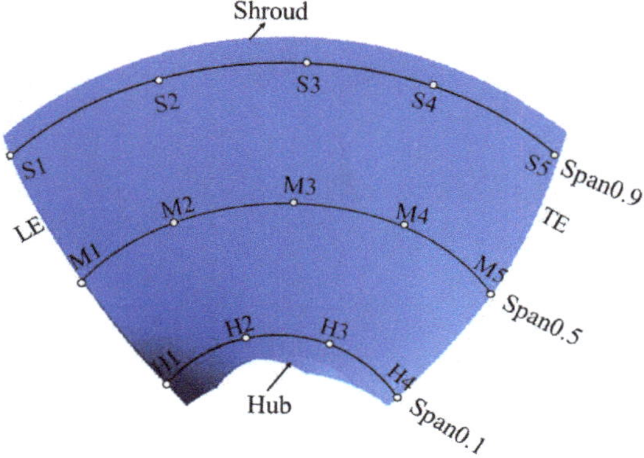

Figure 8. Schematic diagram of monitoring points on blade pressure surface.

As already found, the size of wall roughness has a proportional impact on machine performance. The same observation can be deduced from Figure 9 that shows the relative velocity (V_t) histograms of monitoring points (H3, M4, and S4) aligned radially from hub to shroud on the blade pressure surface. The outstanding observation is that as the relative velocity gets high towards outermost spans, the wall roughness keeps reducing the relative velocity according to its size. As tabulated in Table 3, the relative velocity of roughness $R_a = 0$ µm at H3 is 8.134 m/s, which is 0.87%, 2.75%, 7.15%, 10.14%, 11.23%, 12.3% higher than that of wall roughness $R_a = 0$ µm, 60 µm, 120 µm, 240 µm, 480 µm, and 960 µm, respectively. The relative velocity of M4 roughness $R_a = 0$ µm is 14.04 m/s, which is 2.23%, 4.98%, 8.13%, 11.21%, 13.41%, 15.31% higher than that of $R_a = 0$ µm, 60 µm, 120 µm,

240 μm, 480 μm, and 960 μm, respectively. The relative velocity of S4 roughness $R_a = 0$ μm is 22.62 m/s, which is 0.89%, 3.58%, 6.45%, 10.01%, 12.34%, 14.12% higher than that of $R_a = 0$ μm, 60 μm, 120 μm, 240 μm, 480 μm, 960 μm, respectively. It can be seen that the relative velocity of the impeller blade pressure surface increases continuously along the radial direction. As the wall roughness of the impeller shroud, the friction resistance in the boundary layer increases. This results in energy loss increase and a decrease in the relative velocity of the impeller blade.

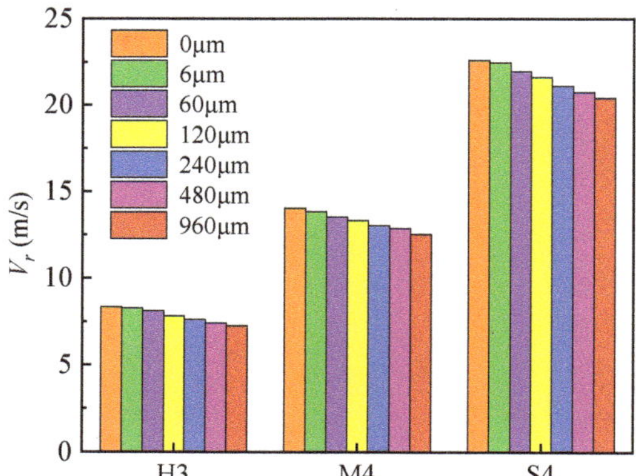

Figure 9. Effect of roughness on relative velocity on blade pressure surface.

Table 3. Relative velocity and percentages with different roughness.

R_a (μm)	H3		M4		S5	
	Vr (m/s)	Relative Deviation (%)	Vr (m/s)	Relative Deviation (%)	Vr (m/s)	Relative Deviation (%)
0	8.13		14.04		22.62	
6	8.06	0.87%	13.73	2.23%	22.42	0.89%
60	7.91	2.75%	13.34	4.98%	21.81	3.58%
120	7.55	7.15%	12.90	8.13%	21.16	6.45%
240	7.31	10.14%	12.47	11.21%	20.36	10.01%
480	7.22	11.23%	12.16	13.41%	19.83	12.34%
960	7.13	12.30%	11.89	15.31%	19.42	14.12%

3.4.2. Effect of Wall Roughness on Impeller Outlet Flow

In order to study the effect of wall roughness on the impeller outlet flow field, the distribution uniformity of the axial velocity component (φ) and velocity-weighted average swirl angle (θ) under different working conditions are analyzed. There is hardly any hydraulic loss when flow state at the impeller outlet is better. That is, when the axial velocity uniformity is around 100% and the velocity-weighted average swirl angle is smaller. φ and θ are calculated according to Equations (9) and (10), respectively.

$$\varphi = \left[1 - \frac{1}{Va}\sqrt{\sum_{i=1}^{n}\left[(Vai - Va)\right]}\right] \times 100\% \tag{9}$$

$$\theta = \sum_{i=1}^{n} Vai\left\{90° - \mathrm{tg}^{-1}\frac{Vti}{Vai}\right\} \Big\backslash \sum_{i=1}^{n} Vai \tag{10}$$

where, φ is the distribution uniformity of axial velocity at the outlet section (%); V_a is the arithmetic average value of the axial velocity of outlet section (m/s); V_{ai} is the axial velocity (m/s) of each computing unit (representing each element) at the outlet section; θ is velocity-weighted average swirl angle (deg) of flow at outlet section; V_{ti} is the transverse velocity (m/s) of each calculation unit in the outlet section; n is the number of computing units on the exit section [30].

Figure 10 shows the curve of uniformity distribution for the axial flow velocity subjected to the variation of roughness under different working conditions. At first glance, the common trend of decreasing axial velocity uniformity with increasing wall roughness can be easily noticed. Under the optimal conditions (Q), the axial velocity distribution uniformity is excellent; i.e., it reaches its highest value, where the roughness-free wall displays a perfect uniformity. However; with the increase of roughness up to 960 µm, the axial velocity uniformity (φ) decreases from 77.2 to 74.3%, which corresponds to an overall downward shift rate of about 2.9%. For small flow rate (0.7 Q), when the roughness increases up to 960 µm, the axial velocity uniformity (φ) decreases from 74 to 68.3%, which is equivalent to the overall downward shift rate of about 5.7%. For large flow rate (1.2 Q), when the wall roughness increases up to 960 µm, the axial velocity distribution uniformity (φ) decreases from 76.4 to 73.9%, which sums up to an overall downward shift rate of about 2.5%.

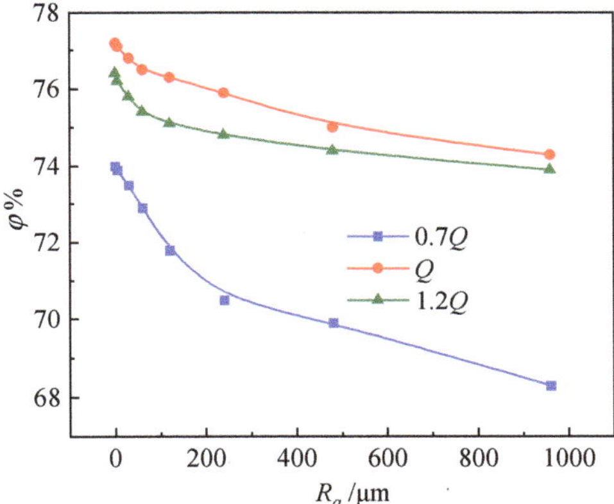

Figure 10. Effect of roughness on axial velocity uniformity.

It can be seen that, under the reverse power generation condition of the slanted axial flow pump device, the increase in wall roughness leads to a decrease in the uniformity of axial velocity distribution. At the small flow rate of $Q = 40 \text{ m}^3/\text{s}$, wall roughness has the most significant effect on uniformity of axial velocity distribution, while the effect of wall roughness on uniformity of axial velocity distribution becomes minimal under the large flow rates. In light of previous findings, the low axial velocity uniformity may be one among several reasons that lead to the low operation efficiency of reverse power generation under the small flow conditions.

Another parameter to evaluate the wall roughness effect on impeller outlet flow is the velocity-weighted average swirl angle. Figure 11 portrays its variation in terms of wall roughness stretch under different working conditions. Under different flow rates, the curve outline of the velocity-weighted average swirl angle closely reproduces the same shape similarity. The common inference is that the velocity-weighted average swirl angle is unambiguously associated to the wall roughness size.

Under the optimal conditions (Q), the velocity-weighted average swirl angle is at its minimum value, no matter the wall roughness. The increase in wall roughness (0–960 µm) at that best efficiency operation results into the slightest increase of the velocity-weighted average swirl angle of 1.66° (8.1° up to 9.76°). Under the small flow rate (0.7 Q), 0–960 µm increase of wall roughness give rise to 8.59–11.25° increase in the velocity weighted average swirl angle (2.66° increment). Under the small flow rate (0.7 Q), the same wall roughness increase brings about 3.46° increases (8.88–12.34°) in the velocity weighted average swirl angle. Hence, the increase in wall roughness leads to the increase in velocity-weighted average swirl angle; and the maximum velocity-weighted average swirl angle is obtained under large flow conditions. With reference to the discussion beforehand, the high velocity-weighted average swirl angle may be one of reasons that lead to the low operational efficiency of reverse power generation under the large flow conditions.

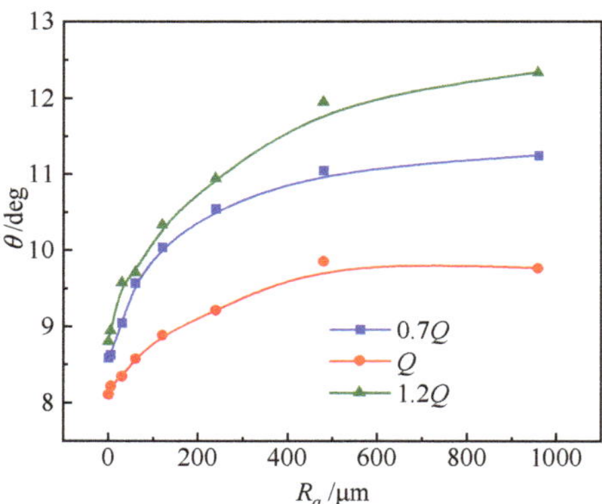

Figure 11. Effect of roughness on velocity-weighted average swirl angle.

3.4.3. Effect of Wall Roughness on Streamline of Outlet Conduit

Streamline at the exit is another aspect to evidence the wall roughness impact on the machine performance. Figure 12 shows the three-dimensional streamlines of the outlet conduit under different flow conditions when the slanted axial flow pump device is used for reverse power generation. Only, three values of wall roughness (R_a = 0 μm, 240 μm, and 960 μm) have been selected for comparative analysis. It can be found that with the increase in wall roughness, the streamlines of the outlet conduit grow disordered.

Under the small flow rates of 0.7 Q, streamlines in the outlet conduit are smooth and less swirling flow appears. This is obviously remarkable within moderate roughness size of R_a = 0 μm and 240 μm. Conversely, backflow phenomena relatively occur in the outlet conduit at higher flow rate (1.2 Q), and they become clearer at a deeper wall roughness (R_a = 960 μm). Under the optimal condition, the flow state in the outlet conduit has minor variation with increasing wall roughness. Thus, the wall roughness has the least influence on the streamline state of the slanted axial flow PAT within such best efficiency operations. Under large flow rates (R_a = 240 μm and 960 μm), the flow state in the outlet conduit becomes worse than that with the smooth wall. With the increase in wall roughness, the flow state in the outlet conduit becomes obviously worse; obvious vortex and backflow phenomena grow quite apparent. In conclusion, swirl flow of the outlet conduit becomes significant when the velocity-weighted average swirl angle at the large flow rate is high with rough surface.

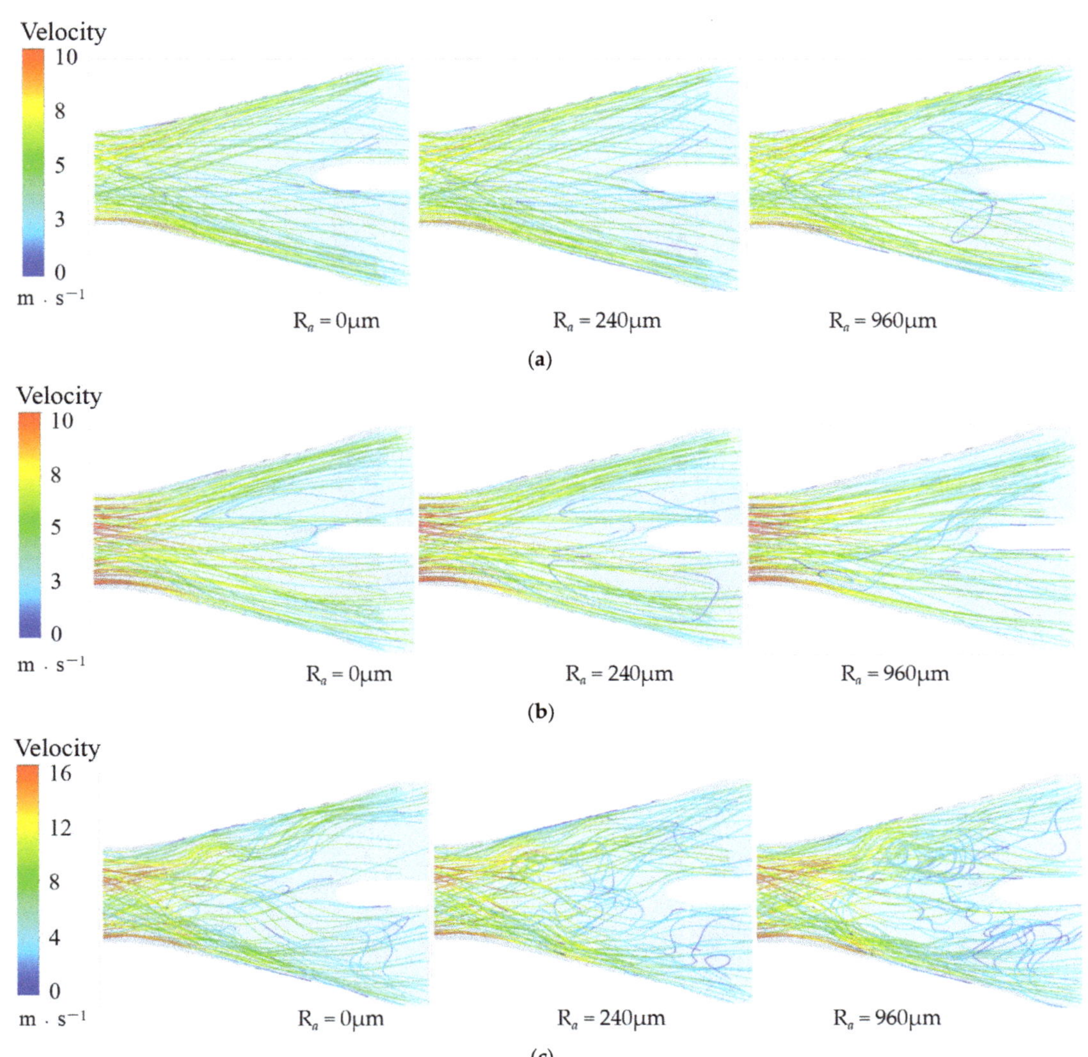

Figure 12. Streamline of the outlet conduit under different flow rates. (**a**) 0.7 Q; (**b**) Q; (**c**) 1.2 Q.

3.4.4. Effect of Wall Roughness on Pressure Distribution over the Blade Pressure Surface

Figure 13 shows the distribution of pressure on blade pressure surface with a changing wall roughness. The pressure on the blade surface obviously varies with a high gradient along the chord direction of the blade, especially at large flow rate. Under the small flow rate, a local low pressure area appears near the leading edge, it increases gradually along the chord direction of the blade, and a local high pressure appears near the impeller shroud of the trailing edge, while there is an opposite variation trend under the larger flow rates.

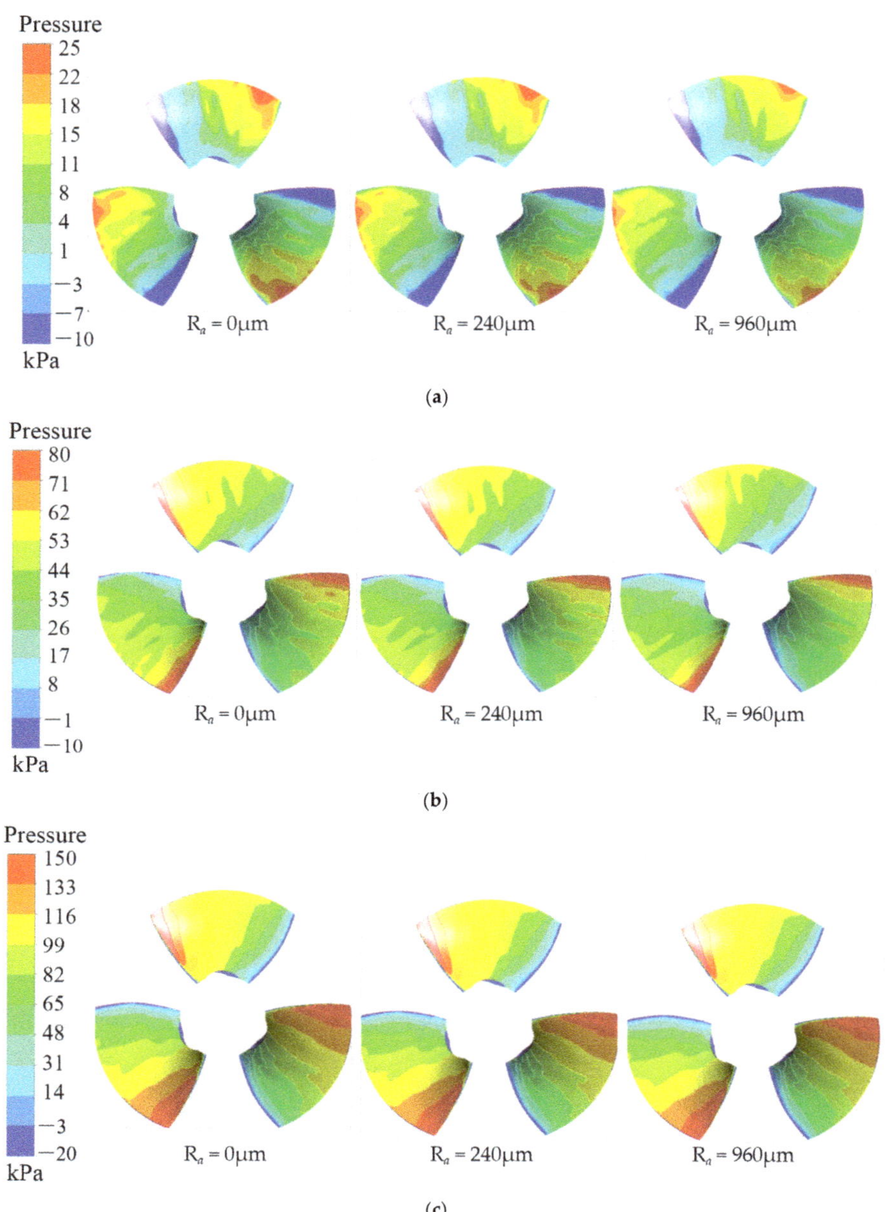

Figure 13. Distribution of pressure on blade pressure surface. (**a**) 0.7 Q; (**b**) Q; (**c**) 1.2 Q.

The negative pressure on the blade is more likely to result in cavitation, which can lead to an increase of wall roughness. At the same flow rate, the negative pressure area becomes larger with the increase of wall roughness. Under the same wall roughness, the negative pressure decreases with the increase of flow rate and the positive pressure increase with the increase in wall roughness. Therefore, the increase of wall roughness easily causes cavitation. In the vicinity of the leading edge, the range of high pressure zone is the largest at the large flow rate, followed by rated flow rate, and the high pressure zone is not obvious at the small flow rate.

Figure 14 shows the variation of pressure at the monitoring point of blade pressure surface with wall roughness under the optimal condition. It can be seen from the figure that with the increase of wall roughness, the pressure decreases gradually, resulting in an increase of energy loss near the boundary layer [31]. In the radial direction (from hub to shroud), the pressure increases gradually. Table 4 shows relative deviation of pressure from 0 μm to 960 μm of monitoring points at different spans. Near the hub, at R_a = 960 μm, the pressure of H1, H2, H3, and H4 is 7752 Pa, 23,318 Pa, 37,201 Pa, and 45,172 Pa, respectively. From Table 4, compared with the smooth wall; the pressure decreases by 11%, 8%, 14%, and 13%, respectively. In the middle of the blade, at R_a = 960 μm, the pressures of M1, M2, M3, M4, and M5 are 119,776 Pa, 54,349 Pa, 37,484 Pa, 34,061 Pa, and 11,626 Pa, respectively, compared with smooth wall, the pressures decrease by 11.4%, 5.2%, 3.6%, 3.2%, and 1.1%, respectively. The pressure of the monitoring point M1 under the smooth wall surface is 135,212 Pa. At R_a = 6 μm, 60 μm, 120 μm, 240 μm, 480 μm, and 960 μm, the pressure of M1 decreases by 1.3%, 2.5%, 4.5%, 7.2%, 8.9%, and 11.4% compared with the smooth wall surface. It indicates that when the roughness is greater than 240 μm, the roughness has a great influence on the pressure near the trailing edge. The pressure of M2, M3, and M4 monitoring points has a similar trend with roughness to that of M1. The minimum pressure value at M5 point under the smooth wall is 13,547 Pa. With the increasing wall roughness, the pressure decreases by 0.36%, 0.12%, 3.5%, 7%, 8.5%, 13.5%, and 18.3%, respectively. It can be seen that when the roughness is less than 60 μm, the roughness has little influence on the pressure near the leading edge. From Figure 12c, at R_a = 960 μm, the pressures of S1, S2, S3, S4, and S5 are 151,730 Pa, 59,123 Pa, 51,258 Pa, 45,943 Pa, and 12,476 Pa, respectively, compared with smooth wall, the pressure decrease by 11%, 14%, 13.2%, 10.8%, and 15%, respectively. It demonstrates that the wall roughness has a great influence on pressure of the blade near the impeller shroud. The results indicate that the boundary layer flow is obviously affected by wall roughness and pressure losses increase significantly due to wall roughness.

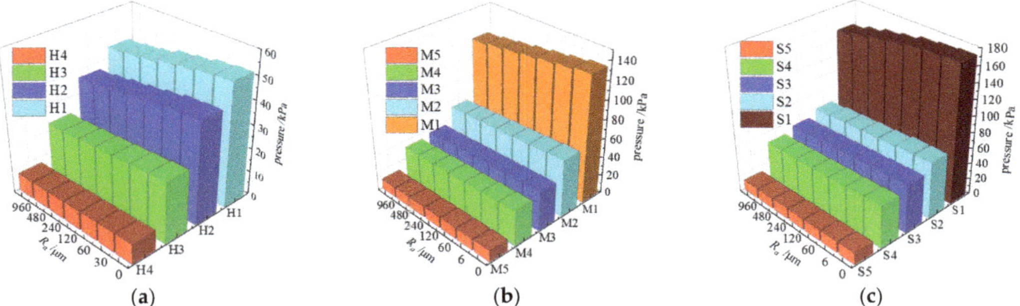

Figure 14. Variation of pressure at the monitoring point with wall roughness. (**a**) span0.1 (**b**) span0.5 (**c**) span0.9.

Table 4. Relative deviation from 0 μm to 960 μm (%).

	Span0.1				Span0.5					Span0.9				
Point	H1	H2	H3	H4	M1	M2	M3	M4	M5	S1	S2	S3	S4	S5
Deviation (%)	11	8	14	13	11	5.2	3.6	3.2	1.1	11	14	13	11	15

3.4.5. Effect of Wall Roughness on Turbulent Kinetic Energy

Turbulent kinetic energy (TKE) is a measure of the turbulence intensity, associated with momentum transport through turbulence eddies in the boundary layer region [32].

The distribution of turbulent kinetic energy on the blade pressure surface is shown in Figure 15. The values of wall roughness of 0 μm, 240 μm, and 960 μm are taken. It can be seen from the figure that, under the same flow condition, the TKE increases with the increase of roughness, especially near the impeller shroud. Under the small flow rate, the TKE near the impeller shroud is obviously higher. Under the optimal condition, when the wall gradually changes from smooth to rough, the TKE is obviously stronger along the radial direction and the distribution range of the TKE increases. The effect of roughness on TKE is more obvious at the large flow rate.

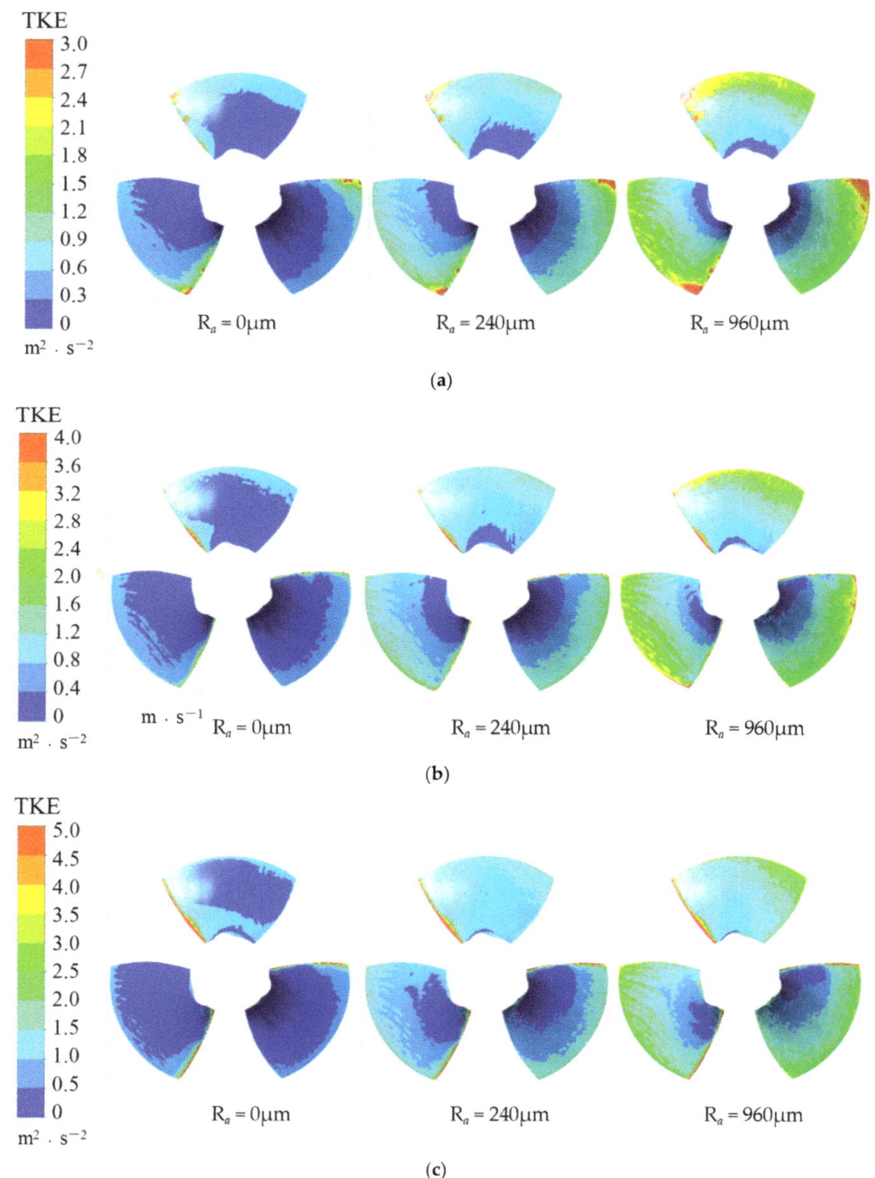

Figure 15. The distribution of TKE on blade pressure surface. (**a**) 0.7 Q; (**b**) Q; (**c**) 1.2 Q.

The distribution of TKE on the blade suction surface is shown in Figure 16. As can it be seen from the figure, the TKE on the suction surface of the blade is larger than that on the pressure surface of the blade under the small flow rate, and the TKE is strengthened near the impeller shroud. Under optimal conditions and large flow conditions, the TKE near the leading edge of the blade increases; this is more obvious under the large flow rate. Under different flow conditions, the distribution range and magnitude of TKE on the blade suction surface increase with the increase in wall roughness.

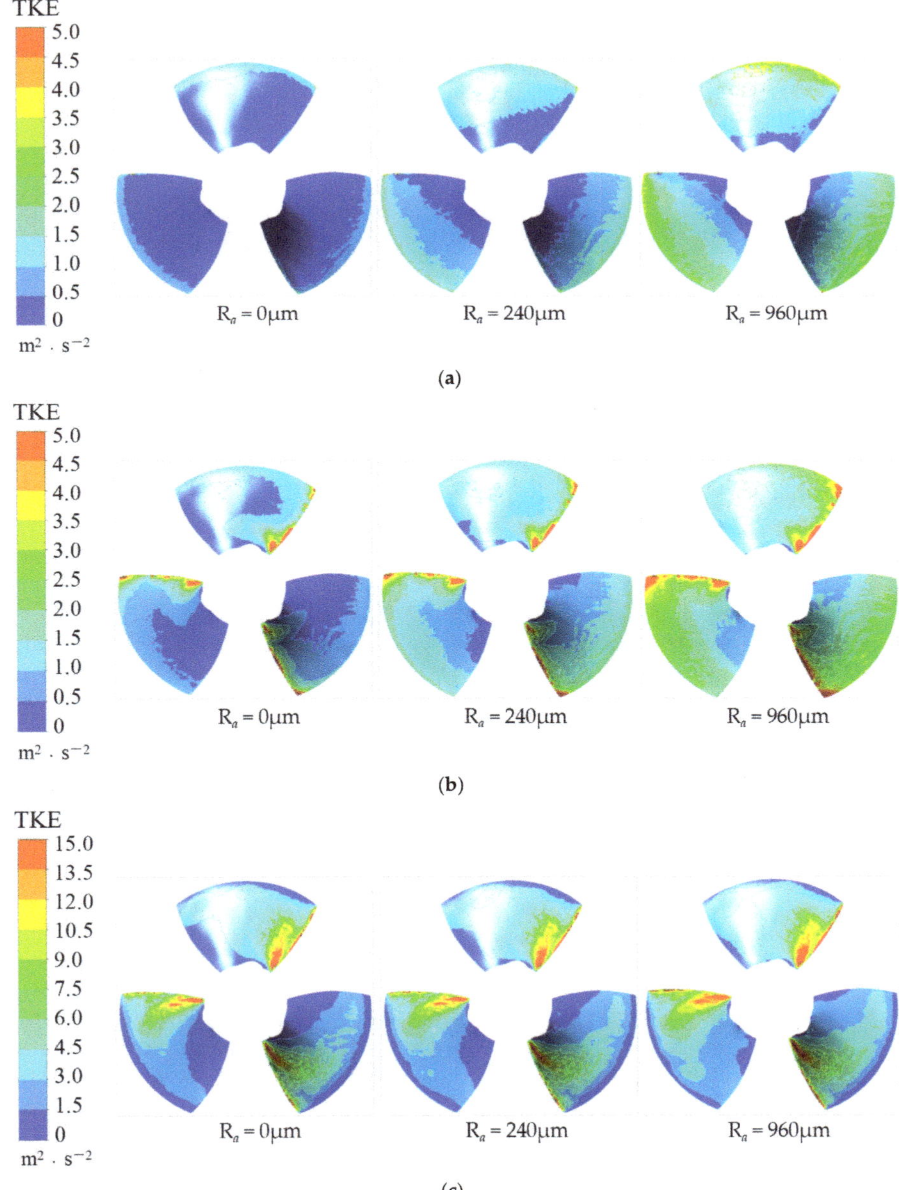

Figure 16. The distribution of TKE on blade suction surface; (**a**) 0.7 Q; (**b**) Q; (**c**) 1.2 Q.

The TKE of blade surface under different working conditions is shown in Figure 17. It can be seen that the TKE gradually increases with the increase in wall roughness, which leads to the increase of energy loss near the boundary layer and is also one of the important reasons for the decrease in efficiency of the slanted axial flow PAT. In each working condition, the TKE of the blade surface increases gradually with the increase in wall roughness, but the variation amplitude is small. When the wall roughness is the same, the TKE of the pressure surface varies little with the flow rate. While the TKE of the suction surface changes obviously with the flow rate and increases significantly with the increase of the flow rate.

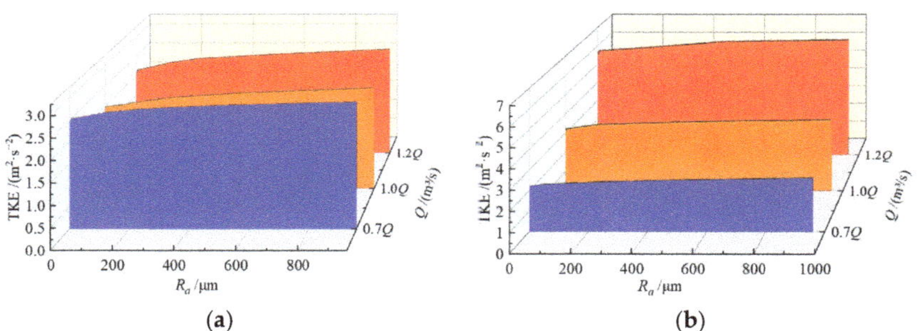

(a)　　　　　　　　　　(b)

Figure 17. TKE of blade surface under different conditions. (**a**) TKE on blade pressure surface (**b**) TKE on blade suction surface.

4. Conclusions

The slanted axial flow PAT operated under turbine mode has been, in this study, numerically simulated. Specifically, the investigation target was the influence of different wall roughness depths on both the external performance and internal flow field characteristics under different working conditions. From narrated findings and pertinent discussion, a number of concluding remarks can be drawn as follows:

(1) Under constant individual flow rates, the gradual deterioration of PAT performance (measured through parameters such as the hydraulic efficiency, head, and shaft power) is conspicuously associated with the increase in wall roughness depth. The later displays a minimum impact on PAT performance only under optimal flow conditions, while for off-design flow conditions, the larger the deviation from the best efficient point, the greater the impact on PAT performance characteristics.

(2) Under turbine operating mode, the increase of wall roughness simultaneously brings about a messy non-uniform distribution of axial velocity and an increase of the velocity-weighted average swirl angle. Furthermore, streamlines within the discharge conduit reflect a disorderly flow pattern, eventually giving rise to backflow structures.

(3) Ultimately, the wall roughness accumulation remarkably triggers the increase of energy losses. This is evidenced by the drop of static pressure on the blade pressure surface and the increase of TKE on the blade. The latter is particularly evident near the impeller shroud. Under the same roughness conditions, the TKE on the blade suction surface proves to be greater than that on the blade pressure surface.

Given the level of PAT performance degradation that can be caused by the machine wall roughness, future research endeavors are encouraged in this field, where the effects of wall roughness can be investigated further and a wider range of machine operating conditions and roughness depths can be considered. This would help monitoring PAT's lengthy operations and assist the timely equipment refurbishment or replacement without highly impairing the machine performance.

Author Contributions: Conceptualization, K.K.; methodology, K.K.; software, Q.Z., Z.X. and J.Z.; validation, K.K. and Q.Z.; formal analysis, K.K., Q.Z. and J.Z.; investigation, K.K. and J.Z.; resources, Y.Z. and H.X.; data curation, Y.Z. and H.X.; writing—original draft preparation, K.K., Q.Z. and Z.X.; writing—review and editing, Y.Z., H.X. and A.M.; visualization, K.K.; supervision, Y.Z. and H.X.; funding acquisition, K.K. All authors have read and agreed to the published version of the manuscript.

Funding: This research was funded by the National Natural Science Foundation of China (52009033), Natural Science Foundation of Jiangsu Province (BK20200509), the Fundamental Research Funds for the Central Universities (B210202066).

Institutional Review Board Statement: Not applicable.

Informed Consent Statement: Not applicable.

Data Availability Statement: All the data in this paper are obtained by physical experiment and numerical simulation, respectively, and the data used to support the findings of this study are available from the corresponding author upon request.

Conflicts of Interest: The authors declare no conflict of interest.

References

1. Yang, F.; Li, Z.; Hu, W.; Liu, C.; Jiang, D.; Liu, D.; Nasr, A. Analysis of flow loss characteristics of slanted axial-flow pump device based on entropy production theory. *R. Soc. Open Sci.* **2022**, *9*, 211208. [CrossRef] [PubMed]
2. Kan, K.; Yang, Z.; Lyu, P.; Zheng, Y.; Shen, L. Numerical study of turbulent flow past a rotating axial-flow pump based on a level-set immersed boundary method. *Renew. Energy* **2021**, *168*, 960–971. [CrossRef]
3. Shi, L.; Zhang, W.; Jiao, H.; Tang, F.; Wang, L.; Sun, D.; Shi, W. Numerical simulation and experimental study on the comparison of the hydraulic characteristics of an axial-flow pump and a full tubular pump. *Renew. Energy* **2020**, *153*, 1455–1464. [CrossRef]
4. Shi, L.; Yuan, Y.; Jiao, H.; Tang, F.; Cheng, L.; Yang, F.; Jin, Y.; Zhu, J. Numerical investigation and experiment on pressure pulsation characteristics in a full tubular pump. *Renew. Energy* **2021**, *163*, 987–1000. [CrossRef]
5. Liu, Y.; Tan, L. Tip clearance on pressure fluctuation intensity and vortex characteristic of a mixed flow pump as turbine at pump mode. *Renew. Energy* **2018**, *129*, 606–615. [CrossRef]
6. Kan, K.; Zhang, Q.; Xu, Z.; Zheng, Y.; Gao, Q.; Shen, L. Energy loss mechanism due to tip leakage flow of axial flow pump as turbine under various operating conditions. *Energy* **2022**, *255*, 124532. [CrossRef]
7. Yang, S.S.; Derakhshan, S.; Kong, F.Y. Theoretical, numerical and experimental prediction of pump as turbine performance. *Renewable. Energy* **2012**, *48*, 507–513. [CrossRef]
8. Jain, S.V.; Patel, R.N. Investigations on pump running in turbine mode: A review of the state-of-the-art. *Renew. Sustain. Energy Rev.* **2014**, *30*, 841–868. [CrossRef]
9. Aldaş, K.; Yapıcı, R. Investigation of effects of scale and surface roughness on efficiency of water jet pumps using CFD. *Eng. Appl. Comput. Fluid Mech.* **2014**, *8*, 14–25. [CrossRef]
10. Han, Y.; Tan, L. Influence of rotating speed on tip leakage vortex in a mixed flow pump as turbine at pump mode. *Renew. Energy* **2020**, *162*, 144–150. [CrossRef]
11. Lin, T.; Zhu, Z.; Li, X.; Li, J.; Lin, Y. Theoretical, experimental, and numerical methods to predict the best efficiency point of centrifugal pump as turbine. *Renew. Energy* **2021**, *168*, 31–44. [CrossRef]
12. Sun-Sheng, Y.; Fan-Yu, K.; Jian-Hui, F.; Ling, X. Numerical research on effects of splitter blades to the influence of pump as turbine. *Int. J. Rotating Mach.* **2012**, *2012*, 123093. [CrossRef]
13. Capurso, T.; Bergamini, L.; Camporeale, S.M.; Fortunato, B.; Torresi, M. CFD analysis of the performance of a novel impeller for a double suction centrifugal pump working as a turbine. In Proceedings of the 13th European Conference on Turbomachinery Fluid dynamics & Thermodynamics. European Turbomachinery Society, Lausanne, Switzerland, 24–28 April 2019.
14. Derakhshan, S.; Mohammadi, B.; Nourbakhsh, A. Efficiency improvement of centrifugal reverse pumps. *J. Fluids Eng.* **2009**, *131*, 021103. [CrossRef]
15. Morabito, A.; Vagnoni, E.; Di Matteo, M.; Hendrick, P. Numerical investigation on the volute cutwater for pumps running in turbine mode. *Renew. Energy* **2021**, *175*, 807–824. [CrossRef]
16. Gülich, J.F. Effect of Reynolds number and surface roughness on the efficiency of centrifugal pumps. *J. Fluids Eng.* **2003**, *125*, 670–679. [CrossRef]
17. Bai, T.; Liu, J.; Zhang, W.; Zou, Z. Effect of surface roughness on the aerodynamic performance of turbine blade cascade. *Propuls. Power Res.* **2014**, *3*, 82–89. [CrossRef]
18. Bellary, S.A.I.; Samad, A. Exit blade angle and roughness effect on centrifugal pump performance. In Proceedings of the Gas Turbine India Conference, Bangalore, Karnataka, 5–6 December 2013; American Society of Mechanical Engineers: New York, NY, USA; pp. 1–10.

19. Deshmukh, D.; Samad, A. CFD-based analysis for finding critical wall roughness on centrifugal pump at design and off-design conditions. *J. Braz. Soc. Mech. Sci. Eng.* **2019**, *41*, 1–18. [CrossRef]
20. Juckelandt, K.; Bleeck, S.; Wurm, F.H. Analysis of losses in centrifugal pumps with low specific speed with smooth and rough walls. In Proceedings of the 11th European Conference on Turbomachinery Fluid dynamics & Thermodynamics, European Turbomachinery society, Madrid, Spain, 23–27 March 2015.
21. He, X.; Jiao, W.; Wang, C.; Cao, W. Influence of surface roughness on the pump performance based on Computational Fluid Dynamics. *IEEE Access* **2019**, *7*, 105331–105341. [CrossRef]
22. Limbach, P.; Skoda, R. Numerical and experimental analysis of cavitating flow in a low specific speed centrifugal pump with different surface roughness. *J. Fluids Eng.* **2017**, *139*, 101201. [CrossRef]
23. Lim, S.E.; Sohn, C.H. CFD analysis of performance change in accordance with inner surface roughness of a double-entry centrifugal pump. *J. Mech. Sci. Technol.* **2018**, *32*, 697–702. [CrossRef]
24. Kan, K.; Xu, Z.; Chen, H.; Xu, H.; Zheng, Y.; Zhou, D.; Muhirwa, A.; Maxime, B. Energy loss mechanisms of transition from pump mode to turbine mode of an axial-flow pump under bidirectional conditions. *Energy* **2022**, *257*, 124630. [CrossRef]
25. Kan, K.; Chen, H.; Zheng, Y.; Zhou, D.; Binama, M.; Dai, J. Transient characteristics during power-off process in a shaft extension tubular pump by using a suitable numerical model. *Renew. Energy* **2021**, *164*, 109–121. [CrossRef]
26. He, X.; Zhang, Y.; Wang, C.; Zhang, C.; Cheng, L.; Chen, K.; Hu, B. Influence of critical wall roughness on the performance of double-channel sewage pump. *Energies* **2020**, *13*, 464. [CrossRef]
27. Shockling, M.A.; Allen, J.J.; Smits, A.J. Roughness effects in turbulent pipe flow. *J. Fluid Mech.* **2006**, *564*, 267–285. [CrossRef]
28. Adams, T.; Grant, C.; Watson, H. A simple algorithm to relate measured surface roughness to equivalent sand-grain roughness. *Int. J. Mech. Eng. Mechatron.* **2012**, *1*, 66–71. [CrossRef]
29. Ji, L.; Li, W.; Shi, W.; Tian, F.; Agarwal, R. Diagnosis of internal energy characteristics of mixed-flow pump within stall region based on entropy production analysis model. *Int. Commun. Heat Mass Transf.* **2020**, *117*, 104784. [CrossRef]
30. Sha, L.; Yang, S.; Kong, X.; Sun, H.; Pei, L. Influence of inlet flow field uniformity on the performance of the nuclear main pump. *IOP Conf. Ser. Mater. Sci. Eng.* **2021**, *1081*, 012024. [CrossRef]
31. Skrzypacz, J.; Bieganowski, M. The influence of micro grooves on the parameters of thecentrifugal pump impeller. *Int. J. Mech. Sci.* **2018**, *144*, 827–835. [CrossRef]
32. Li, W.; Ji, L.; Li, E.; Shi, W.; Agarwal, R.; Zhou, L. Numerical investigation of energy loss mechanism of mixed-flow pump under stall condition. *Renew. Energy* **2021**, *167*, 740–760. [CrossRef]

Article

Roadmap to Profitability for a Speed-Controlled Micro-Hydro Storage System Using Pumps as Turbines

Florian Julian Lugauer [1,2,*], Josef Kainz [1,2], Elena Gehlich [2] and Matthias Gaderer [2]

1 Energy Technology Department, Weihenstephan-Triesdorf University of Applied Sciences, 85354 Freising, Germany; josef.kainz@hswt.de
2 Campus Straubing for Biotechnology and Sustainability, Technical University of Munich, 94315 Straubing, Germany; elena.gehlich@tum.de (E.G.); gaderer@tum.de (M.G.)
* Correspondence: florian.lugauer@hswt.de; Tel.: +49-9421-187-273

Abstract: Storage technologies are an emerging element in the further expansion of renewable energy generation. A decentralized micro-pumped storage power plant can reduce the load on the grid and contribute to the expansion of renewable energies. This paper establishes favorable boundary conditions for the economic operation of a micro-pump storage (MPS) system. The evaluation is performed by means of a custom-built simulation model based on pump and turbine maps which are either given by the manufacturer, calculated according to rules established in studies, or extended using similarity laws. Among other criteria, the technical and economic characteristics regarding micro-pump storage using 11 pumps as turbines controlled by a frequency converter for various generation and load scenarios are evaluated. The economical concept is based on a small company (e.g., a dairy farmer) reducing its electricity consumption from the grid by storing the electricity generated by a photovoltaic system in an MPS using a pump as a turbine. The results show that due to the high specific costs incurred, systems with a nominal output in excess of around 22 kW and with heads beyond approximately 70 m are the most profitable. In the most economical case, a levelized cost of electricity (LCOE) of 29.2 €cents/kWh and total storage efficiency of 42.0% is achieved by optimizing the system for the highest profitability.

Keywords: pump as turbine; micro-hydro pump storage; small-scale hydropower; simulation model; turbine maps

Citation: Lugauer, F.J.; Kainz, J.; Gehlich, E.; Gaderer, M. Roadmap to Profitability for a Speed-Controlled Micro-Hydro Storage System Using Pumps as Turbines. *Sustainability* **2022**, *14*, 653. https://doi.org/10.3390/su14020653

Academic Editors: Mosè Rossi, Massimiliano Renzi, David Štefan and Sebastian Muntean

Received: 2 November 2021
Accepted: 3 January 2022
Published: 7 January 2022

Publisher's Note: MDPI stays neutral with regard to jurisdictional claims in published maps and institutional affiliations.

1. Introduction

Renewable energy generation is often volatile and location-dependent by nature. Its integration into power networks usually requires great flexibility and puts a strain on the grid. To reduce the load on the grid, it must either be expanded, become smarter or storage technologies must be implemented [1]. The requirements for storage technologies are manifold, and no single storage technology fulfills them all. Each energy storage technology has unique characteristics, and the optimal choice depends on the task to be fulfilled [2]. Micro-pump storage systems are one of the storage technologies used for decentralized energy storage and can therefore reduce the load on the network. The pump as turbine (PAT) technology is well-established in a number of water distribution networks for energy recovery [3,4], with a possible payback period of six years [5], by pressure reduction and reducing water loss [6] or for improving network effectiveness [7,8]. It can also be beneficial to enhance the sustainability of such networks [9]. Additionally, studies on PATs have already developed integration strategies for water distribution networks [10,11] or methodologies to select pumps from catalogues [12]. Simulation models were also developed to determine the number and optimal locations of operations as well as the optimal PATs to minimize costs [13]. For instance, a recent publication has shown that the specific decentralized installation of PATs in a water distribution network can be used to charge e-bikes or scooters by reducing the pressure in the network [14]. Something comparable could be achieved by using a PAT in a micro-pump storage (MPS) system, thus reducing the load on the power grid. A publication [15] preceding the present contribution investigated various operating strategies for an MPS system with a pump as a turbine. Therein, speed control using a frequency transformer (FUM) was identified as the best strategy in terms of economy. It has also been proven that the efficiency of pumped hydro storage can be significantly increased by varying the speed with an FUM [16]. The result of a publication on electricity costs in India shows that MPS systems can be economically viable [17]. In contrast, other publications concerned with modelling small hydropower plants [18–23] in combination with photovoltaic (PV) systems [24], wind turbines [25], or both [26], leave open the question of how such a system can be operated economically.

A pilot MPS plant using a PAT was recently implemented in Froyennes, Belgium [27], where the energy produced by a variety of PV systems and a few small wind turbines is stored in the MPS in the event of a local energy surplus and is then withdrawn at a later point in time. Attaining profitability with an MPS system is a challenge. For example, the levelized cost of electricity (LCOE) for storing and retrieving energy in Froyennes is €1.06/kWh [27], which is far above current energy prices. Throughout this paper, this pilot plant is used as an example to illustrate the general boundary conditions that are required for economical operation. For instance, the ratio of length to head (L/H) at the MPS in Froyennes is about 8. Since this ratio is usually much lower at other pumped storage power plants, there is great potential for cost reduction [27]. Depending on the filling level of the reservoir, there is only a maximum head of 11 m, which also brings the potential for improvement. The efficiency of the PAT is 72.7%, but there are now PATs with better efficiency (see Section 2.2.2), which should also lead to a better result for the LCOE. Furthermore, in Froyennes, a 30 kW motor/generator (M/G) was used for a maximum pump power of 17 kW; a better adapted M/G could reduce the investment costs. This publication is intended to identify potential for improvement and therefore focuses on how a MPS can be operated economically and what boundary conditions are necessary for achieving this goal.

In order to be able to evaluate the economic efficiency of such a system, pump and turbine maps are required as input for a simulation model. Since turbine maps, in contrast to pump maps, are usually not publicly available, turbine maps usually have to be measured or calculated. Over the years, quite a few calculation methods have been developed, e.g., by Gülich [28], Alatorre-Frenk [29], Derakhshan [30], Yang [31], and T. Lin [32], and have been compared with experimental measurements [33,34] or with other prediction methods [35]. Prediction models have also been developed, especially for multi-stage PATs [36]. The accuracy of the turbine characteristics obtained depends on the pump mode data available as a basis, and a certain degree of inaccuracy must always be taken into account. Due to the fact that a test bench is very expensive, especially when covering the full performance range considered for this paper, the calculation methods are still necessary. Based upon the given calculated turbine characteristics, a numerical model of the MPS system can be used to simulate its operation and evaluate it according to economic and technical criteria. In particular, a roadmap can be established for illustrating how an MPS system could be profitably operated, which is the major goal pursued by the present paper.

This paper is structured as follows: Section 2 describes the system calculation methods and affinity laws and is followed by a short description of the simulation model employed. In Section 3, the results of the technical and economic efficiency simulations for various generation and load profiles are presented, discussed, and used to sketch the roadmap to profitability for MPS. To cover a larger scope of load and generation scenarios, Section 4 introduces various scaling factors for the load and generation profile which are examined using the most promising pump of Section 3, the KSB 8065200.

2. Materials and Methods

The present section describes the procedure for evaluating the economic viability of an MPS. To this end, the functionality of the MPS, the basic data, the methodology of the simulation model, as well as the calculation and extension of the turbine maps are described.

2.1. Concept and Initial Data

Figure 1 presents a schematic diagram of a speed-controlled micro-pump storage plant that might be used by a small company (e.g., a dairy farmer). A centrifugal pump, which operates both as a pump and a turbine, is located between the lower and the upper reservoirs. The speed-controlled plant is connected to the power grid, to a PV system (specific yield = $1.15\frac{kWh}{kW_p}$) and to the consumer via an electric motor or generator (M/G) and a frequency converter.

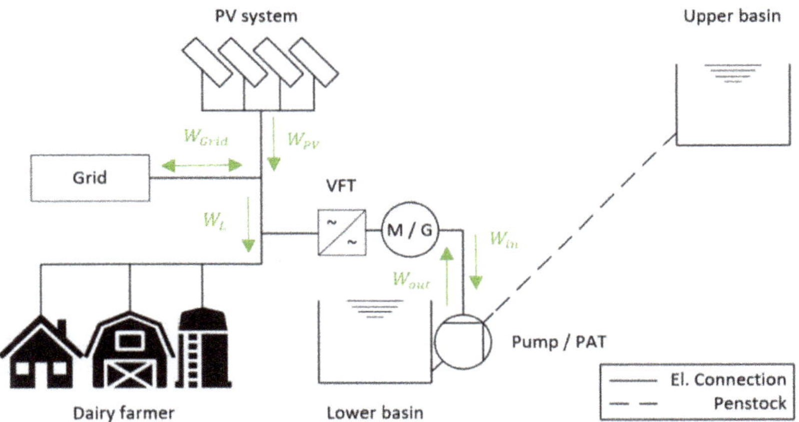

Figure 1. Schema of the speed-controlled MPS including dairy farmer, grid, and PV system, energy demand W_L, grid energy W_{Grid}, total energy input W_{in}, total energy output W_{out}, PV-energy W_{PV} [15].

The functionality of the MPS can be understood as follows. In pump mode, active during periods of low energy consumption and high solar power production, water is pumped into an upper basin, storing potential energy for later use. During times of high power consumption and low PV output, turbine mode is active, with water flowing back into the lower basin and powering the centrifugal pump, which now acts in turbine operation. Thus, the energy demand from the grid can be reduced, and more of the energy generated by the PV system can be used on site.

In the previous publication [15], the optimal location (i.e., optimal head and storage volume) for a given pump was identified for fixed energy production from the PV system and given energy consumption. To investigate the effects of various consumption and generation scenarios, a scaling factor for PV power S_{PV} and energy consumption S_L has now been introduced. The scaling factors of 1.0 are based on the values of the previous publication, i.e., a PV power of 60 kWp, an annual energy production of $W_{PV} = 69.1$ MWh, and an annual energy demand W_L of 43.0 MWh. The ratio of W_{PV}/W_L is therefore around 1.6. For convenience, the relation between the scaling factors and the energy production or demand is shown in Table 1.

Table 1. Energy (per year) produced by the PV system W_{PV}, energy demand W_L, scaling factor S_L (corresponding to W_L) and S_{PV} (corresponding to W_{PV}).

Scaling Factor S_L or S_{PV}	0.5	0.8	1.0	1.2	1.5	2.0	3.0	4.0
W_L [MWh]	22	34	43	52	65	86	129	172
W_{PV} [MWh]	35	55	69	83	103	138	207	276

The penstock was assumed to have a nominal diameter of DN150, with costs of 50 €/m pipe. The slope was assumed to be 15%. One of the two basins was supposed to have already existed. The second, a snow-making style of water reservoir design, was assumed to incur construction costs of 40 €/m^3 [37,38]. In this publication, due to the many different sizes of centrifugal pumps, no individual costs for the electrical installation were taken into account. Because of that, results for the KSB Etanorm 50160D174 deviate slightly from the results from the last study [15]. The investment costs of the PV system were not included in the economic efficiency calculation because this study was based on the assumption that a PV system was already available, e.g., after expiration of the subsidy by the Germany renewable energy act (EEG)

2.2. Map Calculation

The characteristic maps of various centrifugal pumps for turbine and pump operation were required in order to be able to run simulations and compare several centrifugal pumps with each other. While pump maps are provided in full for each specific pump by the respective manufacturer, turbine maps are usually not provided. For this paper, turbine maps that were not provided (or only provided in partial areas) by the manufacturer were either calculated and/or extended via similarity relationships.

2.2.1. Calculation of Turbine Characteristics at Nominal Speed

The calculation of the characteristic curves was based on empirical correlations according to J.F. Gülich [28]. The exact calculation method is explained in detail in Appendix A and briefly explained here: All values (head, flow, speed, power, and efficiency) for the BEP in pump mode are needed for determining the turbine best efficiency point (BEP). The turbine BEP values were calculated from the pump BEP. All required values for the nominal speed characteristic curve were determined from the BEP in turbine operation thus computed. The formulas were based on calculations from eight different correlations using 35 measured pumps, with a specific speed n_q in the range between 12 and 190. Due to the small number of pumps upon which the correlations were based, considerable deviation should be expected when using these calculations, which is why Q and H may be up to 20% different than the calculation predictions [28].

2.2.2. Extension of Turbine Maps

In order to be able to analyze speed control and to determine the optimum head, it is necessary that a certain range of the characteristic map be known. The law of similarity provides the possibility of calculating further characteristic curves for other speeds from a single measured or given speed characteristic curve. The values calculated in Section 2.2.1 (Q, H, P) for nominal speed are extended to create a wider range of the turbine map according to Gülich using similarity relationships [28]. Likewise, maps obtained from the manufacturer covering only a subrange of the full turbine operation map are extended in the same way. In order to calculate an extended turbine map, the head h, the flow Q, and the power for various speeds n are calculated. The similarity laws given in formulas (1)–(3) are based on the respective rated quantities, i.e., speed n_r, and the corresponding flow Q_r, head H_r, and power P_r, from which the turbine behavior for other speeds and operating conditions can be calculated:

$$Q(n) = Q_r \cdot \frac{n}{n_r} \tag{1}$$

$$H(n) = H_r \cdot \left(\frac{n}{n_r}\right)^2 \tag{2}$$

$$P_{T,mech}(n) = P_r \cdot \left(\frac{n}{n_r}\right)^3 \tag{3}$$

One prerequisite for the affinity laws is that the flows in the channels, such as in the impellers, behave similarly kinematically and dynamically. As a consequence, the corresponding inertial and frictional forces behave similarly [28]. In addition, it is assumed that the fluid is incompressible, and no cavitation occurs. When calculating the efficiency with formula (3), there is a deviation, which results from the fact that there are losses that do not increase with the cube of the speed. Among these are the mechanical losses in the bearing and shaft sealing, as well as the disc friction loss [39].

The hydraulic power P_{hydr} is determined for the respective speed n for various points on the speed characteristic curve (formula (3) [40].

$$P_{hydr}(n) = Q_T(n) \cdot \rho \cdot g \cdot H \tag{4}$$

The mechanical efficiency $\eta_{T,mech}$ can now be obtained with [40]:

$$\eta(n)_{T,mech} = \frac{P_{T,mech}(n)}{P_{hydr}(n)} \tag{5}$$

The extended maps are now created by means of interpolation from the associated data of the extended speed curves. There is a set of two maps for the turbine mode. The first map $H_T(Q_T, n)$ represents the dependency of head H_T on flow and rotational speed, while the second map, $P_T(Q_T, n)$ describes the power $P_{T,mech}$.

In order to exemplarily check the accuracy of the calculation and expansion of the turbine maps, the map of the KSB50160174 (see Table 2) was calculated from BEP pump data alone using the method described above. Comparing the calculated turbine maps with measured maps provided by the manufacturer, an efficiency deviation of only one percentage point was seen in the high rpm range. For lower speeds, the difference increased to up to four percentage points. As the pump/turbine often operates close to its nominal speed, the high-speed range is most relevant for us and, therefore, a fairly good overall accuracy can be expected.

Table 2. The centrifugal pumps investigated, sorted by their nominal power P_n at their best efficiency given by the manufacturer $\eta_{P,opt}$, average electrical pump efficiency $\overline{\eta}_{P,el}$, average electrical turbine efficiency $\overline{\eta}_{T,el}$, average electrical turbine efficiency η_{tot}, pump investment I_P and investment for motor and frequency inverter I_{el}. All results here are given for a scaling factor $S_{PV} = S_L = 1.0$.

Pump	P_n	$\eta_{P,opt}$	$\overline{\eta}_{P,el}$	$\overline{\eta}_{T,el}$	η_{tot}	I_P	I_{el}
F065-200A-1102H [1]	11.0 kW	71.1%	64.2%	58.5%	37.5%	3347 €	9323 €
EST 65-160 [2]	15.0 kW	80.9%	67.6%	58.9%	38.7%	4246 €	7343 €
F080-255A-1502H [1]	15.0 kW	75.4%	63.6%	56.0%	35.3%	3419 €	10,475 €
KSB 50160174 [2]	15.0 kW	76.0%	67.7%	62.3%	42.1%	3808 €	11,575 €
F080-255A-1852H [1]	18.5 kW	75.9%	64.7%	57.5%	37.1%	4420 €	12,372 €
F080-255A-2202H [1]	22.0 kW	78.4%	65.9%	60.9%	40.1%	4699 €	14,348 €
F080-240A-2202H [1]	22.0 kW	82.3%	69.3%	53.8%	37.2%	4377 €	13,864 €
F080-330A-2204H [1]	22.0 kW	78.1%	63.2%	56.1%	35.4%	4893 €	15,585 €
KSB 8065200 [3]	30.0 kW	80.9%	65.2%	58.6%	38.2%	2579 €	12,022 €
F080-255A-3002H [1]	30.0 kW	79.2%	63.1%	59.1%	37.2%	4999 €	17,465 €
F080-255A-3702H [1]	37.0 kW	79.3%	62.4%	61.1%	38.1%	4871 €	18,334 €

[1] turbine map calculated, [2] turbine map given by the manufacturer, [3] turbine map partially given by manufacturer and extended via similarity relationships.

2.3. Simulation Model

The simulation model has also been described in detail in [15] and is reproduced for easier access in Appendix B of this work. A short description of the program sequence is provided as follows. After initiating the simulation, all of the necessary data are collected. The main task of the program is to determine the water volume in the storage reservoir. This is performed by calculating the flow rates in pump and turbine operation. To solve this numerically, the flow has to be obtained from characteristic maps which describe, respectively, the head, the efficiency, and the power as a function of the speed and the flow rate. The difference between PV power and power from the load profile results in the possible pump or turbine power depending on flow and speed, taking into account the efficiency of the electrical components. Taking pipe losses into account, the system's characteristic curve is defined to determine the resulting head. The operating point is then determined. The flow rate and speed at the intersection point can then be read directly from the characteristic maps while also taking into account the operating limits for pump or turbine operation.

The simulation model then returns results for every point in the time of the year for all system variables. The annual electricity balance can then be obtained from these results, and a profitability calculation can be performed.

2.4. Finding the Optimal Site for Each PAT and Scaling Factor

The plant simulation model was run within a MATLAB optimizer [15] used to determine an optimal site (as defined by the geodetic head H_{geo} and the volume of the storage V) for each of the PATs as listed in Table 2 as well as the scaling factors provided in Table 1. By using the MATLAB 2019a simulated annealing algorithm [41,42], the economic outcome can be maximized by varying H_{geo} and V. The results for the optimal site (H_{optim}, V_{optim}) for each scaling factor S_{PV} and S_L are obviously specific to the PAT type chosen.

3. Results and Discussion

In this section, the investigated pumps are listed, and the results of the simulation model for the head, total efficiency, average turbine and pump efficiency, and levelized cost of energy (LCOE) are presented and discussed. At first (Section 3), the scaling factors S_{PV} and S_L are set to the same value (i.e., $S_{PV} = S_L$), while in Section 4, S_{PV} and S_L are varied independently. As will be shown later, the specific component costs play an important role, which is why they are discussed in more detail after describing the different pumps investigated.

3.1. Pumps Investigated

The following Table 2 lists 11 suitable pumps, examined with increasing nominal power P_n. The scope of the pumps evaluated is chosen to cover the range from 5 to 50 kW, which defines micro-hydropower stations based on the classification of the Latin American Energy Organization (OKADE) [43].

In addition to the pump efficiency $\eta_{P,opt}$ given by the manufacturer for the BEP, the table also shows the average efficiency for turbine ($\overline{\eta}_{T,el}$) and pump ($\overline{\eta}_{P,el}$) as well as the total efficiency η_{tot} for a scaling factor of 1.0. The values of ($\overline{\eta}_{T,el}$) and ($\overline{\eta}_{P,el}$) are calculated from simulation results for the optimal size and scaling factor using trapezoidal numerical integration [44] with a step size of 0.1 min using the following formulae:

$$\overline{\eta}_{P,el} = \frac{\int Q_P dt \cdot \rho \cdot g \cdot H_{geo}}{W_{in}} \tag{6}$$

$$\overline{\eta}_{T,el} = \frac{W_{out}}{\int Q_T dt \cdot \rho \cdot g \cdot H_{geo}} \tag{7}$$

The total efficiency η_{tot} is defined as the quotient of the total electrical energy output $W_{out} = \int P_{P,el}dt$ and the total electrical energy input $W_{in} = \int P_{T,el}dt$:

$$\eta_{tot} = \frac{W_{out}}{W_{in}} \cdot 100\% \tag{8}$$

As Table 2 shows, the highest pump efficiency at the BEP in pump mode does not indicate the highest total efficiency. Comparing the EST 65-160 pump, with high manufacturer-specified efficiency of 80.9% in pump mode, with the F080-255A-2202 H, which has only a manufacturer-specified pump efficiency of 78.4%, the total efficiency of the F080-255A-2202H exceeded that of the EST 65-160 by about 1.4 percentage points. The better efficiency is mainly based on a better average electrical efficiency in turbine operation of the F080-255A-2202H and may also be related to better efficiency in the part-load range. Comparing the efficiency of the pumps listed in Table 2 with the pump in Froyennes ($\eta_{p,opt} = 72.7\%$) [27], it becomes clear that, with the exception of the F065-200A-1102H, each of the pumps listed in Table 2 has better efficiency. It should therefore be possible to use a PAT with higher efficiency for an MPS.

3.2. Specific Costs

Figure 2 shows the specific prices (€/kW rated power) of the pump including a motor (e.g., F080-255A) [45], asynchronous motor (e.g., VEM PS1R/PS2R) [46], and a frequency converter (e.g., ABB ASC 580) [47]. As expected, the specific prices decreased with the rising power. The drop is especially steep for frequency converters and pumps including motors with an output of less than 22 kW.

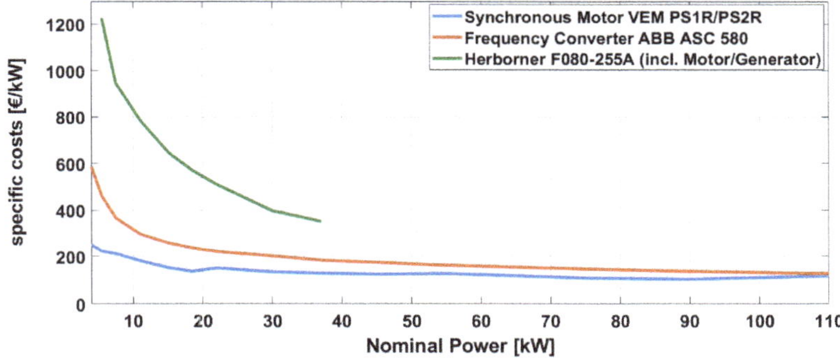

Figure 2. Specific costs on the example of the VEM PS1R/PS2R synchronous motor, the ABB ASC 580 FUM, and the Herborner F080-255A (including motor/generator).

From a nominal output greater than approximately 22 kW, the curve flattens, and the prices per kW nominal power begin to flatten. This indicates that smaller pumped storages with outputs of below 22 kW are difficult to realize economically due to a high specific cost. To illustrate these findings, the results for centrifugal pumps with a nominal output of less than 22 kW are marked with solid lines in a subsequent part of this paper, whereas centrifugal pumps with a nominal output of greater than 22 kW are marked with dash–dot lines.

3.3. Total Efficiency

Plotting the simulated values of the total efficiencies for the various scaling factors (see Figure 3), it becomes clear that for small-scaling factors, the KSB Etanorm 50160D174 pump has the best total efficiency for an output smaller than 22 kW, and it also shows the best total efficiency for scaling factors of less than two. The excellent total efficiency of the KSB 50160D174, especially for smaller scaling factors, is also reflected in the electrical pump and turbine efficiency (see Figures 4 and 5). Furthermore, a clear increase in the efficiency for factors increasing from 0.5 to 1 can be noticed and is related to the increase in the head (see Figure 6). The highest total efficiencies in all of the pumps investigated (efficiency of about 42%) were achieved by the KSB 8065200 and 50160D174, depending on the scaling factors. The KSB8065200's pump and turbine efficiencies (see Figures 4 and 5) were correspondingly rather high.

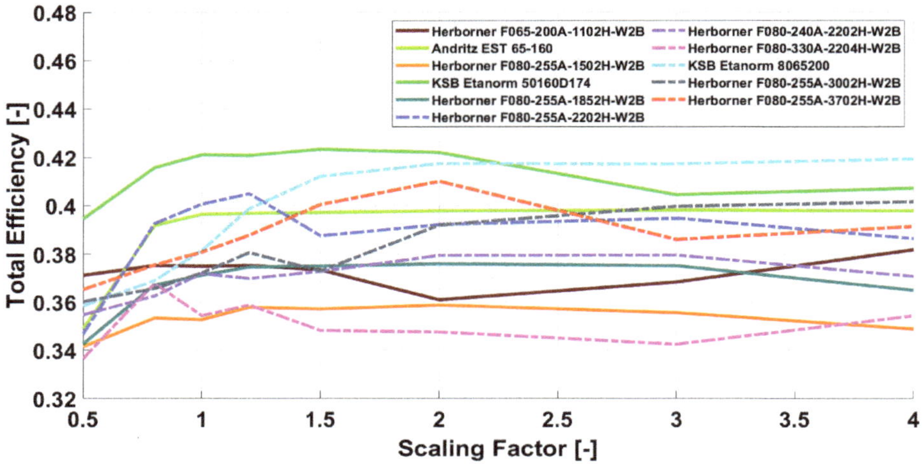

Figure 3. Total efficiency for equal scaling factors $S_{PV} = S_L$ for various centrifugal pumps.

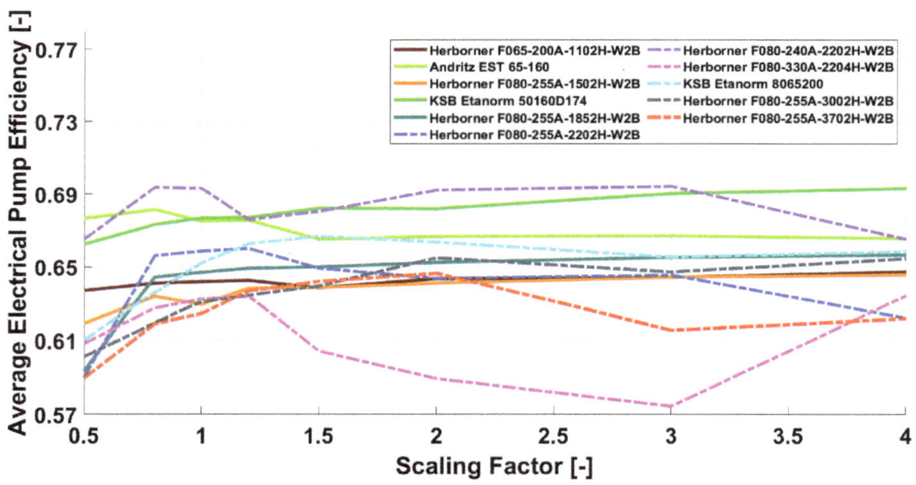

Figure 4. Electrical pump efficiency for equal scaling factors S_{PV} and S_L various centrifugal pumps.

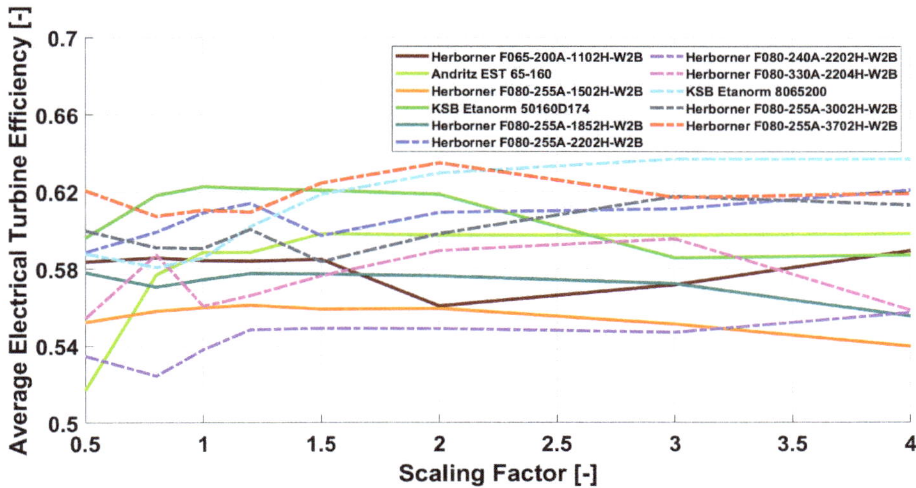

Figure 5. Electrical turbine efficiency for equal scaling factors $S_{PV} = S_L$ for various centrifugal pumps.

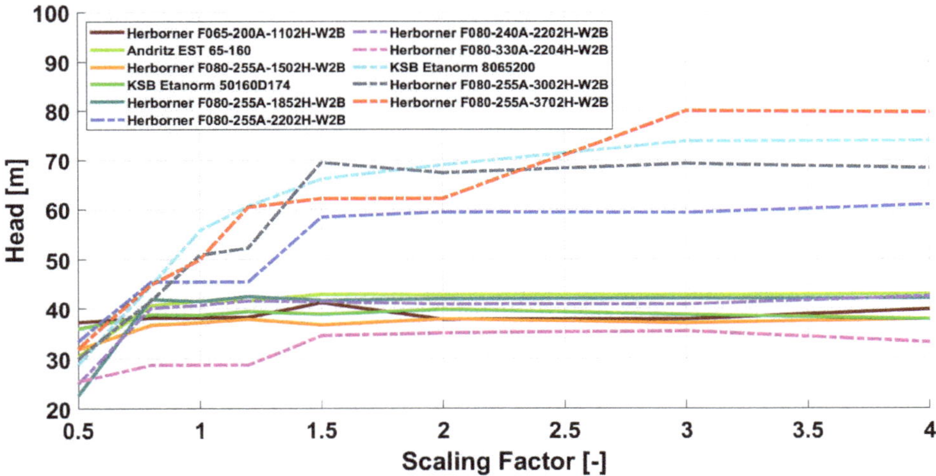

Figure 6. Optimized head for equal scaling factors $S_{PV} = S_L$ for various centrifugal pumps.

3.4. Average Electrical Pump Efficiency

Out of the 11 investigated pumps from Table 2, the F080-240A-2202H-W2B and Andritz EST 65-160 centrifugal pumps showed the best BEP pump efficiency, with more than 80%. The results of the simulation showed a comparably good average electrical pump efficiency (see Equation (6)) of slightly below 70%. This deviation can be explained by the fact that the simulations were based on variable speed operations. As a result, the pump is not always operating at its best point and is also in the partial-load range, where the efficiency is lower than the best point efficiency. In addition, the efficiency losses due to the motor and FUM were also taken into account in calculating the averaged value. The KSB 50160D174 pump had a very good pump efficiency for pumps smaller than 22 kW. For pumps with a power greater than or equal to 22 kW, the pump efficiency showed a trend similar to that of the total efficiency (see Figure 3). The KSB 8065200 and the Herborner F080-240A-2202H showed the best efficiency in this diagram, and the optimum efficiency specified by the manufacturer was also in the upper range at over 80%.

3.5. Average Electrical Turbine Efficiency

The average electrical turbine efficiency of the pumps (see Figure 5) increased between scaling factors 0.5 and 1, which was similar to the behavior observed previously regarding pump efficiency. The pump with the best turbine efficiency varied depending on the scaling factor. Around a scaling factor of one, the KS3 50160D174 pump had an electrical turbine efficiency of 62%, but this dropped to below 60% as the factor increased. In correlation with a (high) pump efficiency, this appliance nevertheless had the best total efficiency. The second-best electrical turbine efficiency for an output smaller than 22 kW was provided by the F080-255A-2202H-W2B.

The turbine efficiencies of pumps of 22 kW and above showed significant variations. Most of the pumps exhibited high electrical turbine efficiencies, with an increasing scaling factor with a peak between 62–64%, thus higher than the turbine efficiencies for the pumps at below 22 kW. For high-scaling factors, the KSB 8065200 clearly stood out compared to the other pumps.

3.6. Head

The optimization process for the simulation model determines the optimal head for each centrifugal pump and for each scaling factor. These are shown in Figure 6. In Figure 6, an increase in the optimal head in the range of factors between 0.5 and 1 can be seen for all heads, which is due to the fact that the best possible head with optimal efficiency can often not be achieved for consumers with low energy demand and production, i.e., for small-scaling factor values. This results in the pump and/or turbine running at reduced efficiency or not at all since the available power was not enough to enter the operation range. Figure 6 thus directly shows which head would be required at the site in order to use the system optimally. For most centrifugal pumps, it can be seen that the head stagnated above a certain scaling factor. The stagnating curves can be related to the limitation of the maximum possible head of the pumps used. Each pump had a maximum head that limits its operating range.

3.7. Levelized Cost of Electricity

The LCOE for the micro-pump storage under consideration herein was calculated using the following formula [48]:

$$LCOE = \frac{I_0 + \sum_{t=1}^{n} \frac{A_t}{(1+i)^t}}{\sum_{t=1}^{n} \frac{W_{t,\ el}}{(1+i)^t}} \tag{9}$$

In year zero $t = 0$, in which the plant is established, all investment costs are considered to be I_0 and no energy $W_{t,\ el}$ is drawn from the pumped storage. For any given year t between one and the final year n, the sum of costs A_t consists of the operating and maintenance costs $O\&M_t$, the possible sale of electricity on the electricity exchange ES_t and reinvestment for replacement purchases RI_t. Since the concept considered is in this case primarily concerned with climate-friendly energy storage, reduction in the grid load and increasing the self-consumption of the produced energy from the PV system, rather than an economically profitable investment, the interest factor i was set to 2%. The Federal Ministry of Finance of Germany publishes the interest rates for economic efficiency studies under document GZ II A 3-H 1012-10/07/0001. The approximate average of the interest factor from the years 2005–2020 in this document also gives a value of 2.0% [49]. The sum of costs A_t is determined for every year and, in the final year, the residual values of the components RW_t are also taken into account and deducted from the costs:

$$A_t = O\&M_t + ES_t + RI_t - RW_t \tag{10}$$

Figure 7 shows the LCOE for all pumps as a function of the scaling factor for the optimal site. Comparing the LCOE of the EST65-160 with that of the F065-200A, which has similarly high total investment costs, the influence of the previously identified significant improvements are clearly evident. For example, the EST65-160 has slightly lower specific costs for the electrical components due to the larger nominal power of 15 kW instead of 11 kW. In addition, the overall efficiency of the EST65-160 is slightly higher (Cf. Figure 5). Altogether, this leads to an LCOE reduction of up to approximately 7 €cent/kWh.

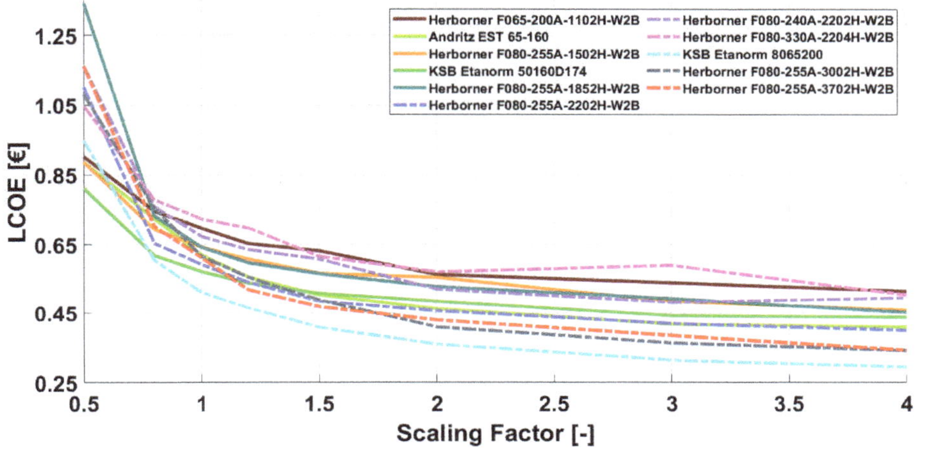

Figure 7. LCOE for equal scaling factors S_{PV} and S_L for various centrifugal pumps.

Within the scope of pumps and boundary conditions investigated, the lowest possible LCOE of slightly less than 29.2 €cents/kWh can be achieved with rising scaling factors by using the KSB 8065200. This value, though at first glance seemingly high, is very promising when compared to the LCOE of around 1.06 €/kWh for a recently built PAT storage in Froyennes, where the high values result from a small head and a long pipeline [27]. Compared to a value of 1.66 €/kWh [50] for a system with a storage tank in buildings with an open water tank on the roof, it is even more encouraging. On the other hand, for a small-scale PV system (>30 kWp) with a lithium-ion battery [51], the LCOE ranges between approximately 5.4 and 17.2 €cents/kWh. The exact value in this rather wide range depends on the irradiation and the assumed PV-battery ratio at an assumed interest rate of $i = 3\%$. For instance, the upper value of 17.2 €cents/kWh was reached for a large battery storage size (capacity in kWh = 80% of PV system power in kW), while the 5.4 €Cent/kWh applied to a small battery storage size (capacity of 20% of the PV system power). As of 1 April 2020, commercial customers in Germany paid 23.03 €cents/kWh of electricity based on a volume-weighted average annual consumption of 50 megawatt-hours [52]. On the same date, household customers in Germany were paying an average of 33.8 €cents/kWh on the basic supply tariff with an annual electricity purchase of between 2500 and 5000 kWh [52]. Given the electricity price for household customers, the system could therefore be profitable. There still remains a gap of around 6.5 cents/kWh in order to reach the electricity prices for commercial customers, which may be closed in the future by further rising electricity prices. So far, and especially in the months of September and October, 2021 shows a sharp increase in the price of electricity on the exchange [53]. It remains to be seen how this will affect end-customer prices. Since the profitability of the pumped storage always depends on the difference between the costs of the energy obtained from the grid and the potential sales revenue on the electricity exchange, strong fluctuations in the exchange and end prices can have a significant impact—both positive and negative—on the profitability of the pumped storage.

A large contribution to the LCOE is made by (specific) investment costs for the FUM, motor, and pump (see Figure 2). As a result, the LCOE results in Figure 7 showed a behavior similar to that observed in Figure 2. As the scaling factors increased, the costs were spread over larger amounts of stored energy, and the specific costs decreased. As shown in Figure 7 of the LCOE, the three pumps (KSB 8065200, F080-255A-3002H-W2B, and F080-255A-3704H-W2B) became the most economical pumps with an increasing scaling factor. Looking at Table 1, it becomes clear that the KSB 8065200 is rather inexpensive for a 30 kW pump. In this case, the motor and pump were procured separately, which offers a cost advantage compared to the purchase of a complete system of the pump, including the motor. Figure 7 also clearly illustrates that, in most cases, pumps with a larger rated power achieve a lower LCOE. Only in areas with a scaling factor of less than one do pumps with a lower nominal power perform better. However, this sector is already rather uneconomical (see Figure 7). Combining these results for the LCOE with the corresponding results for the optimal heads, it becomes clear that these are located in the range of 70–80 m. Compared with the maximum head in Froyennes, which is about 11 m [27], the optimum head found in this study would be up to seven times higher. This is one of the main reasons for the high LCOE in Froyennes.

Finally, it should be noted that for all of the pumps investigated, a choice of geodetic heads suitable for the pump is crucial for the economic efficiency of the system.

4. Simulation Results for Various Scaling Factors Using the Example of the KSB 8065200

In order to cover a larger scope of load and PV generation scenarios, this section will now analyze the impact of various scaling factors for the load and generation profiles using the example of the KSB 8065200. This pump was selected for the extended analysis because it had the best LCOE of all simulated pumps and offered a very good total efficiency. In addition, a large part of the turbine map was provided by the manufacturer, so only a small portion had to be extended using similarity relationships. Therefore, improved accuracy of the results can be expected. The plots (as, e.g., in Figure 8) were created with the MATLAB fit function [54] using cubic interpolation between data points. In Section 3, the scaling factors for PV (S_{PV}) and annual energy demand (S_L) varied but remained equal (i.e., $S_{PV} = S_L$). In order to cover various load scenarios, S_{PV} and S_L are then varied independently.

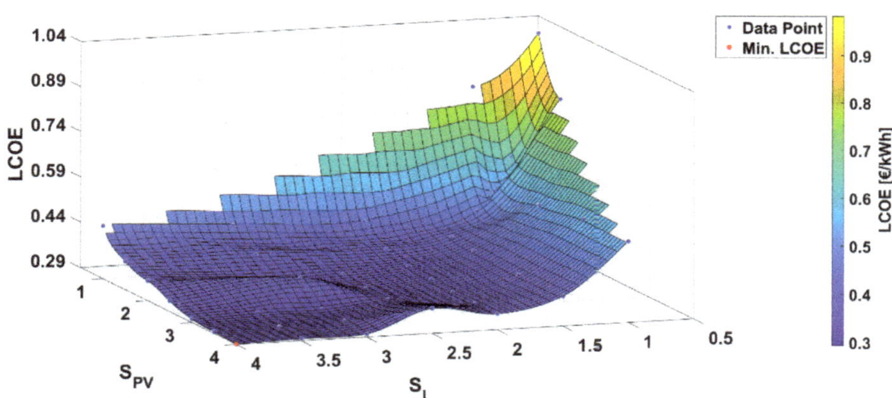

Figure 8. LCOE for various scaling factors S_{PV} and S_L for the KSB 8065200.

As can be seen clearly in Figure 8, the LCOE of the pump decreased if any of the two scaling factors rose. The lowest LCOE can be achieved if both scaling factors are large, as was already demonstrated in Figure 7 for the special case of $S_L = S_{PV}$. Even with a relatively low PV output and high energy consumption, an acceptable LCOE of around

29.2 €cents/kWh can be achieved. Considering the LCOE and the ratio of W_{PV}/W_L, the slightly better results were obtained at a ratio slightly above one. This can be understood easily given that, if the PV power is too small, the storage tank cannot be filled sufficiently. So, there is, of course, no storage content that can be discharged afterwards.

The maximum total efficiency (Figure 9) of 42.6% was reached for $S_{PV} = 4.0$ and $S_L = 4.0$. In addition to the LCOE, Figures 10 and 11 show the electrical pump and turbine efficiencies. Both figures show a low efficiency for scaling factors smaller than one and an increase with rising scaling factors.

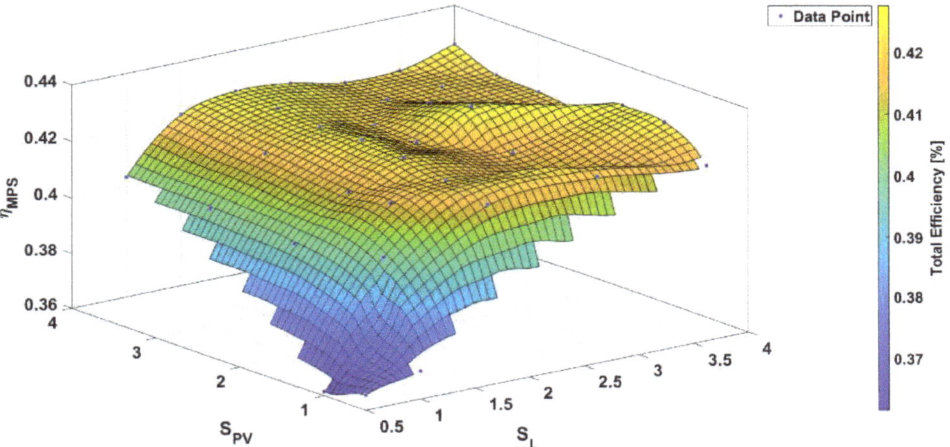

Figure 9. Total efficiency for various scaling factors S_{PV} and S_L for the KSB 8065200.

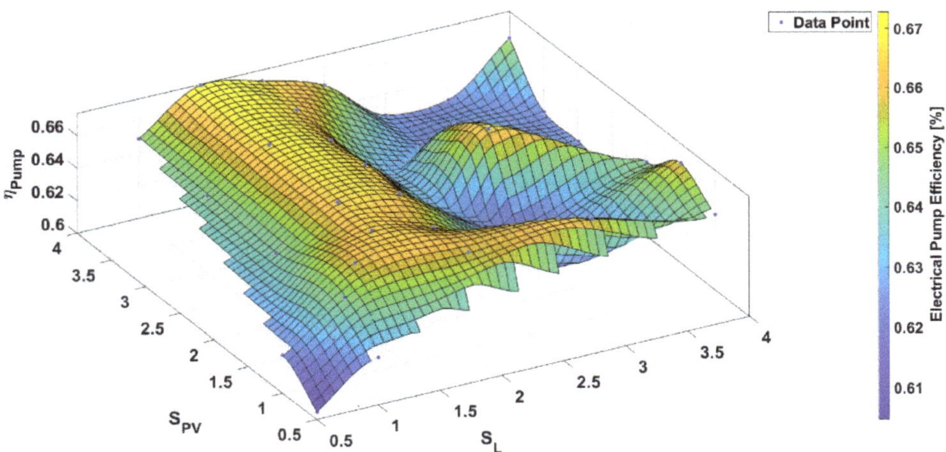

Figure 10. Average electrical pump efficiency for various scaling factors S_{PV} and S_L for the KSB 8065200.

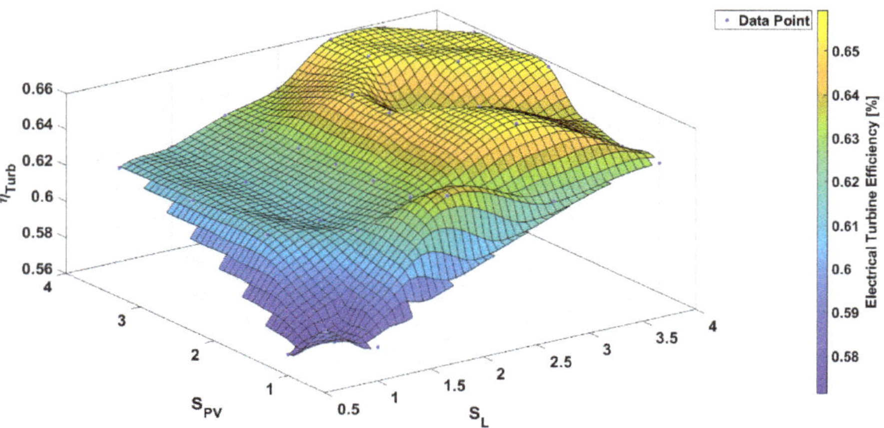

Figure 11. Average electrical turbine efficiency for various scaling factors S_{PV} and S_L for the KSB 8065200.

For $S_{PV} = 2.0$ and $S_L = 3.0$, a local minimum of the electrical pump efficiency (Figure 10) is reached, whereas the electrical turbine efficiency (Figure 11) is at a local maximum. However, it should be noted that the total variation of efficiencies is only around three percentage points. For the factors $S_{PV} = 2.0$ and $S_L = 1.5$, the conditions were reversed, and the turbine reached a minimum and the electrical pump efficiency a maximum.

In summary, it can be concluded that a suitable combination of high pump and turbine efficiencies is of significant importance in obtaining a high level of total efficiency. However, the trade-off is that improving the turbine efficiency often leads to a reduction in pump efficiency or vice versa.

5. Conclusions

Summarizing on the basis of the previous graphical representations of the efficiencies, it can be said that the correlation between pump and turbine efficiency plays a major role in the use of the pumps in PAT operation. Generally, it can be said that the efficiencies for small power plants (i.e., scaling factors of <1) are too low for an economical operation. Moreover, an increase in the efficiency at the scaling factors from 0.5 to 1 can also be seen. A subsequent leveling off, on the other hand, does not follow for every pump, which may be related to increasing or stagnating heads. Based on the total efficiencies (Figure 3), it can be seen that, with increasing factors, the efficiencies of the pumps below 22 kW are more dispersed, showing poorer total efficiencies. In contrast, four of the pumps with a power greater than or equal to 22 kW had a similar curve and a total efficiency of around 40% for high generation and load scenarios. It should also be noted that the PAT with the best total efficiency is not always the most economical pump. High efficiency is a basic prerequisite for profitability, but many other factors such as the head, specific costs, generation, and load scenarios, as well as the efficiency in the partial load range, play a major role. As an operating strategy, speed control using a frequency converter was shown to be the most sensible method of economical operation [15]. Basically, due to the high specific costs, no PAT with an output of less than 22 kW should be selected. As was to be expected, larger pumps were shown to lead to lower specific costs. For pumps with a nominal power of greater than 40 kW, however, the specific costs are barely reduced. Therefore, only minor savings can be expected by increasing the nominal power beyond this amount. The ratio of W_{PV}/W_L, as provided in Table 1, at around 1.6, has turned out to be quite an optimal ratio, depending on the scaling factor. Qualitatively, it can be easily understood that the optimal ratio of W_{PV}/W_L needs to be larger than one since smaller values indicate that the generated energy W_{PV} is not sufficiently larger than the energy demand W_L, which in turn leads to the fact that the storage can often not be filled sufficiently.

Even as it is important to keep in mind that the results are subject to calculation inaccuracies of the turbine maps (see Section 2.2.1), this quantitative examination nevertheless shows that pumped storage with a pump as a turbine can achieve a passable LCOE of around 29.2 €cents/kWh if the nominal power of the pump is sufficiently high and the specific costs can thus be strongly reduced. While the use of PATs in water networks may lead to short payback periods, for example, of six years, the use of a PAT in an MPS faces the challenge of achieving economic operation at all. Nevertheless, if the results obtained in this study are compared to the LCOE of the MPS from Froyennes of around 1.06 €/kWh [27], or a storage tank in buildings with a value of 1.66 €/kWh [50], an LCOE of 29.2 €cents/kWh seems very promising.

In the future, it is planned to further increase the accuracy and reliability of the results reported here by using measured instead of calculated turbine maps for the simulations. In this context, a test bench is currently being set up to measure characteristic maps and create a database that will allow new simulations to be performed. Furthermore, more accurate calculation methods recently published [55] could be used to further increase the reliability of the results.

In summary, it can be said that an MPS system with a pump as a turbine can achieve profitability in the future. However, this will require further cost savings or further increases in electricity prices. Moreover, only a few locations are suitable for the profitable use of micro-pump storage because so many conditions have to be fulfilled. A sufficiently sized PV system, a pre-existing storage basin, sufficient head (around 70 m or more), and appropriate energy consumption are all required. Taken together, this considerably limits the availability of possible locations. The PAT itself should preferably have a nominal power of 22 kW or higher, high efficiency in turbine and pump mode, which preferably extends well into the partial load range. Of course, the PAT must also be adapted to the head, the flow rates, as well as the power of the load and of the PV system. If all these conditions are fulfilled, economic viability may well be possible, but each case has to be examined individually. Based on the results presented in this article, a pre-selection of suitable PATs can be performed for a given prospective MPS site. In a second step, the simulation model can be used to evaluate the economic feasibility for these pumps at the given site in more detail.

The envisioned transition to a more sustainable future with a carbon-free and electricity-based energy system will require storage systems of all sizes and many types. For small and decentralized energy storage, concepts such as those presented in this publication could contribute to the overall solution.

Author Contributions: Conceptualization, F.J.L. and J.K.; methodology, F.J.L.; software, F.J.L. and E.G.; validation, F.J.L., J.K. and M.G.; formal analysis, F.J.L. and E.G.; investigation, F.J.L. and E.G.; resources, F.J.L.; data curation, F.J.L. and E.G.; writing—original draft preparation, F.J.L.; writing—review and editing, J.K., M.G. and E.G.; visualization, F.J.L. and E.G.; supervision, J.K. and M.G.; project administration, J.K.; funding acquisition, J.K. All authors have read and agreed to the published version of the manuscript.

Funding: This research was funded by the Bavarian State Ministry of Science. Grant number: VIII.2-F1116.WE/29/2.

Institutional Review Board Statement: Not applicable.

Informed Consent Statement: Not applicable.

Data Availability Statement: Not applicable.

Acknowledgments: This research was supported by IB Pfeffer, Herborner-Pumpentechnik GmbH & Co KG and KSB Aktiengesellschaft.

Conflicts of Interest: The authors declare that they have no conflict of interest.

Nomenclature

Symbol/Abbreviation	Interpretation
A_t	sum of costs
d_N	nominal pipeline diameter
ES_t	possible energy sales proceeds
f_D	Darcy friction factor
g	gravitational acceleration
H	head
H_{geo}	geodetic head
H_L	head of the idle point
H_m	arithmetic average value of the head
H_N	head at nominal speed
H_{optim}	optimized head
H_{opt}	head at the best efficiency point
H_r	head corresponding to the respective speed
H_T	turbine head
i	discount rate
I_0	investment costs
I_{el}	investment for motor and frequency inverter
I_P	pump investment
kWp	nominal power (PV)
$LCOE$	levelized cost of energy
L_{pipe}	pipeline length
M	torque
n	rpm
n_N	nominal speed
n_q	specific speed
$O\&M_t$	operating and maintenance costs
P_{hydr}	hydraulic power
P_L	power demand (consumer)
P_n	nominal output
P_N	power at nominal speed
P_{max}	maximum power in pump mode
P_P	pump power
$P_{P,el}$	electrical pump power
P_{PV}	PV power
P_r	power corresponding to the respective speed
P_T	turbine power
$P_{T,\,el}$	electrical turbine power
$P_{T,mech}$	mechanical turbine power
P_v	power loss
P_{PV}	PV power
Q	flow rate
Q_L	flow rate of the idle point
Q_m	arithmetic average value of the flow rate
Q_{max}	maximum flow rate
Q_{min}	minimum flow rate
Q_N	flow rate at nominal speed
Q_{opt}	flow rate at the best efficiency point
Q_r	flow rate corresponding to the respective speed
Q_T	turbine flow rate
RI_t	reinvestment costs
RW_t	residual values

Symbol/Abbreviation	Interpretation
S_L	scaling factor for the annual energy demand
S_{PV}	scaling factor for PV power
t	period of time
$\dot{V}$	volume flow
V	storage capacity
V_{optim}	optimized volume of the storage
W_{Grid}	grid energy
W_{in}	total energy input
W_L	annual energy demand
W_{out}	total energy output
W_{PV}	annual energy production of the PV system
Greek letters	**Interpretation**
ζ_{tot}	total dynamic loss coefficient of fittings
η	efficiency
η_{el}	electrical efficiency
$\overline{\eta}_{P,el}$	average electrical pump efficiency
$\overline{\eta}_{T,el}$	average electrical turbine efficiency
η_{opt}	best efficiency point
$\eta_{P,opt}$	best pump efficiency given by the manufacturer
η_q	specific speed
η_T	turbine efficiency
η_{tot}	total efficiency
$\eta_{T,mech}$	mechanical turbine efficiency
ρ	density

Appendix A

This appendix describes a possible procedure for calculating the speed characteristic according to Gülich [28]. The values given by the manufacturer for the flow rate $Q_{P,opt}$, the head $H_{P,opt}$, the nominal speed n_N, and the efficiency η_{opt} at the best efficiency point of pumping operation of the centrifugal pump used serve as the basic data. From this, the necessary parameters in turbine operation are calculated in the next steps, where the index T stands for turbine mode and P for pump mode, respectively.

The first step is the determination of $Q_{T,opt}$, $H_{T,opt}$ and the specific speed n_q at the turbine BEP of operation. There are two variants for the calculation, from which an average value is then formed.

- For the first variant, the BEP in turbine operation is determined from the BEP data of the pump using the following empirical formulas:

$$Q_{T1,opt} = Q_{P,opt} \cdot \frac{1}{\left(\eta_{P,opt}\right)^{0.8}} \tag{A1}$$

$$H_{T1,opt} = H_{P,opt} \cdot \frac{1}{\left(\eta_{P,opt}\right)^{1.2}} \tag{A2}$$

- For the second variant, the BEP in turbine operation is also calculated from the BEP of the pump using another set of empirical formulas:

$$Q_{T2,opt} = Q_{P,opt} \cdot \left(\frac{2.5}{\eta_{P,opt}} - 1.4\right) \tag{A3}$$

$$H_{T2,opt} = H_{P,opt} \cdot \left(\frac{2.4}{\left(\eta_{P,opt}\right)^2} - 1.5\right) \tag{A4}$$

- Gülich states that both methods (Equations (A1)–(A4)) can also be used and an arithmetic mean can be formed from them. When comparing the calculations with the characteristic diagrams given by the manufacturer, the arithmetic mean of the values has proven to be the best method, which is why this is used in our calculations. The the arithmetic average value Q_m and H_m is calculated from both variants as well as the determination of the specific speed of the turbine via the formulas:

$$Q_m = \frac{Q_{T1,opt} + Q_{T2,opt}}{2} \tag{A5}$$

$$H_m = \frac{H_{T1,opt} + H_{T2,opt}}{2} \tag{A6}$$

$$\eta_{qT} = \eta_{qP} \cdot (1.3 \cdot \eta_{P,opt} - 0.3) \tag{A7}$$

- To determine the turbine characteristic curve at the nominal speed n_N, the no-load characteristic curve, which at the same time represents the lower limit of the characteristic diagram. The no-load point $Q_{L,N}$, $H_{L,N}$ are calculated according to Equation (A8). For this, the specific speed of the pump $\eta_{P,q}$ is required.

Determination of the specific speed of the pump:

$$n_{P,q} = n_N \cdot \sqrt{\frac{\frac{Q_{P,opt}}{f_q}}{H_{P,opt}^{0.75}}} \tag{A8}$$

- Calculation of the idle point $Q_{L,N}$, $H_{L,N}$ for the nominal speed n_N:

$$Q_{L,N} = Q_m \cdot \left(0.3 + \frac{\eta_{P,q}}{400}\right) \tag{A9}$$

$$H_{L,N} = H_m \cdot (0.55 - 0.002 \cdot \eta_{qP}) \tag{A10}$$

- For the approximation of the turbine characteristic $H_T = f(Q_T)$ the characteristic curve runs as a parabola through the BEP of the turbine. The head H_T depends on the flow Q_T and is calculated using the following formula:

$$H_T = H_m - \frac{H_m - H_{L,N}}{Q_m^2 - Q_{L,N}^2} - \left(Q_m^2 - Q_T^2\right) \tag{A11}$$

- Calculation of the idle speed characteristic curve (LLK):

The characteristic curve is determined using formula (A12), where $Q_{L,x}$ is an individual volume flow on the no-load characteristic curve. At the no-load point, the torque M is 0 Nm. The idling characteristic connects all points $H(Q)$, which are located at different speeds for M equal to 0 Nm and thus forms a parabola through the coordinate origin.

$$H_L = H_{L,N} \cdot \left(\frac{Q_{L,x}}{Q_{L,N}}\right)^2 \tag{A12}$$

- Calculation of the turbine efficiency $\eta_{T,opt}$ for the nominal speed:

The efficiency is calculated from the quotient of the optimum efficiency and the specific speed of the pump by means of the formula:

$$\eta_{T,opt} = \eta_{P,opt} \cdot \left(1.16 - \frac{\eta_{qP}}{200}\right) \tag{A13}$$

The efficiency curve is then determined from a graph in which $\eta_{T,opt}/\eta_{P,opt}$ is plotted as a function of the ratio of $\frac{Q - Q_L}{Q_{opt} - Q_L}$

- The turbine power can then be calculated using the following formula:

$$P_T = \eta_T \cdot \rho \cdot g \cdot H_T \cdot Q_T \tag{A14}$$

Appendix B

A detailed description of the simulation program sequence is provided as follows. Initially, all required data are gathered by the program. The relationship between the water volume V in the storage reservoir to the pump and the turbine flow through the PAT (Q_P and Q_T, respectively) is given by the following ordinary differential equation:

$$\dot{V} = Q_P - Q_T \tag{A15}$$

It should be noted that no external water inflow or evaporation is considered. To solve this numerically, Q_P and Q_T have to be obtained from characteristic maps. There is a set of two maps each for the turbine and pump operation, respectively. The map $H_T(Q_T, n)$ represents the dependency of head H_T on flow and speed, while the second map $P_T(Q_T, n)$ shows the power (from which the efficiency can also be obtained). Accordingly, pump operation is described by $H_P(Q_P, n)$ and $P_P(Q_P, n)$.

The difference between PV power and power from the load profile $P_{PV-L} = P_{Pv} - P_L$ results in the possible pump $P_P(Q_P, n)$ or turbine $P_T(Q_T, n)$ power depending on the flow and speed and taking into account the efficiency of the electrical components in the drive unit η_{el}:

$$P_{P,mech} = P_{PV-L} \cdot \eta_{el} \tag{A16}$$

$$P_{T,mech} = \frac{-P_{PV-L}}{\eta_{el}} \tag{A17}$$

To determine the resulting head, taking pipe losses h_f into account, the system's characteristic curves $H\left(\dot{V}\right)_P$ and $H\left(\dot{V}\right)_T$ have to be defined. The curves depend on the flow $\dot{V}$, the Darcy friction factor f_D, the dynamic loss coefficient ζ_{tot}, gravitational acceleration g, pipe length L_{pipe} and diameter d_N [56,57]:

$$H_P\left(\dot{V}\right) = H_{geo} + \left[f_D\left(\dot{V}\right) \cdot \frac{L_{pipe}}{d_N} + \zeta_{tot}\right] \frac{\left(\frac{4 \cdot \dot{V}}{3600s/h \cdot \pi \cdot d_N^2}\right)^2}{2 \cdot g} \tag{A18}$$

$$H_T\left(\dot{V}\right) = H_{geo} - \left[f_D\left(\dot{V}\right) \cdot \frac{L_{pipe}}{d_N} + \zeta_{tot}\right] \frac{\left(\frac{4 \cdot \dot{V}}{3600s/h \cdot \pi \cdot d_N^2}\right)^2}{2 \cdot g} \tag{A19}$$

The operating point (Q_i, n) is now determined by numerically solving the following equations:

$$P_i(Q_i, n) = P_{i,mech} \tag{A20}$$

$$H_i(Q_i, n) = H_i\left(\dot{V}\right) \tag{A21}$$

where $i = T$ for turbine mode and $i = P$ for pump mode, respectively.

Graphically, this corresponds to the intersection of the system's characteristic curve $H_i(Q_i)$ with the turbine or pump power isoline corresponding to the power $P_{i,mech}$. The flow rate and speed at the intersection point can then be read directly from the characteristic maps.

The limit values of pump or turbine operation are the lines for maximum n_{max} and minimum speed n_{min}, as well as the line for minimum Q_{min} and maximum flow Q_{max}. It is also important to provide hysteresis to prevent the turbine from shutting down immediately after starting due to a small drop in load request.

In total, the water volume of the storage tank is determined by numerical integration, and the states "storage tank full" and "empty" are defined. The simulation model provides results for every point in the time of year for all system variables, such as storage level, turbine and pump flow, power, and so on. The annual electricity balance can then be obtained from these results and a profitability calculation can be performed.

References

1. Cebulla, F.; Haas, J.; Eichman, J.; Nowak, W.; Mancarella, P. How much electrical energy storage do we need? A synthesis for the U.S., Europe, and Germany. *J. Clean. Prod.* **2018**, *181*, 449–459. [CrossRef]
2. Rohit, A.K.; Devi, K.P.; Rangnekar, S. An overview of energy storage and its importance in Indian renewable energy sector: Part I—Technologies and Comparison. *J. Energy Storage* **2017**, *13*, 10–23. [CrossRef]
3. Ramos, M.H.; Dadfar, A.; Basharat, M.; Adeyeye, K. Inline pumped storage hydropower towards smart and flexible energy recovery in water networks. *Water* **2020**, *12*, 2224. [CrossRef]
4. De Marchis, M.; Fontanazza, C.M.; Freni, G.; Messineo, A.; Milici, B.; Napoli, E.; Notaro, V.; Puleo, V.; Scopa, A. Energy recovery in water distribution networks. Implementation of pumps as turbine in a dynamic numerical model. *Procedia Eng.* **2014**, *70*, 439–448. [CrossRef]
5. Spedaletti, S.; Rossi, M.; Comodi, G.; Salvi, D.; Renzi, M. Energy recovery in gravity adduction pipelines of a water supply system (WSS) for urban areas using Pumps-as-Turbines (PaTs). *Sustain. Energy Technol. Assess.* **2021**, *45*, 101040. [CrossRef]
6. Muhammetoglu, A.; Karadirek, E.; Ozen, O.; Muhammetoglu, H. Full-Scale PAT Application for Energy Production and Pressure Reduction in a Water Distribution Network. *J. Water Resour. Plan. Manag.* **2017**, *143*, 04017040. [CrossRef]
7. Carravetta, A.; Del Giudice, G.; Fecarotta, O.; Ramos, M.H. Pump as Turbine (PAT) design in water distribution network by system effectiveness. *Water* **2013**, *5*, 1211–1225. [CrossRef]
8. Fecarotta, O.; Ramos, H.; Derakhshan, S.; Del Giudice, G.; Carravetta, A. Fine tuning a PAT hydropower plant in a water supply network to improve system effectiveness. *J. Water Resour. Plan. Manag.* **2018**, *144*, 04018038. [CrossRef]
9. Ramos, H.; McNabola, A.; Lopez-Jimenez, P.A.; Perez-Sanchez, M. Smart water management towards future water sustainable networks. *Water* **2020**, *12*, 58. [CrossRef]
10. Carravetta, A.; Del Giudice, G.; Fecarotta, O.; Ramos, H. Energy production in water distribution networks: A PAT design strategy. *Energies* **2013**, *6*, 411–424. [CrossRef]
11. Madeira, C.F.; Fernandes, F.P.J.; Perez-Sanchez, M.; Lopez-Jimenez, P.A.; Ramos, M.H.; Branco, P.J.C. Electro-hydraulic transient regimes in isolated pumps working as turbines with self-excited induction generators. *Energies* **2020**, *13*, 4521. [CrossRef]
12. Stefanizzi, M.; Capurso, T.; Balacco, G.; Binetti, M.; Camporeale, S.M.; Torresi, M. Selection, control and techno-economic feasibility of Pumps as Turbines in Water Distribution Networks. *Renew. Energy* **2020**, *162*, 1292–1306. [CrossRef]
13. Moazeni, F.; Khazaei, J. Optimal energy management of water-energy networks via optimal placement of pumps-as-turbines and demand response through water storage tanks. *Appl. Energy* **2021**, *283*, 116335. [CrossRef]
14. Balacco, G.; Binetti, M.; Caggiani, L.; Ottomanelli, M. A Novel Distributed System of e-Vehicle Charging Stations Based on Pumps as Turbine to Support Sustainable Micromobility. *Sustainability* **2021**, *13*, 1847. [CrossRef]
15. Lugauer, F.J.; Kainz, J.; Gaderer, M. Techno-Economic Efficiency Analysis of Various Operating Strategies for Micro-Hydro Storage Using a Pump as a Turbine. *Energies* **2021**, *14*, 425. [CrossRef]
16. Morabito, A.; Furtado, A.; Gilton, C.; Hendrick, P. Variable speed regulation for pump as turbine in micro pumped hydro energy storage. In Proceedings of the 38th IAHR World Congress, Panama City, Panama, 1–6 September 2019.
17. Anilkumar, T.T.; Simon, S.P.; Padhy, N.P. Residential electricity cost minimization model through open well-pico turbine pumped storage system. *Appl. Energy* **2017**, *195*, 23–35. [CrossRef]
18. Borkowski, D. Analytical model of small hydropower plant working at variable speed. *IEEE Trans. Energy Convers.* **2018**, *33*, 1886–1894. [CrossRef]
19. Barbarelli, S.; Amelio, M.; Florio, G. Predictive model estimating the performances of centrifugal pumps used as turbines. *Energy* **2016**, *107*, 103–121. [CrossRef]
20. Vasudevan, K.R.; Ramachandaramurthy, V.K.; Gomathi, V.; Ekanayake, J.B.; Tiong, S.K. Modelling and simulation of variable speed pico hydel energy storage system for microgrid applications. *J. Energy Storage* **2019**, *24*, 100808.
21. Mohanpurkar, M.; Ouroua, A.; Hovsepian, R.; Luo, Y.; Singh, M.; Muljadi, E.; Gevorgian, V.; Donalek, P. Real-time co-simulation of adjustable-speed pumped storage hydro for transient stability analysis. *Electr. Power Syst. Res.* **2018**, *154*, 276–286. [CrossRef]
22. Schmidt, J.; Kemmetmüller, W.; Kugi, A. Modeling and static optimization of a variable speed pumped storage power plant. *Renew. Energy* **2017**, *111*, 38–51. [CrossRef]
23. Stoppato, A.; Benato, A.; Destro, N.; Mirandola, A. A model for the optimal design and management of a cogenartion system with energy storage. *Energy Build.* **2016**, *124*, 241–247. [CrossRef]
24. Ma, T.; Yang, H.; Lu, L.; Peng, J. Pumped storage-based standalone photovoltaic power generation system: Modeling and techno economic optimization. *Appl. Energy* **2015**, *137*, 649–659. [CrossRef]
25. Yang, W.; Yang, J. Advantage of variable-speed pumped storage plants for mitigating wind power variations: Integrated modelling and performance assessment. *Appl. Energy* **2019**, *237*, 720–732. [CrossRef]
26. Simao, M.; Ramos, H. Hybrid pumped hydro storage energy solutions towards wind and PV integration: Improvement on flexibility, reliability and energy costs. *Water* **2020**, *12*, 2457. [CrossRef]

27. Morabito, A.; Hendrick, P. Pump as turbine applied to micro energy storage and smart water grids: A case study. *Appl. Energy* **2019**, *241*, 567–579. [CrossRef]

28. Gülich, J.F. *Kreiselpumpen*; Springer: Berlin/Heidelberg, Germany, 2013; Volume 4.

29. Alatorre-Frenk, C. Cost Minimization in Microhydro Systems Using Pumps-as-Turbines. Ph.D. Thesis, University of Warwick Coventry, Coventry, England, 1994; pp. 55–113.

30. Derakhshan, S.; Nourbakhsh, A. Experimental study of characteristic curves of centrifugal pumps working as turbine in different specific speeds. *Exp. Therm. Fluid Sci.* **2008**, *32*, 800–807. [CrossRef]

31. Yang, S.; Derakhshan, S.; Kong, F. Theoretical, numerical and experimental prediction of pump as turbine performance. *Renew. Energy* **2012**, *48*, 507–513. [CrossRef]

32. Lin, T.; Zhu, Z.; Li, X.; Li, J.; Lin, Y. Theoretical, experimental, and numerical methods to predict the best efficiency point of centrifugal pump as turbine. *Renew. Energy* **2021**, *168*, 31–44. [CrossRef]

33. Pugliese, F.; De Paola, F.; Fontana, N.; Giugni, M.; Marini, G. Experimental characterization of two Pumps as Turbines. *Renew. Energy* **2016**, *99*, 180–187. [CrossRef]

34. Stefanizzi, M.; Torresi, M.; Fortunato, B.; Camporeale, S.M. Experimental investigation and performance prediction modeling of a single stage centrifugal pump operation as turbine. *Energy Procedia* **2017**, *126*, 589–596. [CrossRef]

35. Renzi, M.; Nigro, A.; Rossi, M. A methodology to forecast the main non-dimensional performance parameters of pumps-as-turbines (PaTs) operating at Best Efficiency Point (BEP). *Renew. Energy* **2020**, *160*, 16–25. [CrossRef]

36. Pugliese, F.; Fontana, N.; Marini, G.; Giugni, M. Experimental assessment of the impact of number of stages on vertical axis multi-stage centrifugal PATs. *Renew. Energy* **2021**, *178*, 891–903. [CrossRef]

37. Beschneiungsanlage Erhält Wasserspeicher. Available online: https://bnn.de/lokales/abb/beschneiungsanlage-erhaelt-waserspeicher (accessed on 30 April 2020).

38. Fahrner, H.; (Nationalpark-Hotel Schliffkopf, Schliffkopf, Germany); Lugauer, F.J.; (TUM Campus Straubing for Biotechnology and Sustainability, Weihenstephan-Triesdorf University of Applied Science, Straubing, Germany). Personal communication, 2020.

39. EDUR-Pumpenfabrik Eduard Redlien GbmH & Co. KG. Die Kennlinien. Available online: www.treffpunkt-kaelte.de/wp-content/uploads/2017/komponenten/pumpen/edur/kennlinien_2.pdf (accessed on 24 August 2021).

40. KSB AG Pump Affinity Laws. Available online: https://www.ksb.com/centrifugal-pump-lexicon/pump-affinity-laws/191832 (accessed on 17 August 2021).

41. Kirkpatrick, S.; Gelatt, C.D., Jr.; Vecchi, M.P. Optimization by simulated annealing. *Science* **1983**, *220*, 671–680. [CrossRef] [PubMed]

42. Mathworks. Global Optimization Toolbox: Simulated Annealing (r2019a). Available online: https://de.mathworks.com/help/gads/simulannealbnd.html (accessed on 28 January 2020).

43. Nourbakhsh, A.; Derakhshan, S.; Javidpour, E.; Riasi, A. Centrifugal & axial pumps used as turbines in small hydropower stations. In Proceedings of the Hidroenergia 2010 Intenational Congress on Small Hydropower International Conference and Exhibition on Small Hydropower, Lausanne, Switzerland, 26–28 October 2009; pp. 16–19.

44. Mathworks. Trapezoidal Numerical Integration (r2021a). Available online: https://de.mathworks.com/help/matlab/ref/trapz.html (accessed on 1 September 2021).

45. Herborner Pumpentechnik GmbH & Co. KG. Price List 2019. Available online: https://www.herborner-pumpen.de/en/downloads/ (accessed on 20 May 2020).

46. VEM GmbH. Price List 2021: Drive Technology. Available online: www.vem-group.com/fileadmin/content/pdf/Download/Kataloge/Preisliste/2021_VEM_Preisliste_Antriebstechnik.pdf (accessed on 30 June 2021).

47. Ghv Vertriebs-Gmbh für Antribstechnik und Automation. Prices for Frequency Converters. Personal Communication, 2017.

48. Branker, K.; Pathak, M.J.M.; Pearce, J.M. A review of solar photovoltaic levelized cost of electricity. *Renew. Sustain. Energy Rev.* **2011**, *15*, 4470–4482. [CrossRef]

49. German Federal Ministry of Finance. Kalkulationszinssätze für Wirtschaftlichkeitsuntersuchungen GZ II A 3—H 1012-10/07/0001. 2020. Available online: https://www.bundesfinanzministerium.de/Content/DE/Standardartikel/Themen/Oeffentliche_Finanzen/Bundeshaushalt/personalkostensaetze-2020-anl.pdf (accessed on 20 November 2021).

50. de Oliveira e Silva, G.; Hendrick, P. Pumped hydro energy storage in buildings. *Appl. Energy* **2016**, *179*, 1242–1250. [CrossRef]

51. Kost, C. Study: Levelized Cost of Electricity-Renewable Energy Techologies. June 2021. Available online: https://www.ise.fraunhofer.de/en/publications/studies/cost-of-electricity.html (accessed on 17 August 2021).

52. Bundesnetzagentur; Bundeskartellamt. *Bundesnetzagentur—Monitoringbericht 2020.* 2020, pp. 269, 272, 282. Available online: https://bundesnetzagentur.de/SharedDocs/Mediathek/Berichte/2020/Monitoringbericht_Energie2020.pdf (accessed on 26 August 2021).

53. Bundesnetzagentur: Market—Day-Ahead Prices. Available online: https://www.smard.de/page/home/marktdaten/78?marketDataAttributes=%7B%22resolution%22:%22day%22,%22region%22:%22DE-LU%22,%22from%22:1626300000000,%22to%22:1634248799999,%22moduleIds%22:%5B8004169%5D,%22selectedCategory%22:null,%22activeChart%22:true,%22style%22:%22color%22%7D (accessed on 20 October 2021).

54. Mathworks. Fit Curve or Surface to Data (r2021a). Available online: https://de.mathworks.com/help/curvefit/fit.html (accessed on 7 September 2021).

55. Macías Ávila, C.A.; Sánchez-Romero, F.-J.; López-Jiménez, P.A.; Pérez-Sánchez, M. Definition of the Operational Curves by Modification of the Affinity Laws to Improve the Simulation of PATs. *Water* **2021**, *13*, 1880. [CrossRef]

56. KSB Aktiengesellscahft. *Selecting Centrifugal Pumps*; KSB Aktiengesellschaft: Frankenthal, Germany, 2005; Volume 4, pp. 19–31.
57. Menny, K. *Strömungsmaschinen*; Teubner: Wiesbaden, Germany, 1985; ISBN 978-3-519-06317-9.

 *sustainability*

Article

Development and Numerical Performance Analysis of a Micro Turbine in a Tap-Water Pipeline

Huixiang Chen [1,2,3], **Kan Kan** [3,4,*], **Haolan Wang** [5], **Maxime Binama** [3], **Yuan Zheng** [3,4] **and Hui Xu** [1,3]

1   College of Agricultural Science and Engineering, Hohai University, Nanjing 211100, China; chenhuixiang@hhu.edu.cn (H.C.); hxu@hhu.edu.cn (H.X.)
2   Key Laboratory of Fluid Machinery and Engineering, Xihua University, Chengdu 610039, China
3   College of Water Conservancy and Hydropower Engineering, Hohai University, Nanjing 210098, China; binalax05@yahoo.fr (M.B.); zhengyuan@hhu.edu.cn (Y.Z.)
4   College of Energy and Electrical Engineering, Hohai University, Nanjing 211100, China
5   Rugao Banjing Town Water Service Station, Rugao 226500, China; whlgongzuo322@126.com
*   Correspondence: kankan@hhu.edu.cn; Tel.: +86-151-5186-2390

Abstract: The induction faucet has been widely used in public due to its advantages of convenience, sanitation, water, and electricity saving. To solve the problem of environmental pollution caused by dry batteries used in induction faucets, a suitable micro pipe mixed-flow turbine installed in a tap-water system with only 15 mm in diameter, that uses the pipeline water pressure to generate electricity for the induction faucet was designed and developed, based on computational fluid dynamics (CFD) and model tests. According to the specific speed, a preliminary design of each flow component of the turbine was first produced. Then, using the multi-objective orthogonal optimization method, the optimum test schemes were determined, and the influence of various test factors on the turbine's hydraulic performance was revealed. Under the design flow rate, the turbine's power output and efficiency were 6.40 W and 87.13%, respectively, which were 34.45% and 4.99% higher than those of the preliminary scheme. Both the power output and efficiency of the optimized turbine met the design requirements. Numerical and model test results showed good agreement, where the deviation in turbine power output predictions was below 5% under large flow condition. Model test results also showed that the turbine can be started as long as the inlet flow is greater than 0.14 kg/s. Overall, the micro-pipe turbine designed in this paper exploits the (mostly wasted) water kinetic energy in induction faucets for power production, contributing to environmental pollution reduction and realizing energy conservation.

Keywords: tap-water pipeline; micro-pipe turbine; selection design; multi-objective optimization; orthogonal test; model test

Citation: Chen, H.; Kan, K.; Wang, H.; Binama, M.; Zheng, Y.; Xu, H. Development and Numerical Performance Analysis of a Micro Turbine in a Tap-Water Pipeline. *Sustainability* **2021**, *13*, 10755. https://doi.org/10.3390/su131910755

Academic Editor: M. Sergio Campobasso

Received: 2 August 2021
Accepted: 24 September 2021
Published: 28 September 2021

Publisher's Note: MDPI stays neutral with regard to jurisdictional claims in published maps and institutional affiliations.

1. Introduction

Water is essential for life and plays an irreplaceable role in the human production and proper functioning of the Earth's ecosystems [1]. Water quality and quantity are major environmental problems, and important factors affecting human health and safety [2]. With the current global progress in science and technology there has been an improvement in people's awareness of energy conservation and environmental protection and the induction faucet has been become widely used in various public places. This has come as a result of several advantages being linked to its use, such as convenience, sanitation, and water and electricity saving. However, many sensors in the induction faucet have relied on battery power supply due to the fact that their locations make it hard to connect to local utility grid power [3]. Generally speaking, the batteries in an induction faucet need to be replaced every two to three months. Extensive use of an induction faucet will require a large number of batteries, which will cause pollution to the environment and soil if the batteries are not properly handled [2,4]. However, the environmental pollution from batteries usage can be solved by changing the power supply mode of sensors in the induction faucet, which would make the induction faucet more energy-saving and environmentally friendly.

Therefore, this has become the main challenge for many researchers. McNabola et al. [5] indicated that water supply is a core service on which civilized society depends, and which involves considerable energy consumption and economic cost. They showed that a break pressure tank installed in water supply networks to resolve the high flow pressure provides the opportunity to recover energy from networks through a hydropower turbine system. Moreover, they proved that the above approach showed preliminary technical and economic feasibility, so as to minimize energy consumption. Eventually, more and more investigators turned their attention to the utilization of micro-hydropower in water systems to take advantage of the excess energy in the pipelines. In a recently conducted study [6], a micro-pipe turbine (MPT) installed in a tap-water pipeline was gradually developed and utilized. Its working principle is as follows: this micro-pipe turbine is designed to be driven by water kinetic energy, where, through the connecting shaft, the MPT drives the generator to produce usable electrical power. Then the generator charges the battery of the sensor after sending out an AC electric signal, thus providing power for the detection equipment of the water supply pipeline (WSP). Finally, the conversion of fluid kinetic energy, mechanical energy, and electric energy is realized. The entire MPT system is closed in a rectangular box, only leaving out the inlet and outlet pipes. The box is fully installed in the tap-water pipeline connected to the induction tap, and the bottom of the box is fixed to the wall or floor. The water flows through the MPT to drive the generator, and the electrical energy is stored in the rechargeable battery to realize the constant power supply for the induction faucet. In the production process of the MPT, the engineering plastic polyamide material would be selected for its advantages such as the low cost, long service life, small volume, and light weight. 3D printing technology has been used to manufacture MPT, which is convenient and cost-effective. The excellent features of this MPT system are, among others, the associated convenient installation, small size, simple production process, stable power supply, long service life and replacement cycles, environment-friendliness, and energy conservation ability.

Some investigations have been done to study the hydro energy harvest in the water pipelines, and several attempts have been made to select or design proper turbines [7,8]. For example, Du et al. [9] investigated the impact of the runner inlet arc angle on the performance of an inline cross-flow turbine in urban water mains. Sinagra et al. [10] suggested a cross-flow with positive outlet pressure (a power recovery system), which could provide potential energy consumption service with additional significant hydropower production. Generally, in line with the placement mode of the main shaft, micro hydro turbines can be divided into horizontal axis and vertical axis types, which are derived from the wind turbines [11]. Sakti et al. [12] studied a horizontal-axis type in-pipe hydro-turbine by using the SolidWorks simulation method, where the best combination of deflector design and blade curvature angle was realized. Li et al. [13] designed an ultra-low head turbine, which was similar to a marine and hydrokinetics turbine, but their research focus was on the performance of composite runner blades. Several mini bulb propeller turbines which also can be installed inside water pipelines have been developed [14]. Ferro et al. [15] designed the rotor blades of a mini hydraulic turbine using a quasi-three-dimensional method and the results validated the design hypothesis of small flow velocity through the rotor. Samora et al. [16] optimized and tested a new five-blade tubular propeller turbine with an 85 mm diameter for pipe inline installation. Study results showed that this turbine has potential for further development since the efficiencies were around 60%. Compared to the horizontal axis type, vertical axis type of pipeline turbines have attracted more attention due to linked advantages of flow direction holding, simple structure, small space, and easy installation. Chen et al. [3] developed a novel vertical-axis water turbine for hydropower harness inside water pipelines and found that the rotor with a hollow structure combined with an eye-shaped slanted block in a pipeline could achieve the maximum power as compared with other types of vertical-axis rotors. The vertical-axis-type micro-turbines mainly consist of two sub-types. One is the lift-type and the other is the drag-type. The lift type hydro turbine, also known as the Darrieus type, was developed for big water pipelines (pipe diameters larger than 25 inch) by Lucid Energy [17]. Langroudi et al. [18] conducted new spherical lift-based in-pipe turbine modelling and parameterization research, and the power coefficient, the probability of cavitation, and the pressure coefficient obtained by using a reasonable numerical methods were validated with some experiments. Oladosu et al. [19] generated a pipeline lift-based spherical turbine with NACA 0020 airfoil cross-section, which showed appropriate energy extraction in water distribution pipelines with diameters of 250 mm. Though this lift-type MPT exhibits good energy performance after reasonable design, the start-up performance is still a big problem due to the dynamically stalled blade characteristics [20,21]. The drag-type turbine, sometimes named the Savonius turbine, has also been successfully developed and utilized on different occasions. Payambarpour et al. [22,23] performed a parametric study on a 3D modified Savonius turbine with a deflector in a 100 mm water pipe by using both experimental and numerical methods, where they found that the relationship between the flow rate and aspect ratio depended on the blockage coefficient. Chen et al. [24] designed a drag-type vertical-axis hydro-turbine for pico-hydropower in water pipelines and conducted an optimization study based on previous design scheme. The proposed turbine showed good performance at the velocity rage from 1.5 m/s to 3.5 m/s and the rotating speed range from 100 rpm to 1000 rpm. Yang et al. [25] compared the start-up hydraulic performances of a lift-type and a drag-type hydro-turbine. The results showed that the lift-type turbine gives better hydraulic characteristics in terms of tip speed ratio and power coefficient than those of the drag-type turbine, while the drag-type turbine displays better start-up performance than the lift-type.

Compared to other turbines, energy recovery from water supply systems using pumps as turbines (PATs) presents many advantages such as the marketable availability, low investment cost, and easy installation and maintenance. This has increasingly attracted the attention from many field players and researchers, where PATs have been considered to be an effective way to generate power in water pipelines. Different studies have been carried out to verify the feasibility of PAT technology in water pipelines, using computational fluid dynamics (CFD) techniques [26–28]. Fontana et al. [29] investigated the water distribution system in one district of Naples (Italy) to assess the hydropower generation potential using the PATs, where it was found that the PATs could achieve a good balance between hydropower recovery and head reduction. Du et al. [30] numerically and experimentally studied the performance of a PAT, which was selected through the application of empirical equations, based on the given working conditions in one typical high-rise building. The results verified that the PATs are a good way for avoiding excessive water-head reduction at higher flow rates. Kramer et al. [31] investigated a PAT device both in the laboratory and in the field to improve the economic profitability of energy recovery plants at low installed capacities (around 15 kW). Buono et al. [32] used 3D CFD modeling methodology to study a PAT installed in a hydraulic urban network. The maximum efficiency was found to be 66.3%. De Marchis et al. [33] developed a hydrodynamic model of a PAT, which was applied in the Palermo district network (Italy). Their analysis indicated that a really attractive capital payback period could be achieved if PATs are installed in the pipes located close the water supply node. Pérez-Sánchez et al. [34] defined a new approach to estimating the PAT best efficiency point and predicting its energy performance curve, the accuracy of which was verified using an experimental database. However, due to the fixed geometrical characteristics and inner flow space, the PAT performance in terms of head reduction was easily affected by the variation of flow discharge. To solve this problem, Carravetta et al. [35] designed a PAT installation scheme, which ensured that PAT operating conditions stayed around their best efficiency point, where at the same time, PAT's water pressure reduction task was effectively performed. For PATs, the fundamental blade design only considers its pump mode of operation (normal hydraulic pump operation). Therefore, PAT flow dynamics under reverse operating mode do not conform to the conventional turbine design concept. Meanwhile, due to PAT's lack of guide vanes (GV), it is difficult to form an appropriate in-flow velocity at the runner inlet, resulting in a relatively low efficiency.

Due to their comparatively big sizes, existing MPT designs don't fit the power generation task within micro-pipes (pipe diameter less than 15 mm) with very small flows. Moreover, their operational efficiency has been less than 65% for most of the so-far-tested pipe turbines, which indicates the associated low extraction rate of water-contained energy within water pipelines. In this paper, a high-efficiency and stable mixed-flow turbine was applied to a micro tap-water-pipe power generation system that was only 15 mm in diameter. This turbine uses the water pressure in the WSP to kick off its rotational motion, and generates the electrical power to feed the induction faucet batteries. The main goal of this study was to design and develop an efficient micro turbine that specifically met two performance requirements, namely low-flow-rates start-up and having enough output power. To this end, the design process is shown in Figure 1 and the structure of the paper is as follows: In Section 1, which is the introductory section, the research background to the present study is discussed. In Section 2, the preliminary design of the MPT is explained. Then the hydraulic design and optimization of the concerned turbine, based on orthogonal optimization method and numerical simulation, are discussed in Section 3. In the same section, these methods are verified through experimental testing. Finally, Section 4 contains the study's main concluding remarks.

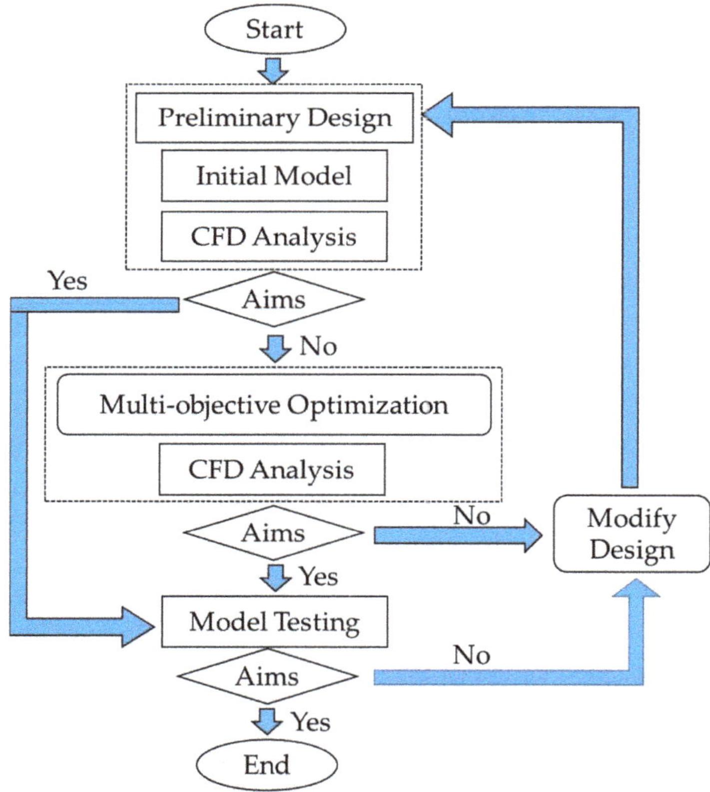

Figure 1. Design process of the MPT.

2. Preliminary Design of the MPT

2.1. Determination of Basic Parameters

2.1.1. Design Head

The pressure at the end of the urban water supply network is 0.14 Mpa, that is, the water head is 14 m according to the requirements of the national water supply pressure service code of China. However, the kinetic hydraulic power inside WSP is mainly used to drive the water flow, and only superfluous head or small part of water energy can be used by turbines to generate electricity. Considering the pressure loss of water flow in the pipeline, the head is taken as 3 m.

2.1.2. Design Discharge

On the basis of the actual specification and size of the WSP, the diameter of the four-branch pipe is 15 mm, and the flow rate is set at 1.5 m/s [36], so the discharge Q can be calculated by the following equation:

$$Q = \frac{\pi}{4}d^2\rho u \tag{1}$$

where d is the pipe diameter, ρ is the water flow density (1000 kg/m^3), and u is the water flow velocity. Based on the known conditions, the design flow rate can be calculated as 0.265 kg/s.

2.1.3. Design Power Output

In order to offer continuous and stable power for the monitoring system in induction faucet, a rechargeable storage battery is installed between the control circuit and the induction faucet. The electric energy emitted by the generator first charges the battery, and then the battery supplies power to the induction faucet, since the power consumption of the induction faucet is less than 6 W.

2.1.4. Design Rotational Speed

The operating voltage of the rechargeable battery is 4 V, and the output current of the generator needs to reach 1.5 A in order to meet the power supply requirements. Due to the small size of the MPT and the high speed rotation driven by the water flow, a micro-sized high-speed generator independently developed by a company in Zhejiang Province, China was adopted. When the speed of the generator is 3500 r/min, the output current is 1.5 A. Thus, the design speed of the MPT is 3500 r/min.

2.2. Determination of Turbine Type

The specific speed n_s is an important characteristic parameter of water turbine classification, and its definition can be expressed as follows [37]:

$$n_s = \frac{n\sqrt{P}}{H^{5/4}} \tag{2}$$

where n, P, and H are the rotational speed, power output, and water head of the MPT, respectively.

According to the above basic design parameters of MPT, the calculated specific speed is about 50 m·kW, which is a low specific speed. Generally, micro hydro turbines can be classified into two general types based on their working principle: (1) impulse turbines used for high heads of water and low flow rates and (2) reaction turbines normally employed for heads below about 450 m and moderate or high flow rates. Meanwhile, according to the different types of reaction turbine applicable speed range, the Francis turbine was selected. Finite element studies have been widely used in structural analysis [38–40]. Many studies show that Francis turbines not only perform well, but also have a stable and reliable structure [41,42]. Moreover, the flow in WSPs is usually unstable, so the energy performance of axial flow hydraulic machinery is obviously affected by the tip leakage flow [43], and the mixed flow structure has better stability. A Francis turbine consists of spiral casing, water guiding mechanism, runner, and draft tube.

2.3. Initial Modelling

For initial models (Figure 2), the diameter of runner was 30 mm, a curved-shape upper crown and lower ring were adopted, the number of runner blades was preliminarily selected as 13, and the blade thickness was 1mm. The distribution circular diameter of GV was 33 mm, the number of GV was temporarily 12, and negative curvature blade profile was initially chosen. The volute form employed a combination of circular and elliptical sections with an wrap angle of 345°. The draft tube adopted a straight, tapered shape with a cone angle of 12° and the length and outlet diameter of draft tube outlet were both 15 mm, which was limited by the diameter of the tap-water pipeline.

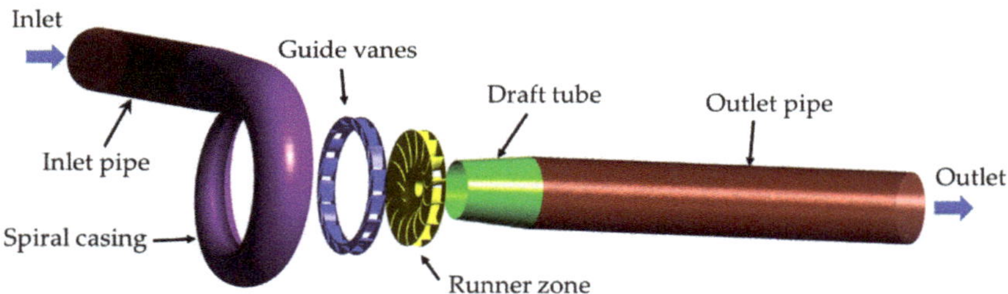

Figure 2. MPT computational domain and components.

2.4. Numerical Simulation of the Initial MPT

2.4.1. Governing Equations

The water flow in the WTP is an incompressible viscous and turbulent fluid, so the continuity and the quasi-steady-state Reynolds-Averaged Navier–Stokes equations can be expressed as follows [44]:

$$\frac{\partial u_i}{\partial x_i} = 0 \tag{3}$$

$$\rho u_j \frac{\partial}{\partial x_j}(u_i) = \rho F_i - \frac{\partial p}{\partial x_i} + \mu \frac{\partial^2 u_i}{\partial x_j \partial x_j} \tag{4}$$

where u_i and u_j are the velocity components in different direction, $i, j = 1,2,3$, and to simplify the equation form, Einstein summation convention were used for the subscript i and j; x_i and x_j are the Cartesian coordinate components, $i, j = 1,2,3$; ρ is the density of fluid, F_i is the mass force, p is the pressure, and μ is kinematic viscosity.

The finite volume method was adopted to discretize the controlling system of equations [45]. The Menter's shear stress transport $k - \omega$ turbulence model was chosen to close the governing equations due to its effective prediction for cases involving flow separation under adverse pressure gradient conditions [46,47]. To obtain an accurate calculation, the second-order center-difference scheme for the pressure term was employed and the second-order upwind scheme and implicit solution were applied for momentum item, turbulent dissipation rate, and turbulent kinetic energy. The semi-implicit method for pressure-linked equations-consistent method was adopted for the coupling of velocity and pressure of turbulent flow.

2.4.2. Grid Partition and Boundary Condition

The CFD calculation domain consists of six parts: the inlet passage, the guide vanes, the spiral casing, the runner, the draft tube, and the outlet passage. The grid partition method was applied for grid generation, and the interfaces were defined to connect different parts. Generally, the quality and quantity of the grid have a major impact on the accuracy of numerical simulation results [48]. The ANSYS-ICEM meshing software is used to generate the grid. Considering the complexity of turbine's geometry, the unstructured mesh was generated throughout the whole MPT model due to its superiority in terms of fast-generation speed and good adaptability for most complex structures [49]. The local mesh-refinement method was adopted for runner and the guide vanes, and the fluid regions were directly divided in volume grids.

In this paper, 10 layers of grids were defined in the boundary layer around the guide vanes and runner blades wall to reduce the $y+$ value near the wall region, and finally the $y+$ region was 8~60, since the main concern of this research was the turbine's power output. In order to reduce the influence of grid number on the simulation results, five grid schemes were tested (Table 1). From the table, when the grid number of the whole model came to 4.0×10^6 (S3), the relative variation ratio of turbine's efficiency was no more than 0.1%. After weighing computing resources and grid computing accuracy, the grid-division S4 was chosen, where the total grid number of flow components was 4.7×10^6, and the minimum quality was no less than 0.3.

Table 1. Mesh independency verification.

Scheme	Grid Number (10^6)	Efficiency (%)
S1	2.8	81.64
S2	3.5	81.88
S3	4.0	82.97
S4	4.7	82.99
S5	5.3	83.01

In addition, the mass flow inlet with a value of $Q = 0.265$ kg/s was chosen as the inlet and the pressure outlet was applied to outlet for MPT. The interfaces were used for information transmission between the calculation areas. The runner zone was simulated by the moving reference frames method with the rotational speed of 3500 r/min. A no-slip boundary condition was imposed to the solid walls, and standard wall functions used for the near-wall region [42]. The convergence criteria of the residuals at each iteration were below a typical criterion of 10^{-5}.

2.4.3. CFD Results of the Initial MPT

Table 2 shows the calculation results of the initial MPT model. The power output and efficiency of the initial MPT model were 4.76 W and 82.99%, respectively, which were close to the design requirements but did not yet meet them. So optimization should be carried out to reach the design goal, where the power output of 6 W and an 85% efficiency are the target.

Table 2. CFD results of the initial MPT model.

Parameters	Value
Head/m	2.22
Runner torque/N·m	0.013
Power output/W	4.76
Efficiency/%	82.99
Hydraulic loss of spiral casing/m	0.064
Hydraulic loss of guide vanes/m	0.128
Hydraulic loss of draft tube/m	0.037

3. Hydraulic Performance Improvement of the MPT

3.1. Orthogonal Experiment Design

The orthogonal experiment is also known as a multi-factor preferred design. When there are multiple investigated factors and multiple values for each parameter, the orthogonal experiment method is very suitable for exploring the influence among the factors. Through comprehensive design, integrated comparison, and statistical analysis, this method can give out better results. Due to the characteristics of balance and dispersion by balanced sampling in the range of factors for the orthogonal table, each test scheme has strong representative, so the tests can achieve the purpose of the experiment. Based on the structural parameters discussed above, the orthogonal experiment method was adopted to analyze the influence of the parameters on the WPT's hydraulic performance and to explore the best combination of them by extremum difference analysis.

The purpose of the orthogonal experiment was to improve the power output P and efficiency η of the MPT, so the power output and efficiency were selected as the test evaluation index. Blade number z_b (factor A), guide vane profile G (factor B), guide vane number z_g (factor C), and guide vane opening (GVO) α (factor D) were selected for the orthogonal experimental study, and three levels were taken for each factor. The three levels of the blade numbers were 11, 13, and 15, respectively. The three levels of the guide vane type were negative curvature profile (B1), symmetric curvature profile (B2), and positive curvature profile (B3). The three levels of guide vane number were 12, 14, and 16, respectively. And the three levels of guide vane opening were $23°$, $25°$, and $27°$, respectively. The factors and levels are shown in Table 3.

Table 3. Factors and levels of orthogonal experiment design.

Factors		Levels of Factor (t)		
		$r = 1$	$r = 2$	$r = 3$
A	z_b	11	13	15
B	G	B_1	B_2	B_3
C	z_g	12	14	16
D	α	$23°$	$25°$	$27°$

The orthogonal layout is represented by $L_l(t^c)$. In this layout, L stands for orthogonal layout, l is the number of experiments, t is the level of factor, and c is the number of columns (the maximum factors). So the orthogonal experiment design adopted a $L_9(3^4)$ layout (Table 4) in this study. The $L_9(3^4)$ design was carried out pursuant to the orthogonal principal in order to examine the influences of the four factors on the power output and efficiency at the design condition of the MPT.

Table 4. $L_9(3^4)$ orthogonal experiment schemes.

Trail No.	Parameter Combination				Indicators	
	A (z_b)	B (G)	C (z_g)	D (α)	P (W)	η (%)
1	11	B_1	12	$23°$	4.32	77.99
2	11	B_2	14	$25°$	4.70	79.29
3	11	B_3	16	$27°$	5.81	76.93
4	13	B_1	14	$27°$	4.42	82.89
5	13	B_2	16	$23°$	7.76	84.79
6	13	B_3	12	$25°$	3.98	76.00
7	15	B_1	16	$25°$	6.40	87.13
8	15	B_2	12	$27°$	4.02	82.38
9	15	B_3	14	$23°$	5.11	74.54

According to the table, only trail No. 5 and 7 achieved the required power output at the design discharge point. Therefore, the superior scheme was as follows: the optimal scheme in terms of power output was trail No. 5 ($A_2B_2C_3D_1$, named Scheme (a)) and the optimal scheme in terms of efficiency was trail No. 7 ($A_3B_1C_3D_2$, named Scheme (b)).

3.2. Extremum Difference Analysis

The extremum difference value (R) can represent the level of a factor's influence on the indicators. The larger the R value, the greater the influence a certain factor imposes on the indicator. The analysis of extremum difference requires the K_r and K_{ravg} values. These can be defined as follows:

$$K_r = \sum_{k=1}^{m} Y_k \tag{5}$$

$$K_{ravg} = \frac{1}{m} K_r \tag{6}$$

where r is the number of levels, m is total test times for each level, K_r is the sum of test results when number of the level is r, Y_k is the kth index value, and K_{ravg} is the arithmetic mean value of K_r.

The analysis of different R values is displayed in Tables 5 and 6. According to the table, the impact of each factor level on the two test indicators can be obtained.

Table 5. Extremum difference analysis of test results for MPT's power output P (W).

Parameter	A (z_b)	B (G)	C (z_g)	D ($\alpha/°$)
K_1	14.83	15.14	12.32	17.19
K_2	16.16	16.48	14.23	15.08
K_3	15.53	14.90	19.97	14.25
K_{1avg}	4.94	5.05	4.11	5.73
K_{2avg}	5.39	5.49	4.74	5.03
K_{3avg}	5.18	4.97	6.66	4.75
R	0.45	0.52	2.55	0.98

Table 6. Extremum difference analysis of test results for MPT's efficiency η (%).

Parameter	A (z_b)	B (G)	C (z_g)	D ($\alpha/°$)
K_1	234.21	248.01	236.37	237.32
K_2	243.68	246.46	236.72	242.42
K_3	244.05	227.47	248.85	242.20
K_{1avg}	78.07	82.67	78.79	79.11
K_{2avg}	81.23	82.15	78.91	80.81
K_{3avg}	81.35	75.82	82.95	80.73
R	3.28	6.85	4.16	1.70

Based on the extremum difference analysis, the influence degree of each factor on the two evaluation indicators was different. To intuitively demonstrate the influence of each factor on the evaluation index, we took the factors as the horizontal coordinate and each evaluation index as the vertical coordinate, and the horizontal index relationship is shown in Figure 3.

With the increase of z_b, the power output first increased and then decreased, the efficiency continuously rose to a certain value then slowed down. On one hand, as the z_b increased, the extrusion effect of the blades on the water flow increased, causing the climb of the head and the turbine's power output. On the other hand, after z_b increased, the total area of the blade increased, and the pressure difference between the front and back of the blade decreased, resulting in the drop of turbine's power output. The common action between the blades' extrusion effect on the water flow and the blades' surface pressure difference affected the power output and efficiency of the turbine.

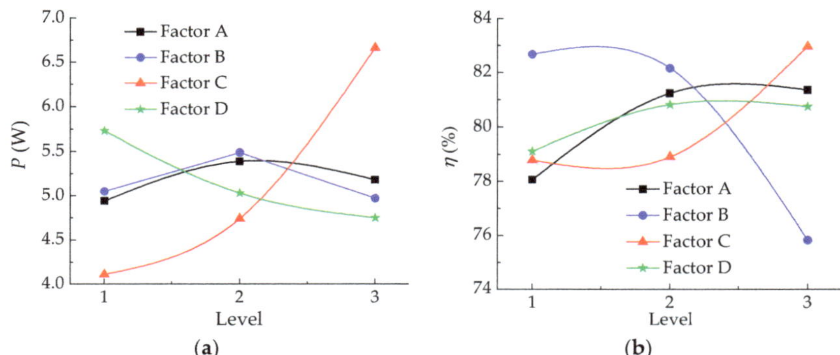

Figure 3. Relationship between levels of each factor and evaluation indicators, (**a**) output index and (**b**) efficiency index.

In terms of the guide vane profile (factor B), the positive curvature guide vane had the minimum power output and efficiency, so this type is not suitable for this MPT. The symmetrical guide vane had the largest power output because of large volume, which enhanced the extrusion effect on the water flow, leading to large head and power output. The negative curve guide vane had the highest efficiency and a higher utilization rate of water flow energy.

As z_g increased, both the MPT's efficiency and the power output increased. The increase of z_g within a certain range was conducive to guiding the water flow into the rotating runner and forming a certain circulation amount, which then improved the MPT's power output and efficiency.

With the increase of α, the power output continuously decreased, and the efficiency increased first and then decreased. At the same flow rate condition, the water head rose and the power output dropped after the α rose. The excessively small or large GVO led to a large angle of attack at the inlet of the guide vane, resulting in flow separation and greater hydraulic loss. The large GVO did not allow the water flowing through the guide mechanism to form circulation before the runner, reducing the power output and efficiency of the MPT. Therefore, the proper GVO should be determined in the design of the turbine to improve the power output and efficiency.

From the extremum difference analysis, the superior test scheme of power output and efficiency index were $A_2B_2C_3D_1$ (named Scheme (c)) and $A_3B_1C_3D_2$ (named Scheme (d)), respectively. According to the calculation of extremum difference value R, the significant order of four test factors on the test index was: C > D > B > A for power output and B > C > A > D for efficiency.

3.3. Optimized Scheme Determination

The two evaluation indicators of this orthogonal test are of the equal importance. According to the test schemes (a) and (b) obtained from the intuitive analysis above and the test schemes (c) and (d) obtained from the extremum difference analysis of each index, the comprehensive frequency analysis of the corresponding levels of the four factors A, B, C and D was carried out, as shown in Table 7.

Table 7. Factors A, B, C, and D correspond to different horizontal frequency tables.

Factor Level	A_1	A_2	A_3	B_1	B_2	B_3	C_1	C_2	C_3	D_1	D_2	D_3
Frequency	0	0.5	0.5	0.5	0.5	0	0	0	1	0.5	0.5	0

From Table 6, both schemes $A_2B_2C_3D_1$ and $A_3B_1C_3D_2$ were superior according to the integrated frequency analysis method. However, the calculated water head of scheme $A_2B_2C_3D_1$ was 3.52 m, while the calculated head of scheme $A_3B_1C_3D_2$ was 2.85 m, which was closer to the design water head of 3m. Furthermore, the efficiency of scheme $A_3B_1C_3D_2$ was the highest when the power output reached 6 W, so the optimal test scheme was initially determined to be $A_3B_1C_3D_2$. The results of the orthogonal experiment showed that the ideal design model was as follows: the number of runner blades $z_b = 15$, the guide vane type is negative curvature profile, the number of guide vanes $z_g = 16$, and the GVO $\alpha = 25°$. After CFD verification, the MPT's power output and efficiency at design condition were 6.4 W and 87.13%, respectively.

3.4. Analysis of the Optimized Scheme

To verify the optimization results, the average internal flow streamlines in the guide vane and runner zone were extracted at the design discharge, as shown in Figure 4. The streamlines in the blade channel at the three spans all presented some obvious vortices on the front and back of the blades, resulting in disorderly flow pattern and large hydraulic loss before optimization. The vortex in blade channel of the optimized model is significantly reduced, and only small vortices appear on the back of the individual blades. The velocity distribution is relatively uniform, and the streamlines are overall smooth and the flow pattern is good. After optimization, the flow velocity on blades increased.

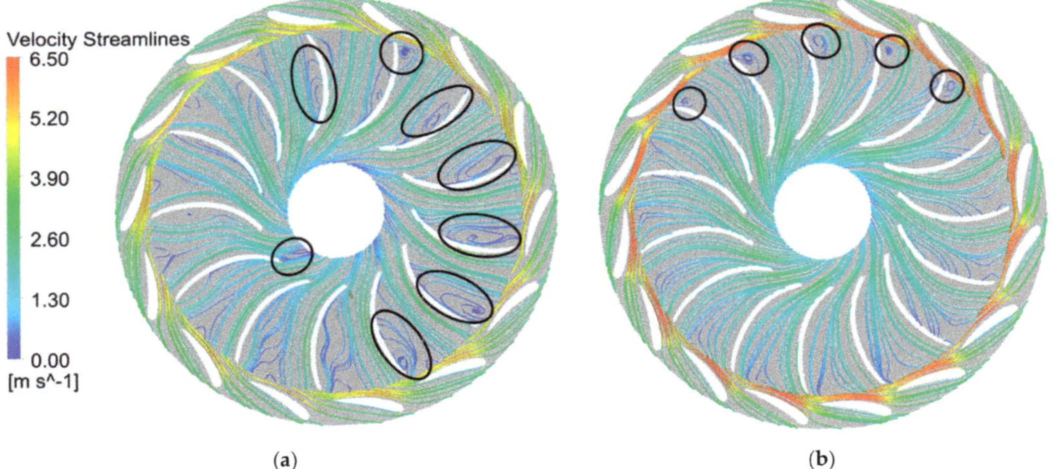

Figure 4. Streamline distribution in guide vane and runner zone at design flow rate, (**a**) before optimization and (**b**) after optimization.

Figure 5 presents the pressure contour of the blade surface under design flow rate. The blade surface pressure decreases uniformly along the inlet edge to the outlet edge of the blade, and there are obvious negative pressure areas near the lower ring of the blade. Obviously, the negative pressure zone on the optimized blade surface is significantly reduced (Figure 5b). Local high pressure appears at the back surface of the blade before the optimization, while the high pressure area basically disappeared after optimization. Moreover, the pressure difference on the two sides of the blade increases significantly after optimization, which improves the working capacity of the blade and the power output of the MPT, so the optimization effect is better.

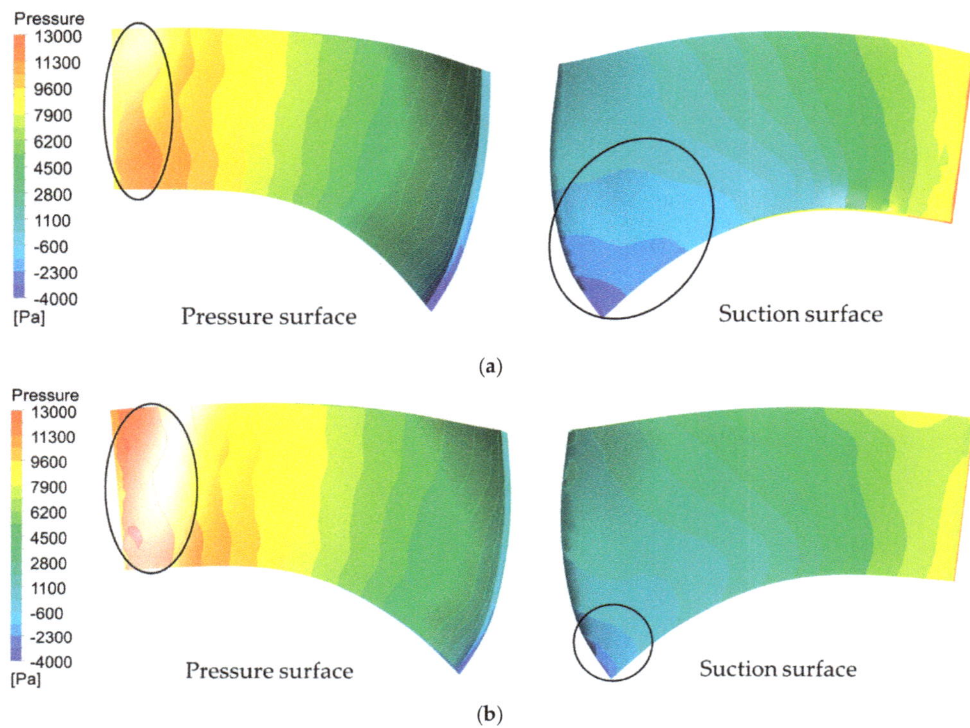

Figure 5. Blade contour on blade surface at design flow rate, (**a**) before optimization and (**b**) after optimization.

4. Experimental Validation and Analysis

4.1. Test Bench Set-Up

The model MPT was made according to the 1:1 model scale (Figure 6), in order to reduce the difficulty of production and financial cost. The spiral casing and draft tube were made of engineering plastic polyamide material and produced by 3D printing technology.

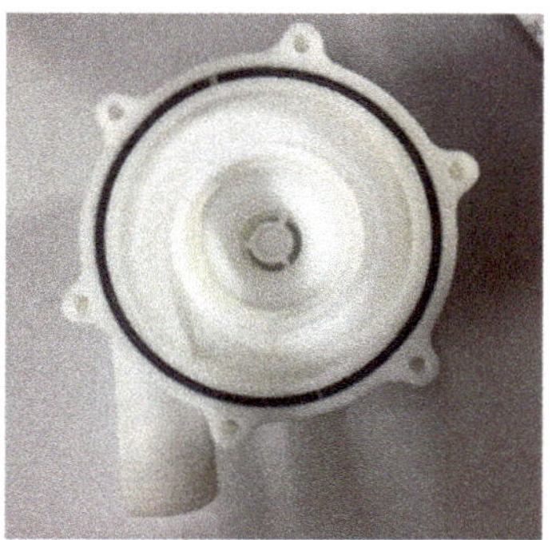
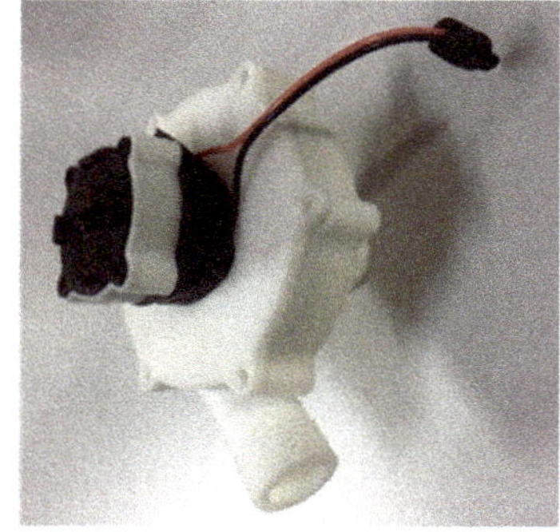

Figure 6. Photo of the physical WPT model.

The MPT experimental test was carried out on a flow test machine, which was independently developed by a company in Zhejiang province, China (Figure 7). The spiral casing inlet was installed at the water pipe outlet on the test bench, and the inlet flow value was input through the flow test machine. After the test instrument was started, the multimeter was used to measure the current and calculate the power. The measuring current range of the multimeter was 2 mA–20 mA–200 mA–20 A, and the measurement accuracy was ±1.8%.

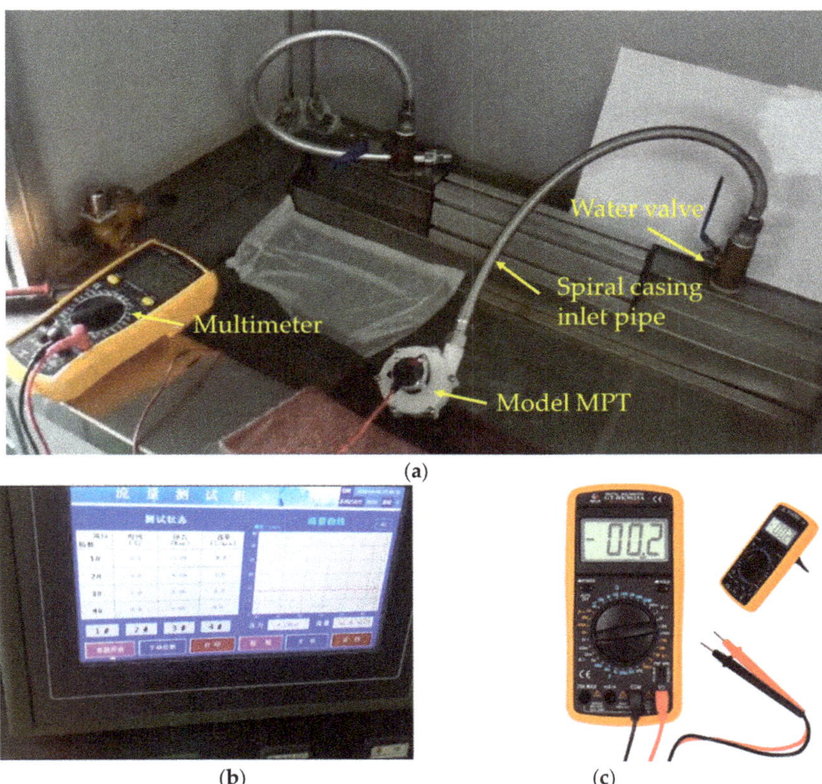

(a)

(b)　　　　(c)

Figure 7. Model test bench of MPT, (**a**) photo of the test bench, (**b**) flow test machine, and (**c**) multimeter.

4.2. Analysis of the Experimental Results

Table 8 shows the results of comparison between numerical simulation results and experimental test results. During the test, the power output of the turbine under different flow conditions was measured by changing the input flow value of the flow tester. Since the flow velocity in the tap-water pipeline is generally 1~3 m/s, the measured flow velocities in the experiment were 1 m/s, 1.5 m/s, 2 m/s, 2.5 m/s, and 3 m/s, and the corresponding inlet mass flow rates were 0.177 kg/s, 0.265 kg/s, 0.353 kg/s, 0.442 kg/s, and 0.53 kg/s, respectively. With the experimental results of the flow rate of 0.8 m/s (0.14 kg/s), the MPT could still start, but the power output was only 0.44 W. Multiple experimental results showed that the turbine could start when the inlet flow was greater than 0.14 kg/s.

Table 8. Comparison of numerical simulation results and experimental results under different working conditions.

Working Condition	Q (kg/s)	P (W)		Error (%)
		Num.	Exp.	
1	0.177	2.37	2.14	9.7
2	0.265	6.40	6.04	5.6
3	0.353	12.56	11.89	5.3
4	0.442	20.91	19.95	4.6
5	0.530	31.09	29.86	4.0

According to Table 8, the power output measured by the model test was lower than the obtained value through numerical simulation. The main reason for this is that the index measured in the experiments was the actual power of the MPT after the generator conversion, while the one obtained through numerical simulation was the power output of the MPT without considering the generator conversion. Also, the MPT model made during the experiment used a plastic material inducing friction between the runner and the runner chamber, and this was not considered in numerical simulation. Moreover, the size of MPT was very small, and the blade and guide vane actually made for model testing deviated from the numerical simulation. There were also some measurement errors, especially when measuring the current with an electric meter. In addition, under the small flow rate condition, the flow pattern was poor, which led to large fluctuations of measured values. This also considerably contributed to the appearance of large errors between the test and numerical predictions. Overall, the error value was within 10% under low flow conditions, while it reduced to within 5% under large flow conditions. The generated power met the design target under design flow conditions.

5. Conclusions

In this study, an MPT used to generate electric power for induction faucet sensors within a 15 mm pipeline was developed. The fundamentally expected MPT performance characteristics were the power output larger than 6 W when the pipe water velocity is 1.5 m/s and the turbine efficiency of no less than 85%. The initial design was done by using the 3D CFD method, where multi-objective optimization through orthogonal experiments was later applied to improve its hydraulic performance. Finally, the optimized version of the MPT model was experimentally tested.

After simulating and comparing a variety of schemes, the optimal MPT was achieved with the output power of 6.40 W at the design flow rate, representing an increase of 34.45% as compared to the preliminary design scheme. In the same respect, the optimal efficiency was 87.13%, representing an increase of 4.99% as compared to the preliminary design scheme. The feasibility of combining the numerical simulation technology and comprehensive frequency analysis method through the multi-objective orthogonal optimization test for MPT studies, has been verified. For the optimized turbine model, the inter-blade vortex and negative pressure area on the blade surface were significantly reduced, and the high-pressure area on the back of the blade basically disappeared, which improved the power capacity of the blades, collectively leading to the improved MPT power output and efficiency. A good and effective optimization was therefore achieved.

According to the model test results, the MPT-generated energy reached 6 W at the design flow rate, which meets the design requirements. The numerical simulation error as compared to test data was within 10%. Under large flow conditions, the error was around 5%, indicating that numerical simulation outcomes can offer a good guidance for micro turbine design even though the difference between the simulation and experimental results is fairly large. The experimental measurement showed that the MPT can start when the inlet flow rate is greater than 0.14 kg/s.

So far, the literature on the application of mixed-flow turbines for energy generation task within water supply pipelines with extra-small diameters (less than 15 mm) is still limited. This study creatively applies the mixed flow water turbine to the micro tap-water pipeline, changing the induction faucet power supply system from the battery-dependent one to a hydropower-fed one. This technology not only gives a long-term solution to energy saving issues within these systems, but also contributes to the environmental pollution control by eliminating the usage of batteries. The kind of MPT design presented in this article pays more attention to the turbine's efficiency and ensures that the output power meets the power requirements of induction faucet sensors.

Future studies could also consider runner blade profile and the frame line of flow components, which may also have a large impact on the machine performance. In the future, more factors can be selected to explore a more complete and faster optimization method, so as to more comprehensively reflect the impact of various design parameters on the turbine performance. Furthermore, a WPT for smaller discharge can be designed to achieve the power demand when the flow velocity in the pipeline is less than 1.5 m/s.

Author Contributions: Conceptualization, K.K. and H.C.; methodology, H.C. and H.W.; software, H.C. and H.W.; validation, Y.Z., K.K. and M.B.; formal analysis, H.C. and H.W.; investigation, K.K.; resources, Y.Z. and H.X.; data curation, H.W., Y.Z. and H.X.; writing—original draft preparation, H.C. and K.K.; writing—review and editing, Y.Z., M.B. and H.X.; visualization, K.K.; supervision, Y.Z. and H.X.; funding acquisition, H.C. All authors have read and agreed to the published version of the manuscript.

Funding: This research was funded by the National Natural Science Foundation of China (52006053), the Natural Science Foundation of Jiangsu Province (BK20200508), the China Postdoctoral Science Foundation (2021M690876), and the Open Research Subject of Key Laboratory of Fluid Machinery and Engineering (Xihua University), Sichuan Province (grant number LTJX2021-001).

Institutional Review Board Statement: Not applicable.

Informed Consent Statement: Not applicable.

Data Availability Statement: Not applicable.

Acknowledgments: The computational work was supported by High Performance Computing Platform, Hohai University. The support of Hohai University, China is also gratefully acknowledged.

Conflicts of Interest: The authors declare no conflict of interest.

Nomenclature

CFD	Computational fluid dynamics
GV	Guide vane
GVO	Guide vane opening
MPT	Micro-pipe turbine
WSP	Water supply pipeline
c	Number of columns
d	Pipe diameter
F_i	Mass force
H	Water head
K_r	Sum of test results when number of the level is R
K_{ravg}	Arithmetic mean value of K_r
l	Number of experiments
m	The total test times of each level
n	Rotational speed
n_s	Specific speed
P	Power output
p	Pressure
Q	Discharge
R	Extremum difference value
r	Number of levels
t	The level of factor

Y_k	The kth index value
u	Water flow velocity
u_i	Velocity component in i direction
u_j	Velocity component in j direction
z_b	Blade number
z_g	Guide vane number
ρ	Water flow density
μ	Kinematic viscosity
η	Efficiency
α	Guide vane opening

References

1. Water and Sustainable Development. Available online: https://www.un.org/waterforlifedecade/water_and_sustainable_development.shtml (accessed on 27 September 2021).
2. Schwarzenbach, R.P.; Egli, T.; Hofstetter, T.B.; von Gunten, U.; Wehrli, B. Global Water Pollution and Human Health. *Annu. Rev. Environ. Resour.* **2010**, *35*, 109–136. [CrossRef]
3. Chen, J.; Yang, H.X.; Liu, C.P.; Lau, C.H.; Lo, M. A novel vertical axis water turbine for power generation from water pipelines. *Energy* **2013**, *54*, 184–193. [CrossRef]
4. Semu, E.; Singh, B.R.; Selmer-Olsen, A.R. Mercury pollution of effluent, air, and soil near a battery factory in Tanzania. *Water Air Soil Pollut.* **1986**, *27*, 141–146. [CrossRef]
5. McNabola, A.; Coughlan, P.; Williams, A.P. The technical and economic feasibility of energy recovery in water supply networks. *Renew. Energy Power Qual. J.* **2011**, *1*, 1123–1127. [CrossRef]
6. Aqel, M.O.A.; Issa, A.; Qasem, E.; El-Khatib, W. Hydroelectric Generation from Water Pipelines of Buildings. In Proceedings of the 2018 International Conference on Promising Electronic Technologies (ICPET), Deir El-Balah, Palestine, 3–4 October 2018; IEEE: New York, NY, USA, 2018; pp. 63–68. [CrossRef]
7. Zhou, D.; Deng, Z. Ultra-low-head hydroelectric technology: A review. *Renew. Sustain. Energy Rev.* **2017**, *78*, 23–30. [CrossRef]
8. Campbell, R.J. Small Hydro and Low-Head Hydro Power Technologies and Prospects. 2010. Available online: https://www.researchgate.net/profile/Ammar-Kamel/project/Small-dams-for-electric-power-production/attachment/57bd6c690 8ae98f5947f2f9a/AS:398544616869890@1472031849894/download/Small_hydro_and_Low-head_hydro_power.pdf?context= ProjectUpdatesLog (accessed on 27 September 2021).
9. Du, J.; Shen, Z.; Yang, H. Numerical study on the impact of runner inlet arc angle on the performance of inline cross-flow turbine used in urban water mains. *Energy* **2018**, *158*, 228–237. [CrossRef]
10. Sinagra, M.; Sammartano, V.; Morreale, G.; Tucciarelli, T. A New Device for Pressure Control and Energy Recovery in Water Distribution Networks. *Water* **2017**, *9*, 309. [CrossRef]
11. Dhadwad, A.; Balekar, A.; Nagrale, P. Literature review on blade design of hydro-microturbines. *Int. J. Sci. Eng. Res.* **2014**, *5*, 72–75.
12. Sakti, A.; Prasetyo, A.; Tjahjana, D.; Hadi, S. The horizontal axis type of savonius water turbine in pipe using SolidWork simulation. In Proceedings of the 4th International Conference on Industrial, Mechanical, Electrical, and Chemical Engineering, Solo, Indonesia, 9–11 October 2018.
13. Li, H.D.; Zhou, D.Q.; Martinez, J.J.; Deng, Z.D.; Johnson, K.I.; Westman, M.P. Design and performance of composite runner blades for ultra low head turbines. *Renew. Energy* **2019**, *132*, 1280–1289. [CrossRef]
14. Saftner, D.A.; Hryciw, R.D.; Green, R.A.; Lynch, J.P.; Michalowski, R.L. The use of wireless sensors in geotechnical field applications. In Proceedings of the 15th annual Great Lakes Geotechnical/Geoenvironmental Conference, Indianapolis, IN, USA, 9 May 2008.
15. Ferro, L.M.C.; Gato, L.M.C.; Falcao, A.F.O. Design of the rotor blades of a mi Water and sustainable development ni hydraulic bulb-turbine. *Renew. Energy* **2011**, *36*, 2395–2403. [CrossRef]
16. Samora, I.; Hasmatuchi, V.; Munch-Alligne, C.; Franca, M.J.; Schleiss, A.J.; Ramos, H.M. Experimental characterization of a five blade tubular propeller turbine for pipe inline installation. *Renew. Energy* **2016**, *95*, 356–366. [CrossRef]
17. Lucid Energy Inc. How Drinking Water Pipes Can Also Deliver Electric Power. Available online: http://www.pbs.org/newshour/bb/drinking-water-pipes-can-also-deliver-electric-power (accessed on 27 September 2021).
18. Langroudi, A.T.; Afifi, F.Z.; Nobari, A.H.; Najafi, A.F. Modeling and numerical investigation on multi-objective design improvement of a novel cross-flow lift-based turbine for in-pipe hydro energy harvesting applications. *Energy Convers. Manag.* **2020**, *203*, 15. [CrossRef]
19. Oladosu, T.L.; Koya, O.A. Numerical analysis of lift-based in-pipe turbine for predicting hydropower harnessing potential in selected water distribution networks for waterlines optimization. *Eng. Sci. Technol.-Int. J.-Jestech* **2018**, *21*, 672–678. [CrossRef]
20. Mcalister, K.W.; Carr, L.W.; Mccroskey, W.J. *Dynamic Stall Experiments on the NACA 0012 Airfoil*; NASA Technical Paper 1100; NASA Ames Research Center: Moffett Field, CA, USA, 1978.
21. Carr, L.W.; Mcalister, K.W.; Mccroskey, W.J. *Analysis of the Development of Dynamic Stall Based on Oscillating Airfoil Experiments*; NACA Technical Paper D-8382; NASA Ames Research Center: Moffett Field, CA, USA, 1977.

22. Payambarpour, S.A.; Najafi, A.F.; Magagnato, F. Investigation of deflector geometry and turbine aspect ratio effect on 3D modified in-pipe hydro Savonius turbine: Parametric study. *Renew. Energy* **2020**, *148*, 44–59. [CrossRef]
23. Payambarpour, S.A.; Najafi, A.F. Experimental and numerical investigations on a new developed Savonius turbine for in-pipe hydro application. *Proc. Inst. Mech. Eng. Part A-J. Power Energy* **2020**, *234*, 195–210. [CrossRef]
24. Chen, J.; Lu, W.; Hu, Z.; Lei, Y.; Yang, M. Numerical studies on the performance of a drag-type vertical axis water turbine for water pipeline. *J. Renew. Sustain. Energy* **2018**, *10*, 44503. [CrossRef]
25. Yang, W.; Hou, Y.; Jia, H.; Liu, B.; Xiao, R. Lift-type and drag-type hydro turbine with vertical axis for power generation from water pipelines. *Energy* **2019**, *188*, 116070. [CrossRef]
26. Rossi, M.; Comodi, G.; Piacente, N.; Renzi, M. Effects of viscosity on the performance of Hydraulic Power Recovery Turbines (HPRTs) by the means of Computational Fluid Dynamics (CFD) simulations. *Energy Procedia* **2018**, *148*, 170–177. [CrossRef]
27. Sekar, K.; Kumar, N.; Vasanthakumar, P. Centrifugal pump as turbine for micro-hydro power scheme in rural areas of India: A review. *Int. J. Chem. Tech. Res.* **2017**, *10*, 106–109.
28. Capurso, T.; Bergamini, L.; Camporeale, S.; Fortunato, B.; Torresi, M. CFD analysis of the performance of a novel impeller for a double suction centrifugal pump working as a turbine. In Proceedings of the 12th European Conference on Turbomachinery Fluid Dynamics & Thermodynamics ETC13, Lausanne, Switzerland, 8–12 April 2019. [CrossRef]
29. Fontana, N.; Giugni, M.; Portolano, D. Losses Reduction and Energy Production in Water-Distribution Networks. *J. Water Resour. Plan. Manag.* **2012**, *138*, 237–244. [CrossRef]
30. Du, J.Y.; Yang, H.X.; Shen, Z.C.; Chen, J. Micro hydro power generation from water supply system in high rise buildings using pump as turbines. *Energy* **2017**, *137*, 431–440. [CrossRef]
31. Kramer, M.; Terheiden, K.; Wieprecht, S. Pumps as turbines for efficient energy recovery in water supply networks. *Renewable Energy* **2018**, *122*, 17–25. [CrossRef]
32. Buono, D.; Frosina, E.; Mazzone, A.; Cesaro, U.; Senatore, A. Study of a Pump as Turbine for a Hydraulic Urban Network Using a Tridimensional CFD Modeling Methodology. *Energy Procedia* **2015**, *82*, 201–208. [CrossRef]
33. De Marchis, M.; Fontanazza, C.M.; Freni, G.; Messineo, A.; Milici, B.; Napoli, E.; Notaro, V.; Puleo, V.; Scopa, A. Energy Recovery in Water Distribution Networks. Implementation of Pumps as Turbine in a Dynamic Numerical Model. *Procedia Eng.* **2014**, *70*, 439–448. [CrossRef]
34. Pérez-Sánchez, M.; Sánchez-Romero, F.J.; Ramos, H.M.; López-Jiménez, P.A. Improved Planning of Energy Recovery in Water Systems Using a New Analytic Approach to PAT Performance Curves. *Water* **2020**, *12*, 468. [CrossRef]
35. Carravetta, A.; del Giudice, G.; Fecarotta, O.; Ramos, H.M. PAT Design Strategy for Energy Recovery in Water Distribution Networks by Electrical Regulation. *Energies* **2013**, *6*, 411–424. [CrossRef]
36. Yeung, A.T.C. Asset management of drainage facilities using advanced technologies. In Proceedings of the 2014 Drainage Services Department International Conference (DSDIC 2014), Hong Kong, China, 12–14 November 2014.
37. Turbine Selection on the Basis of Specific Speed. Available online: https://www.britannica.com/technology/turbine/Turbine-selection-on-the-basis-of-specific-speed (accessed on 27 September 2021).
38. Chu, W.; Jiang, T.; Ho, P.S. Effect of Wiring Density and Pillar Structure on Chip Packaging Interaction for Mixed-Signal Cu Low k Chips. *IEEE Trans. Device Mater. Reliab.* **2021**, *21*, 290–296. [CrossRef]
39. Chu, W.; Spinella, L.; Shirley, D.R.; Ho, P.S. Effects of Wiring Density and Pillar Structure on Chip Package Interaction for Advanced Cu Low-k Chips. In Proceedings of the 2020 IEEE International Reliability Physics Symposium (IRPS), Dallas, TX, USA, 28 April–30 May 2020; pp. 1–4.
40. Kawase, Y.; Yamaguchi, T.; Masuda, T. Magnetic field analysis of micro turbine generator using finite element method. In Proceedings of the 10th International Symposium on Electromagnetic Field, ISEF 2001, Krakow, Poland, 20–22 September 2001; pp. 237–241.
41. Zhu, D.; Tao, R.; Xiao, R.F.; Pan, L.T. Solving the runner blade crack problem for a Francis hydro-turbine operating under condition-complexity. *Renew. Energy* **2020**, *149*, 298–320. [CrossRef]
42. Chen, Z.; Singh, P.M.; Choi, Y.D. Structural Safety Evaluation of a Francis Hydro Turbine Model at Normal Operation Condition. *KSFM J. Fluid Mach.* **2017**, *20*, 12–18. [CrossRef]
43. Kan, K.; Yang, Z.; Lyu, P.; Zheng, Y.; Shen, L. Numerical study of turbulent flow past a rotating axial-flow pump based on a level-set immersed boundary method. *Renew. Energy* **2021**, *168*, 960–971. [CrossRef]
44. Zhou, D.; Chen, H.; Zheng, Y.; Kan, K.; Yu, A.; Binama, M. Development and Numerical Performance Analysis of a Pump Directly Driven by a Hydrokinetic Turbine. *Energies* **2019**, *12*, 4264. [CrossRef]
45. Kversteeg, H.K.; Malalasekera, W. *An Introduction to Computational Fluid Dynamics: The Finite Volume Method*; Wiley: New York, NY, USA, 1995; pp. 4–30.
46. Menter, F.; Rumsey, C. Assessment of two-equation models for transonic flows. In Proceedings of the Fluid Dynamics Conference, Colorado Spring, CO, USA, 20–23 June 1994; pp. 1994–2343.
47. Kan, K.; Chen, H.; Zheng, Y.; Zhou, D.; Binama, M.; Dai, J. Transient characteristics during power-off process in a shaft extension tubular pump by using a suitable numerical model. *Renew. Energy* **2021**, *164*, 109–121. [CrossRef]

48. Devals, C.; Vu, T.C.; Zhang, Y.; Dompierre, J.; Guibault, F. Mesh convergence study for hydraulic turbine draft-tube. In Proceedings of the 28thIAHR Symposium on Hydraulic Machinery and Systems, Grenoble, France, 4–8 July 2016; Volume 49.
49. Chen, H.; Zhou, D.; Kan, K.; Guo, J.; Zheng, Y.; Binama, M.; Xu, Z.; Feng, J. Transient characteristics during the co-closing guide vanes and runner blades of a bulb turbine in load rejection process. *Renew. Energy* **2021**, *165*, 28–41. [CrossRef]

Article

Hydrostatic Pressure Wheel for Regulation of Open Channel Networks and for the Energy Supply of Isolated Sites

Ludovic Cassan [1,*], Guilhem Dellinger [2], Pascal Maussion [3] and Nicolas Dellinger [2]

[1]   Institut de Mécanique des Fluides, Toulouse INP, Université de Toulouse, Allée du Professeur Camille Soula, 31400 Toulouse, France
[2]   Laboratoire des Sciences de l'Ingénieur, de l'Informatique et de l'Imagerie (ICUBE), ENGEES, 67000 Strasbourg, France; guilhem.dellinger@engees.unistra.fr (G.D.); dellinger@unistra.fr (N.D.)
[3]   LAPLACE, Toulouse INP, Université de Toulouse, 2 Rue Camichel, 31071 Toulouse, France; pascal.maussion@toulouse-inp.fr
*   Correspondence: lcassan@imft.fr

Abstract: The Hydrostatic Pressure Wheel is an innovative solution to regulate flow discharges and waters heights in open channel networks. Indeed, they can maintain a water depth while producing energy for supplying sensors and a regulation system. To prove the feasibility of this solution, a complete model of water depth–discharge–rotational speed relationship has been elaborated. The latter takes into account the different energy losses present in the turbine. Experimental measurements achieved in IMFT laboratory allowed to calibrate the coefficients of head losses relevant for a large range of operating conditions. Once the model had been validated, an extrapolation to a real case showed the possibility of maintaining upstream water level but also of being able to produce sufficient energy for supplying in energy isolated sites. The solution thus makes it possible to satisfy primary energy needs while respecting the principles of frugal innovation: simplicity, robustness, reduced environmental impact.

Keywords: renewable energy; water wheel; low tech; frugal innovation; experimental model; theoretical model

Citation: Cassan, L.; Dellinger, G.; Maussion, P.; Dellinger, N. Hydrostatic Pressure Wheel for Regulation of Open Channel Networks and for the Energy Supply of Isolated Sites. *Sustainability* **2021**, *13*, 9532. https://doi.org/10.3390/su13179532

Academic Editors: Mosè Rossi, Massimiliano Renzi, David Štefan and Sebastian Muntean

Received: 6 July 2021
Accepted: 18 August 2021
Published: 24 August 2021

Publisher's Note: MDPI stays neutral with regard to jurisdictional claims in published maps and institutional affiliations.

1. Introduction

Irrigation systems have a lot of weir structures in order to change the water elevation, water velocities, etc. These "small" drops represent a non-negligible hydroelectric potential [1]. Nevertheless, the role of the turbine in this study is a little bit different. Its function is not only to produce electricity but also to serve as a water level controller. Indeed, the regulation of water height and flow discharge in irrigation systems is also crucial to save the water resources and, as mentioned before, requires many regulation systems (sluice gates, spillways, weirs, etc). The fact is that these systems can be very isolated and therefore difficult to power, especially in remote areas. To overcome this problem, a turbine could regulate the water level automatically by adjusting its rotational speed. Part of the energy supplied by the turbine would then be used to power the sensors and the control system needed for this regulation and data transmission. Moreover, in the context of isolated areas, the rest of the energy provided by the turbine could also be used for supplying the local electricity grid. Indeed, the water wheels are also a good example of the low-tech concept for energy supply [2]. Therefore, we will also discuss the power performance in the context of isolated communities in developing countries in line with the frugal innovation concept. Although equipping a low head in a large river may involve some environmental issues, power generation with low-head coupled with solar energy and storage devices is fully relevant if no other source is available (isolated zone).

As noted above, irrigation systems are constructed with weirs that can be exploited. Typically, these structures have head differences between 0.5 and 3 m and have a low flow rate. Turbines suitable for this kind of hydraulic sites are the Kaplan turbine, the Archimedean Screw Turbine (AST) and the various types of water wheels [3]. It is also possible to add the Very Low Head (VLH) turbine to this list. For obvious economical and ecological reasons, the turbine chosen must be inexpensive, robust, fish-friendly and provide the sediment continuity. The Kaplan and VLH turbine reach a better hydraulic efficiency than the other two, but they require complex control elements and are much more expensive [4]. The AST turbine has almost the same efficiency as a well-designed water wheel but is slightly more expensive and much more complex to build. Finally, the choice was made to consider water wheels.

By combining the different types of water wheels, they are able to exploit water head from 0.5 to 12 m [5]. The three main types of water wheels are the overshot wheels, the breastshot wheels and the undershot wheels. It is also possible to speak about stream water wheels but they are designed to exploit the flow velocity and not the head difference [6]. For the overshot wheels, the water enters from above and the wheel is rotated by the water weight. These turbines are used for heads between 2.5 and 12 m and can reach efficiencies of about 80% [4,7]. In the case of the breastshot wheels, the water enters at approximately the same height than the turbine's rotation axis. This type of wheel works for head differences between 1.5 and 4 m and can achieve efficiencies between 60% and 70% [8–11]. The last main type of water wheel is the undershot wheel. In this case, the flow enters below the axis of rotation. This type of turbine is adapted for very low head differences between 0.5 and 2.5 m and can reaches efficiencies of 80% [12,13]. Regarding the head differences in the context of irrigation systems, undershot wheels seem to be the most suitable. Specifically, the Hydrostatic Pressure Wheel (HPW) developed by [14] is chosen in order to achieve the objectives presented above. The wheel is composed of radial blades and is driven to rotate by the hydrostatic pressure exerted by the flow on the blades. In addition, this turbine is chosen because it presents a very simple design, is robust, inexpensive and fish-friendly [14].

As a reminder of the general context, the stated objectives are to replace the regulation structures of irrigation system that require manual operation and maintenance with a turbine. The latter will adjust the water level automatically by changing its rotational speed and, at the same time, will produce some energy for local consumption. In order to properly design the system according to the site characteristics, a theoretical model that links hydraulic conditions, geometrical parameters and turbine performance is required. Currently, there is only one theoretical model, proposed by [14]. However, this model is not able to predict directly the water level upstream to the wheel and the flow discharge through it. In irrigation systems, due to the modulation of water demand, it is essential to control water levels and flows throughout the network. Therefore, to control these parameters with water wheels, an improved model is needed and is presented in this article.

The paper is structured as follows. In Section 2, a new theoretical model is established based on the [14] model. Special attention is paid to the consideration of various energy losses (gap leakage, drag forces, etc.) as they are very significant. The experimental setup installed in the IMFT hydraulic laboratory and used for providing experimental data on the HPW turbine is presented in Section 3. After calibrating the loss models with the experimental results, the ability of the theoretical model to accurately reproduce the wheel performance, the upstream water levels and the flow discharges absorbing by the wheel are discussed in Section 4. Finally, a discussion about the possibilities to use these turbines for the electrification of an isolated village in a low-tech context is analyzed in Section 5.

2. Operating Principle and Theoretical Models

2.1. Operating Principle

As shown in Figure 1, the HPW developed in [14] consists of a wheel made up of radial blades longer than the inlet water height. This turbine converts the potential energy of a fluid into mechanical energy thanks to rotation induced by the fluid pressure exerted on its blades. This mechanical energy can then be transformed into electrical energy thanks to a generator. This last transformation will not be discussed here although it is a crucial point for power generation [14]. The mechanical power supplied by the wheel is given by:

$$P = P_{hyd}\,\eta = \rho g Q \Delta H \eta \tag{1}$$

where P_{hyd} is the available hydraulic power, ρ is the density of water, Q is the flow rate, g is the acceleration of gravity, ΔH is the hydraulic head difference and η—the hydraulic efficiency of the turbine. The efficiency depends on the different losses present in the turbine which are mainly due to the flow leakages and the drag forces. The latter present when the blades enter into the flow and when they leave it. Thus, minimizing these losses is a key issue in order to increase the turbine performance.

Figure 1. 3D view of a HPW turbine.

To use the wheel effectively as a regulation and energy production system, it is necessary to determine the influence of the wheel on the upstream water level and to predict its energy recovery performance. For this purpose, a theoretical model is established in the following part. The latter is able to determine the upstream water level and the performance of the turbine according to the geometrical parameters of the wheel and the flow conditions.

2.2. Theoretical Model

As presented in Figure 2, the HPW turbine has radial blades longer than the inlet water level. This is why this type of wheel is dedicated to very low heads between 0.3 and 1 m [14]. The main geometrical parameters of this turbine are the radius of the wheel r, the width L and the number of blades N. These parameters are shown in Figure 2. It should be noted that each blade is defined by a plane that contains the axis of rotation. The mechanical power is obtained by the displacement of a vertical blade submitted to a force induced by the flow pressure more important on the upstream face than on the downstream side. The model is derived by considering force balance on a blade. Compared to the previous study [14], the theory is rewritten here with a global efficiency formula η incorporating leakages and turbulence disturbance. The theory presented here will help upscale the wheel in a real scale application. Furthermore, as discussed earlier, an integrated formula for mechanical power as a function of upstream water depth and rotational speed is needed to develop efficient control of water level. To establish the model, a vertical blade with a finite radius r and a width L is considered. The water level upstream to the blade is equal to d_1 and the downstream one is equal to d_2. The distance between the blade and the bed is denoted d_l (Figure 2).

The total flow rate Q can be decomposed in two parts:

$$Q = Q_w + Q_l \tag{2}$$

with Q_w being the flow through the surface bounded by the height d_1 and the width L that actually contributes to wheel rotation, and Q_l representing the flow leakage that appears around the wheel. It may be noted that to optimize the wheel performance, it is necessary to limit the value of Q_l.

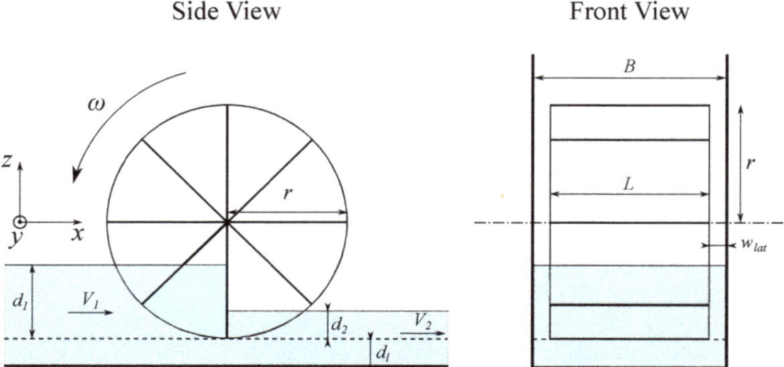

Figure 2. Schematic view of the HPW machine.

The flow leakage Q_l can be decomposed in two leakages. The first one is $Q_{l,1}$, which occurs between the blades and the bottom and, the second one—$Q_{l,2}$, which occurs between the wheel and the two side walls. Both leakages can be determined by using the Torricelli equation and are a function of the head difference $(d_1 - d_2)$, the wheel width L and the gaps between the wheel and the canal. To determine the first leakage $Q_{l,1}$ occurring between the blades and the bottom, the gap d_l is not used directly. Instead, an equivalent opening w representing the surface opening for flow under the wheel is introduced. This value has to be calibrated for each wheel geometry for the following reasons. Firstly, unlike sluice gate, the discharge formula is written here without considering the vertical contraction of the flow stream. Indeed, it is difficult to measure and evaluate accurately this value knowing that it evolves with the submergence d_2/d_1 [15,16]. Secondly, the wheel cannot be modeled like an orifice because the blades are moving and then the opening depends on the radial position of the wheel. Therefore w represents an averaged value of the space between blade and bed during a rotation. Eventually, to determine the leakage $Q_{l,2}$, the gap w_{lat} between the wheel an the sidewall is used. As a consequence, the portion of the discharge Q_l that was not flowing into the wheel can by expressed as:

$$Q_l = \sqrt{2g(d_1 - d_2)}\,wL + 2\sqrt{2g(d_1 - d_2)}\,d_1 w_{lat} \tag{3}$$

The water flowing into the wheel Q_w can be determined by assuming that when a blade is in the vertical position in the fluid, the water and the blade have the same velocity locally. The submerged length of the blade is equal to d_1, as represented in the Figure 2. The averaged velocity of the flow v_a is then given by the integration of the horizontal component of the velocity of the submerged part of the blade:

$$v_a = \frac{1}{d_1}\int_r^{r-d1} z\omega\,dz = \omega r\left(1 - \frac{1}{2r/d_1}\right) \tag{4}$$

The flow Q_w is then equal to:

$$Q_w = Ld_1 v_a = Ld_1 \omega r\left(1 - \frac{1}{2r/d_1}\right) \tag{5}$$

For scaling purposes, the following geometric aspect ratios are defined:

- water depths ratio: $X = d_2/d_1$;
- vertical gap ratio: $Y = d_l/d_1$;
- equivalent vertical gap ratio: $W = w/d_1$;
- horizontal gap ratio: $W_l = w_{lat}/B$;
- blade length ratio: $D = r/d_1$;
- wheel width ratio: $Z = L/B$.

The continuity equation provides the following relationship:

$$
\begin{aligned}
Q &= Q_w + Q_l \\
&= v_a d_1 L + \sqrt{2g(d_1 - d_2)}wL + 2\sqrt{2g(d_1 - d_2)}d_1 w_{lat} \\
&= L d_1 \omega r \left(1 - \frac{1}{2r/d_1}\right) + \sqrt{gd_1}\sqrt{2(1 - X)}\,(wL + 2d_1 w_{lat})
\end{aligned}
\tag{6}
$$

The ratio $R_Q = Q_w/Q$ between the water flowing through the wheel and the total discharge is defined and equal to:

$$
R_Q = \left(1 - \frac{[WZ + W_l]\sqrt{2(1 - X)}}{F_r}\right)\frac{1}{1 - \frac{1}{2D}}
\tag{7}
$$

where $F_r = Q/(Bd_1\sqrt{gd_1})$ is the Froude number calculated in the upstream part of the canal. R_Q is then the ratio of leakage ($R_Q = 1$ for no leakage).

The mechanical power recovered by the wheel is determined by multiplying the torque presents on the wheel axis by the rotational speed of the turbine. The torque can be decomposed in two parts: the drive torque induced by the pressure exerted by the water on the blades and, the resisting torque induced by the drag forces. The latter are mainly present when a blade enters or leaves the flow. To simplify the model, all drag forces acting on the wheel are modeled with a single overall drag force F_{drag} that would be applied to the blade tip with a direction orthogonal to the blade surface. The drag forces were assumed to depend primarily on the rotational speed of the wheel, the flow velocity, both inlet and outlet water levels and turbine width. The question remained as to what velocity and area should be used to effectively model all of the drag forces acting on the blades. After several trials, it appears that the best way to model an overall drag force F_{drag} is to use the blade tip velocity with the area given by the difference in water depth times the width of the wheel. The drag force F_{drag} is then equal to:

$$
F_{drag} = \frac{1}{2}\rho(d_1 - d_2)L(r\omega)^2 C_d
\tag{8}
$$

with C_d a drag coefficient. The resistant torque induced by the drag force is then given by drag force F_{drag} multiplied by the wheel radius r. The power P_d is then given by the resisting torque multiplied by the rotation speed of the wheel ω:

$$
P_d = F_{drag}.r\omega = \frac{1}{2}\rho C_d(r\omega)^3 L(d_1 - d_2) = \frac{1}{2}\rho grwLd_1(1 - X)C_d F_\omega^2
\tag{9}
$$

where $F_\omega = r\omega/\sqrt{gd_1}$.

To analyze the wheel performance, it is necessary to give the expression of the hydraulic efficiency η. As it can be seen in Equation (1), this efficiency corresponds to the ratio between the mechanical power recovered by the wheel P and the available hydraulic power P_{hyd}. To determine the hydraulic head ΔH that appears in the expression of P_{hyd}, it is assumed that the kinetic energy is computed from a zone of uniform flow in the upstream and downstream part of the canal. This means that the energy losses at the inlet and at the outlet of the wheel are taken into account in the efficiency. Considering a flat bottom near the wheel, the expression of ΔH is then equal to:

$$\Delta H = d_1 - d_2 + \frac{Q^2}{2gB^2}\left(\frac{1}{(d_1+d_l)^2} - \frac{1}{(d_2+d_l)^2}\right) \tag{10}$$

The mechanical power P provided by the wheel corresponds to the available hydraulic power calculated with the flow Q_w minus the drag loss P_d:

$$P = \rho g Q_w \Delta H - P_d = \frac{1}{2}\rho g Q d_1 \left((1+X) - \frac{1}{3D}\left(1+X^2+X\right) - C_d F_\omega^2\right)(1-X)R_Q \tag{11}$$

Finally, combining the Equations (7), (9) and (10); the wheel efficiency η is equal to:

$$\eta = \frac{1}{2}\frac{(1+X) - \frac{1}{3D}\left(1+X^2+X\right) - C_d F_\omega^2}{1 - \frac{1}{2}F_r^2\frac{1+X+2Y}{(1+Y)^2(X+Y)^2}}R_Q \tag{12}$$

Except the drag term, the Equation (12) is similar to the formula proposed by [14]. Here all corrections are given in the same equation which provides a better view of the dimensionless significant numbers. These independent numbers are $X, Y, Z, C_d, D, F_r, \omega$. The number R_Q, F_ω and η can be deduced from others. For a given rotational speed ω, the Froude number provides the relationship between flow rate and water depth. The coefficient C_d is considered constant for a given wheel shape even though a Reynolds dependence could be expected. This analysis tends to prove that a Froude similarity could be applied for the wheel design.

3. Experimental Setup

Experimental data are necessary to calibrate and validate the theoretical model established previously. In addition, experimental measurements are needed to investigate the performance of the HPW turbine experimentally. An experimental setup was therefore installed in the hydraulic laboratory of the Toulouse Institute of Fluid Mechanics (IMFT). This device allows to test the efficiency and the mechanical behavior of the turbine for different geometrical and hydraulic parameters. The wheel geometries considered in this study are as simple as possible in order to obtain low-cost and low-maintenance turbines. These characteristics are essential in a context of isolated sites where it is difficult to replace a broken component with a new one. Moreover, plane blades seem to be more suitable for controlling the flow through the wheel even at low speeds by stopping it completely if necessary. The blades are attached to the side disks. They are slightly shorter than the radius which allows the air and water pressures to be balanced between the blades.

Eventually, the geometrical parameters of the tested wheel and the flow conditions are shown in Table 1.

Table 1. Geometrical and hydraulic parameters of the experimental device.

Geometrical Parameters	Outer radius—r (m)	0.4
	Material	PVC
	Width—L (m)	0.2
	Number of blades	8
	Channel width—B (m)	0.4
	Vertical gap—d_l (m)	0.025 and 0.05
	Horizontal gap—w_{lat} (m)	0.002
Flow conditions	Flow rate—Q (m^3 s^{-1})	0.005 … 0.025

The wheel is installed in a 4 m long, 0.4 m wide and 0.4 m high open channel. Aluminum walls of 0.9 m width are placed on each side of the wheel to conduct the flow into the turbine. The bed and the wall of the channel are in glass. The slope of the bed is chosen to be zero. Water is brought at the inlet of the experimental device by a centrifugal pump. The flow rate is measured with a electromagnetic flowmeter (IFS 4000/6, Krohne, Holland) with a given measuring accuracy of $\pm 0.5\%$. The water then flows through the wheel and leads to its rotation. The water then exits at the outlet. All water levels are measured with dial gauges. The torque provided by the wheel C_h is balanced by an overall brake torque, which consists of two terms: a friction term C_f induced by the friction in the device (rotational guidance) and a brake torque C_b induced and controlled by a Prony brake. The motion equation is defined as follows:

$$J\frac{d\omega}{dt} = C_h - C_f - C_b \tag{13}$$

with J—the inertia of the rotating parts. A torque meter is coupled directly in line, between the wheel axis (accuracy of 1 mN m) and the Prony brake. This latter provides the value of C_b; in steady state the equation above becomes:

$$C_b = C_h - C_f \tag{14}$$

Note that the instantaneous values of torque and rotational speed are averaged over 30 s, a sufficient value considering the frequency of the dynamic phenomenon. C_f was determined for different values of ω without water in the channel. It appears that this torque increases linearly with the speed and its value remains close to zero when $\omega = 0$. The rotational speed of the turbine is measured with the incremental coder incorporated in the torque meter (T21WN, HBM, Germany). Finally, the mechanical power provided by the wheel is determined by multiplying the torque by its rotational speed. A schematic diagram of the whole device is shown in Figure 3.

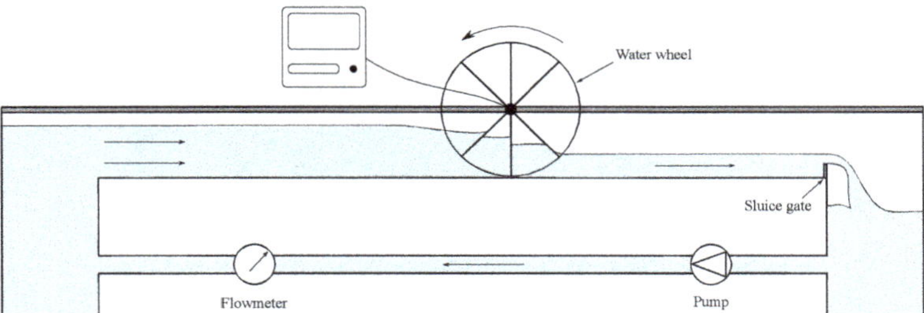

Figure 3. Schema of the experimental device with the HPW wheel without a removable bottom.

In the experiments, two configurations were tested. In the first, called C_1, there is a vertical gap d_l of 0.025 m between the blade and the bed of the channel. This configuration was chosen to determine the value of the equivalent gap w in the flow leakage equation. In the second configuration, named C_2, the wheel axis is placed at a higher position. The gap d_l is then equal to 0.05 m. Nevertheless, a block of aluminum (0.4 m × 0.045 m × 0.9 m) is placed just below the wheel, as shown in Figure 4. This block can be considered as a simplified shape of the curved shroud proposed in [14]. In this case, the gap between the block and the blade of the wheel is equal to 0.01 m. This important gap leads to a strong flow leakage. However, the objective of this study is not to obtain the best hydraulic efficiencies but to be able to correctly quantify this leakage and its influence on the performance. Moreover, this configuration represents the case of nonideal implementation that can be found in isolated areas. Lastly, the wheel axis being positioned higher in the C_2 configuration than in the C_1 one, it appears that the downstream water level was under the wheel for some operating points. A photo of the experimental device with the HPW turbine can be seen in Figure 5.

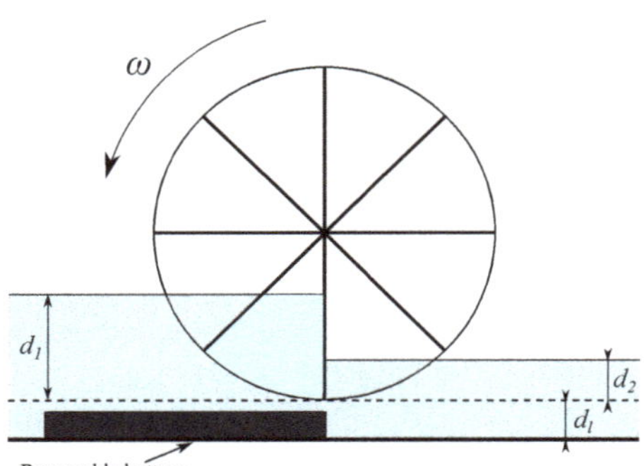

Figure 4. Schematic view of the HPW with the removable bottom in the configuration C_2.

Figure 5. Photo of the experimental device with the HPW wheel (configuration (C_1)).

4. Results

4.1. Model Calibration

Flow leakage is the main energy loss in HPW wheels. In order to accurately determine the wheel performance, it is necessary for the theoretical model to best reproduce this loss. As can be seen in Equation (3), the flow leakage Q_l is a function of an equivalent opening w. This opening must then be calibrated with experimental data. To do so, the experimental and theoretical values of R_Q are compared for several operating points, i.e., for various hydraulic (flow rate, water levels, etc.) and geometrical parameters (configurations C_1 and C_2). The theoretical values of the ratio R_Q are calculated from Equation (7). The latter involves the constant $W = w/d_1$ and the Froude number F_r. The first one must be calibrated and the second one is measured experimentally. The other constants in Equation (7) are known. The experimental values of $R_Q = Q_w/Q$ are determined by calculating Q_w from Equation (5) and by measuring the flow rate Q. Indeed, as demonstrated above, the flow Q_w is a function of known geometrical parameters and of the rotational speed, which is measured experimentally.

Figure 6 exposes the comparison between modeled and measured values of R_Q. It can be seen that the theoretical model reproduces the measured data well. The model is more accurate in configuration C_1 for which the maximum deviation is 10%. This can the explained by the fact that the block of aluminum placed behind the wheel in configuration C_2 perturbs the flow pattern, especially for free-flow experiments ($X = 0$). The calibrated values of w, the lateral gap w_{lat} and the vertical gap d_l are shown in Table 2 for the C_1 and C_2 configurations. For the C_1 configuration, the value of w is greater than the gap d_l between the channel bottom and a vertical blade. This can be explained by the fact that the gap increases when the considered blade rotates and is not vertical anymore. For the C_2 configuration, the value of w is lower than d_l because of the presence of the block under the wheel. Interestingly, although the d_l deviation in the C_1 and C_2 configurations is different, the calibrated values of w are both very similar. This could be due to the fact that the wheel has only eight blades. Most of the leakage then flows between the blades. It is expected that the value of d_l should have more influence on the wheel performance if it has more blades.

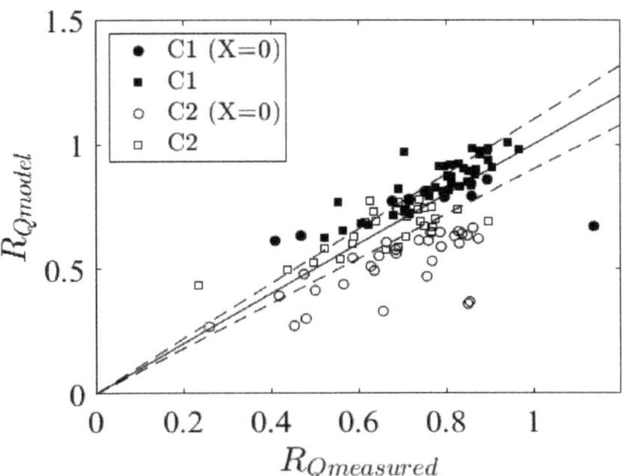

Figure 6. Comparison between modeled and experimental values of $R_Q = Q_w/Q$ for configuration C_1 (without removable bottom) and C_2 (with removable bottom). The solid line represents the perfect agreement and the dashed lines represent the gap of 10%.

Table 2. Measured values of the lateral w_{lat} and vertical gap d_l, calibrated value of the equivalent opening w and of drag coefficient C_d for configurations C_1 and C_2.

	d_l (m)	w_{lat} (m)	w (m)	C_d
C_1	0.02	0.002	0.026	4
C_2	0.05	0.002	0.025	4–10

The power loss P_d induced by the drag force and the associated turbulent dissipation is given by the Equation (9). In this equation, the drag coefficient C_d must be calibrated with experimental data. To do this, special attention is paid to the drag term $C_d F_\omega^2$ presents in Equation (11). The experimental value of this drag term is obtained by measuring experimentally the other terms (P, Q, ω, d_1 and d_2) present in Equation (11) and by using the values of w calibrated previously (Table 2). Figure 7 exposes the evolution of the drag term as a function of F_ω for the experimental configurations C_1 and C_2. It can be shown that the dissipation depends on the square of F_ω and can be expressed as the proposed modeling $C_d F_\omega^2$. As shown in Table 2, it was possible to deduce from experiments a constant $C_d = 4$ for configurations C_1 and C_2 when F_ω is greater than 0.3. For configuration C_2, the agreement is worse at low F_ω and the value of C_d is higher ($C_d = 10$). This is explained by the flow disturbance induced by the block placed under the wheel. It can be noted that for very low values of downstream water level (i.e., for $X = 0$), the flow is almost similar to a free flow over a weir. It differs from the assumption made to establish the theoretical analysis and an additional dissipation appears then. Although this behavior is partially taken into account by the term $(d_1 - d_2)L$ representing the drag area, this correction is not sufficient. To better understand the flow pattern, 3D numerical simulations are planned for the future.

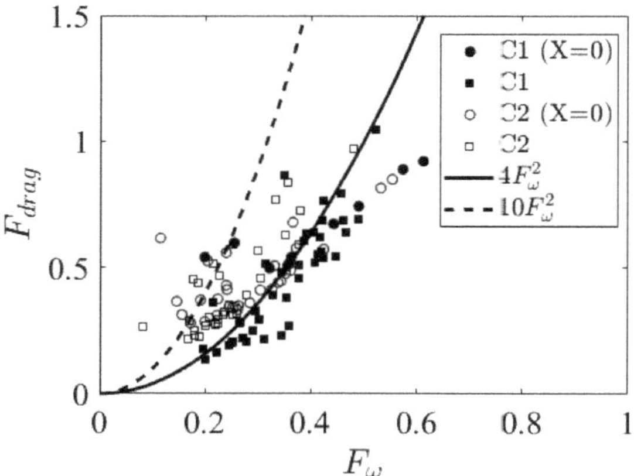

Figure 7. Evolution of the drag term $C_d F_\omega^2$ depending on F_w for configurations C_1 and C_2.

4.2. Power and Efficiency

Figures 8 and 9 show the comparison between experimental and modeled values of the inlet water level d_1 for the C_1 and C_2 configurations, respectively. The theoretical values of d_1 are determined using the calibrated values of w and C_d (Table 2). As expected, it can be shown that the value of d_1 decreases when the rotational speed increases because the flow capacity of the wheel increases with the rotational speed. Figure 8 shows that the model can reproduce the experimental values in configuration C_1 accurately. For the C_2 configuration, the theoretical values of d_1 are slightly higher than experimental values (Figure 9). Again, this difference can be explained by the flow disturbance induced by the block. Nevertheless, these results show that the model is capable of predicting the water level upstream of the wheel. This ability is very important for the control of water levels and flows in irrigation networks.

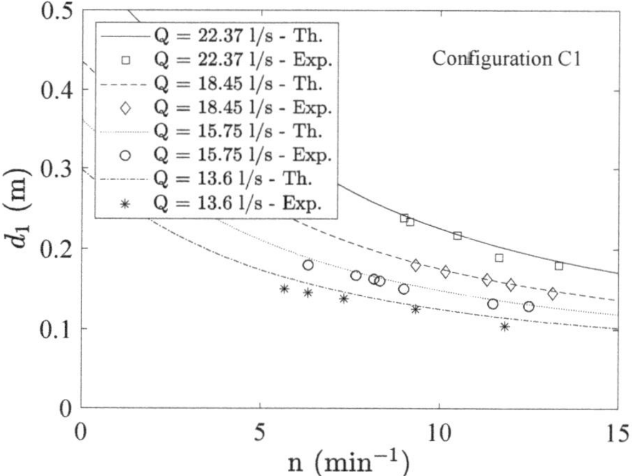

Figure 8. Comparison between modeled and experimental values of the inlet water level d_1 for the configurations C_1.

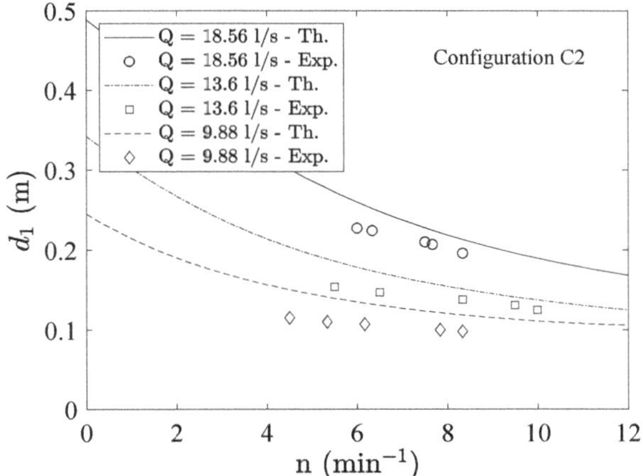

Figure 9. Comparison between modeled and experimental values of the inlet water level d_1 for the configurations C_2.

Figures 10 and 11 show the evolution of the mechanical power supplied by the wheel as a function of the rotational speed for the C_1 and C_2 configurations, respectively. The theoretical power evolution is determined by the Equation (11). It should be noted that for a given value of d_1 and Q, the rotational speed can be directly deduced from Equation (7). Then at a fixed rotational speed, a single value of d_1 corresponds to one discharge if d_2 is constant. It can be seen that the evolution of power is bell-shaped. Thus, for low rotational speed, the principal loss comes from flow leakages. Conversely, for too-high rotational speeds, the loss induced by turbulence and friction become predominant. As expected, there is an optimal rotational speed of the wheel that provides the best wheel performance. Comparison between theoretical and experimental values exposes that the model is able to determine the power delivered by the turbine, especially for the C_1 configuration (blockless configuration).

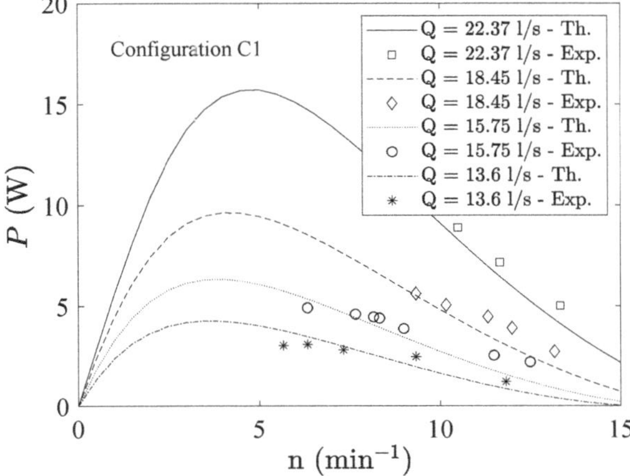

Figure 10. Theoretical and experimental values of mechanical power depending on the rotational speed for different flow discharges Q and for configuration C_1.

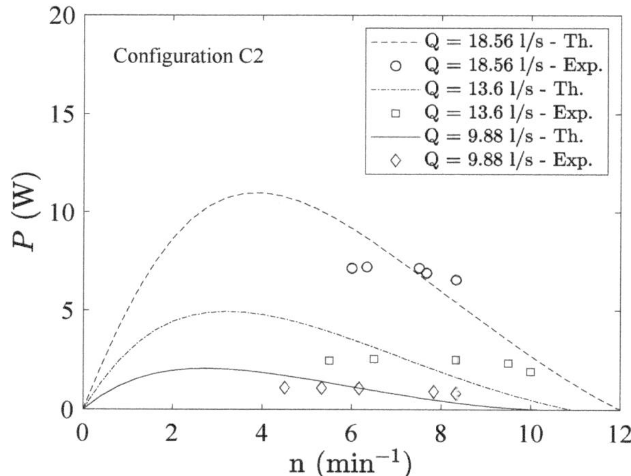

Figure 11. Theoretical and experimental values of mechanical power depending on the rotational speed for different flow discharges Q and for configuration C_2.

Figures 12 and 13 expose the evolution of the hydraulic efficiency η as a function of the rotation speed $n = 60/(2\pi)\omega$ for different flow rates. The theoretical efficiency is calculated from Equation (12). The experimental values are obtained by dividing the measured mechanical power by the hydraulic power according to the flow rate, d_1 and d_2. As for the mechanical power, a bell-shaped evolution of the efficiency can be shown. Comparing the modeled and experimental efficiencies for the C_1 configuration, it can be noticed that the model gives slightly higher efficiency values. The optimal rotational speeds of the wheel (i.e., the speed that gives the higher efficiency) given by the model are also close to the experimental ones for the C_1 configuration. For the C_2 configuration, the modeled efficiency values are much higher than experimental values because of the block under the wheel. When n tends to 0, the efficiency also tends to 0 because the leakage flow rate becomes equal to the total flow rate ($R_Q \approx 0$). Indeed, the wheel behaves like a sluice gate and no mechanical power is obtained.

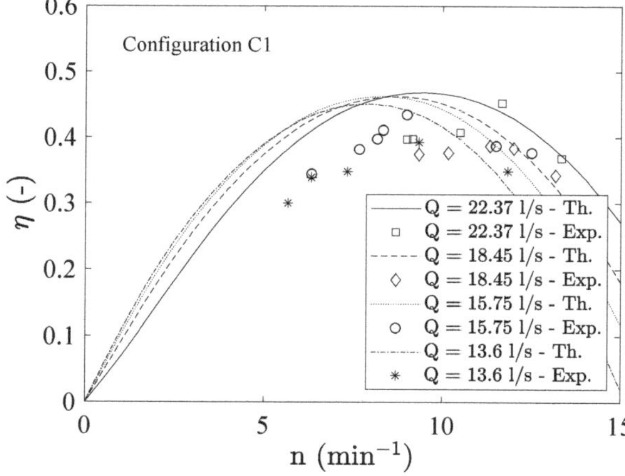

Figure 12. Theoretical and experimental values of hydraulic efficiency depending on the rotational speed for different flow discharges Q and for the configuration C_1.

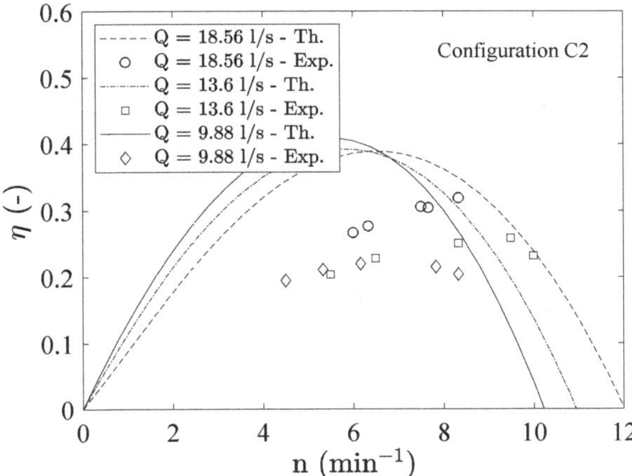

Figure 13. Theoretical and experimental values of hydraulic efficiency depending on the rotational speed for different flow discharges Q and for the configuration C_2.

Regarding the experimental efficiency, it can be seen that the values are much lower than those presented in [14]. Indeed, while efficiencies higher than 80% are measured in [14], maximal values in this study are approximately of 45% (in the C_1 configuration), possibly because of the low number of blades—8 blades here and 12 in [14]. The loss by flow leakage is therefore much more important here. It can also be noticed that the C_2 configuration does not improve the HPW performance. The rectangular shape of the block appears to be too "simple" compared to curved blocks. Indeed, the leakage reduction is not significant with a rectangular shape of the removable bottom and the block-induced flow disturbances that dissipate energy.

5. Discussion

In this section, the previously established theoretical model is used to evaluate the ability of the hydraulic pressure wheel to achieve the two initial goals, i.e., to control the water level and to produce electricity in irrigation systems. To ensure the validity of the model, both low and high rotational speeds of the wheel are avoided. Indeed, very low velocities induce significant leakage that could be misrepresented by the model. Conversely, for too high rotational speeds, energy losses due to turbulence would not be modeled correctly. For the rest of the study, the number of blades and the distance between the blade and the bottom d_l are respectively taken as equal to 8 and 0.025 m. These are the same values as in the experimental setup presented earlier. To approximate the equivalent opening w, it will be assumed that w increases linearly with the wheel diameter.

5.1. Regulation of Water Levels

Firstly, the use of HPW in irrigation networks is considered in conjunction with the installation of a weir, as shown in Figure 14. It is then possible to limit the size of the device while ensuring sufficient spillage in case of too-important flows. The main advantage of using this turbine in an irrigation context is the fact that it is possible to supply energy locally. This can be used to control water levels (upstream and downstream), flow rates, etc. To do this, it is planned that the wheel supplies in electricity sensors and a regulation system that will adapt the speed of the turbine according to the desired values. In order to have a live coverage of the network, this energy can also be used for supplying wireless data transmission in addition to solar energy and storage devices. Other interests can be found in the use of HPW machines. For example, by ensuring an adequate design of the wheel, with tip blades close to the bed bottom, it is possible to ensure free flow of sediments while avoiding silting. For anthrcpized natural systems, the water wheel can also ensure the free circulation of natural species present in the hydraulic system [17,18]. Eventually, HPW can also avoid the problem of floats usually present in the recirculation zone upstream of the gates.

To analyze the operation of the HPW system with respect to the water control objective, the example of a medium-sized network is chosen. The dimensionless Equations (7) and (12) ensure that the wheel behavior can be applied to a full-scale configuration. In the following, the case of a wheel of 2 m in diameter and 1 m in width is then considered. The dimensions of the device are shown in Table 3.

Figure 14. Example of sluice gate with a weir in ar. irrigation system.

Table 3. Geometrical parameters of the Hydraulic Pressure Wheel used for the discussion.

Geometrical Parameters		
	Outer radius—r (m)	1
	Width—L (m)	1
	Number cf blades	8
	Vertical gap—d_l (m)	0.025
	Horizontal gap—w_{lat} (m)	0.02
	Equivalent opening—w (m)	0.075

Figure 15 exposes the inlet water level d_1 as a function of flow rate Q for different values of rotational speed ω and outlet water level d_2. It can be observed that a HPW machine of 2 m in diameter can regulate d_1 by changing ω up to about 0.8 m^3/s. It can also be noted that, for a given flow rate, the value of d_1 is very sensitive to the rotational speed. A wide range of flow rates (between 0.2 and 0.8 m^3/s) is covered for rotational speeds varying only by a factor of 3. This method is advantageous for coupling devices with already existing alternators. The influence of the downstream water level d_2 on the flow rate Q is almost zero. This can be explained by the fact that the slight deviation comes from the flow leakage.

Figure 16 shows the mechanical power delivered by the wheel P as a function of flow rate Q for different values of rotational speed ω and outlet water level d_2. As expected, it can be seen that the power increases with the flow rate Q. Moreover, the power increases as d_2 decreases. This is because the head H and thus the available hydraulic power P_{hyd} increases as d_2 decreases for a given value of d_1. For the entire range of flow and speed, a few hundreds of watts can be produced. This production is sufficient to power a control and transmission system. It should be remember that, for this application, achieving maximum efficiency is not the main objective. However, as long as the ration $X = d_1/d_2$ is greater than 0.5, the efficiency remains greater than 30%. This value ensures a sufficient electrical production since the remaining power is also of the order of several hundred watts.

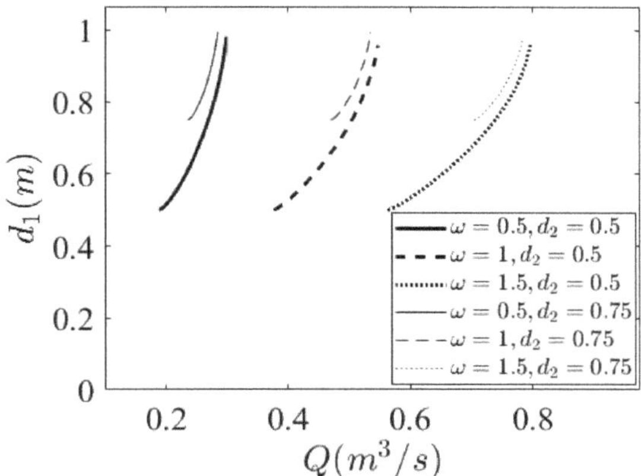

Figure 15. Simulated inlet water level estimated depending on the flow rate for different values of rotational speed and outlet water level.

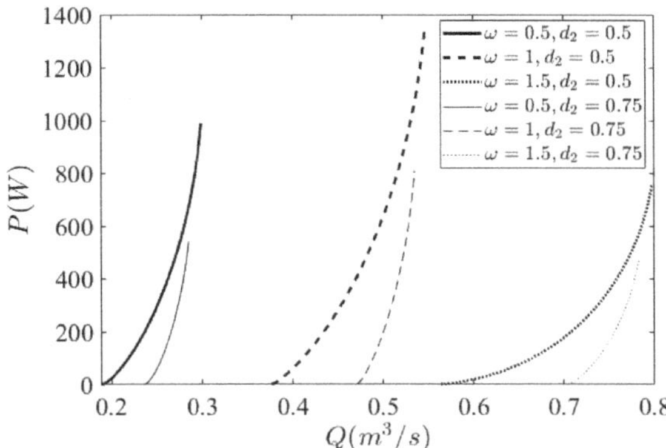

Figure 16. Simulated mechanical power estimates of the real-scale wheel depending on the flow rate for different values of rotational speed and outlet water level.

5.2. Electric Production for Isolated Site

Initially, the HPW wheel presented here is designed to produce electricity for low heads and low flows. For various environmental reasons, it seems judicious to limit this application to particular cases which benefit the most from the advantages of this system. These are its simplicity and robustness. Thus, run-of-river installations should be avoided to limit impacts and management of flood events. On the other hand, the diversion of high flows requires significant civil engineering costs. The interest is then focused on cases with low flows diverted (a few m^3/s) or small river courses where the wheel can be removed during floods. This context corresponds well to the electrification of isolated villages where even a few kilowatts can bring an important development aid [19].

As before, it is assumed that the HPW is installed in a 1 m channel (i.e., $L = 1$ m). It is then possible to work by unit width as one might do with a hill chart for a turbine. The other geometrical characteristics of the wheel are the same as those presented in Table 3. The outlet water level is taken equal to $d_2 = 0.7$ m.

In Figures 17 and 18, the operating diagram of the HPW is drawn in the P, Q plans. The value of the inlet water level is limited by the wheel size ($d_1 < r$) and the rotational speed by $F_\omega = r\omega/\sqrt{g_d 1} < 1$. The evolution of the rotational speed depending on the flow discharge is nearly linear. The flow leakage has an influence only for low power and then for low flow discharges. This is explained by low value of d_l chosen in this example and by the fact that the ratio Q_l/Q decreases as the rotational speed of the wheel increases. Despite the low efficiency values, which are generally lower than 50%, the results show that a wheel with a radius of one meter can provide 600 to 800 W per unit width. Again, these low efficiency values could come from the fact that the wheel has only 8 blades. Increasing the number of blades will increase the performance of the wheel (experimental data in [14]). It can be noted that the same powers can be obtained with the same upstream level. This is due to the fact that the efficiencies are higher for low rotational speeds. Indeed, for high rotational speeds, a lot of energy is dissipated by turbulence. Eventually, the maximum mechanical power is reached for $d_1 = r$, i.e., when the value of $X = d_2/d_1$ is minimal.

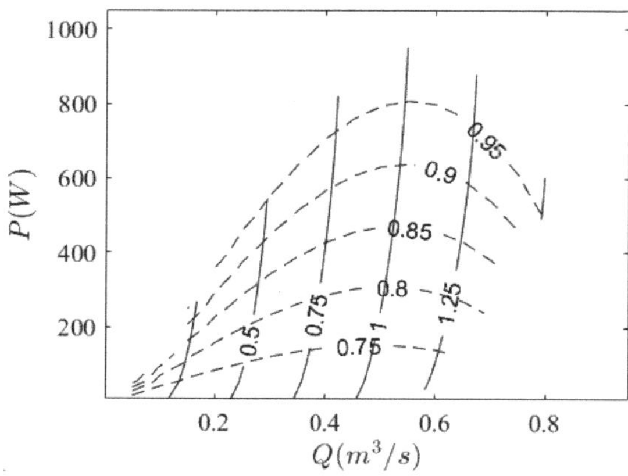

Figure 17. Iso-contours of rotational speed ω (solid line) and of upstream water level d_1 (dashed line) for the HPW with $d_2 = 0.7$ m.

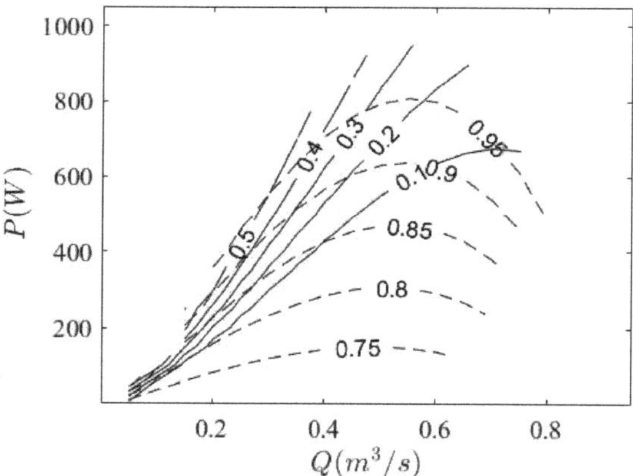

Figure 18. Iso-contours of efficiency η (solid line) and of upstream water level d_1 (dashed line) for the HPW with $d_2 = 0.7$ m.

We recall that the results presented above are quite similar to those presented in [14] but with a different definition and calibration of dissipation terms. The contribution of this study concerns the quantification of flow losses, which is useful to properly design a "low cost" micro-power plant. Indeed, in a context of isolated areas where wheels can be manufactured directly on site, the bank heights are not always adapted to an optimal installation (i.e., $d_l = 0$). Moreover, the presence of large space under the wheel is an advantage to limit the environmental impacts of the wheel (sediment transport and fish displacement). In the case of the previous wheel and for $h_u = 0.9$ m, $h_d = 0.8$ m, the influence of $d_l = w$ is studied. It appears that the recoverable power drops quickly when d_l increases (Figure 19). For low flows, a few centimeters are enough to cancel the interest of the wheel. Nevertheless, for large flows (between 0.6 and 0.8 m³/s), a consistent production of electricity can be maintained with values of d_l close to 0.5 r. The minimum value of d_l for a sedimentary and fish passage is of the order of 0.3 to 0.4 m [20]. These values provide sufficient production for the highest flows.

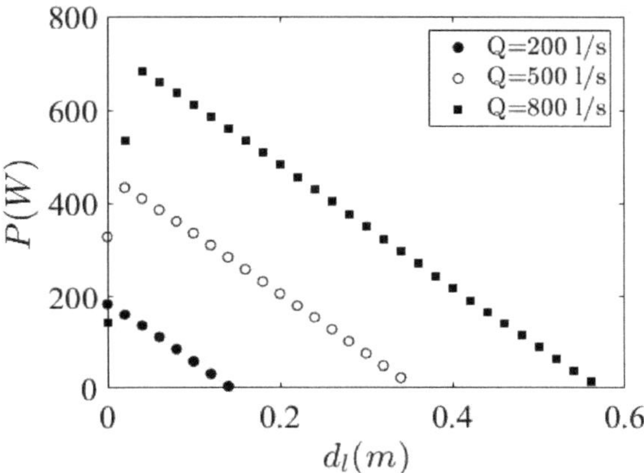

Figure 19. Influence of the gap between the wheel and the bottom d_l on the mechanical power provided by the wheel.

6. Conclusions

Irrigation systems feature many hydraulic structures to change water elevations or flow velocities. These must be controlled according to the hydraulic conditions and the demand. This study was then focused on the use of a turbine to control these physical parameters by modifying its rotational speed. Moreover, this turbine could also be used to exploit the hydraulic energy present in this irrigation network. With controlled electricity storage, this energy production could be used to power sensors and provide electricity in the cases of isolated sites.

To achieve these objectives, the choice was made to use the Hydrostatic Pressure Wheel (HPW) developed by the authors of [21]. Indeed, although this technology does not have the best hydraulic performance, the HPW is simple, robust and low-cost. In order to properly design the wheel, a theoretical model relating the performance of the turbine to the hydraulic conditions is necessary. Therefore, this paper presents an improved theoretical model based on the one presented in [14]. An important novelty resides in the ability of the model to predict the inlet water level. This is particularly important in an irrigation network control context. Moreover, all energy loss terms are taken into account and modeled.

The model was then calibrated and evaluated with experimental data. The latter were provided by an experimental device installed in the IMFT laboratory. The comparison between modeled and measured values have shown that the model is able to reproduce the experimental results in a large range of hydraulic conditions. However discrepancies appear when the operating conditions are far from the hydrostatic functioning due to insufficient downstream water depth.

The extrapolation of the model to a real case has given indications on the installation of this type of turbine in irrigation systems. First, it was shown that a wheel with dimensions compatible with usual irrigation networks can produce between 100 and 1000 W per unit meter width. This power range is sufficient to power a remote control system. Moreover, an HPW's wheels can regulate the water level over a wide range of flow rates by adjusting its rotation speed. Eventually, electricity production with a HPW can be significant even if there are high flow leakages due to implementation difficulties. Nevertheless, the mechanical power can drop quickly when the gap between the wheel and the bottom is too large. The turbine then no longer works with hydrostatic forces but recovers the kinetic energy.

Author Contributions: Conceptualization: L.C.; experimentations: L.C.; analysis: L.C., G.D., P.M. and N.D.; original draft preparation: L.C. and G.D.; review and editing: L.C., G.D., P.M. and N.D.; supervision: L.C. All authors have read and agreed to the published version of the manuscript.

Funding: This research received no external funding.

Informed Consent Statement: Not applicable.

Data Availability Statement: Not applicable.

Conflicts of Interest: The authors declare no conflict of interest.

References

1. Butera, I.; Balestra, R. Estimation of the hydropower potential of irrigation networks. *Renew. Sustain. Energy Rev.* **2015**, *48*, 140–151. [CrossRef]
2. Kim, B.; Azzaro-Pantel, C.; Pietrzak-David, M.; Maussion, P. Life cycle assessment for a solar energy system based on reuse components for developing countries. *J. Clean. Prod.* **2019**, *208*, 1459–1468. [CrossRef]
3. Williamson, S.J.; Stark, B.H.; Booker, J.D. Low head pico hydro turbine selection using a multi-criteria analysis. *Renew. Energy* **2014**, *61*, 43–50. [CrossRef]
4. Quaranta, E.; Revelli, R. Gravity water wheels as a micro hydropower energy source: A review based on historic data, design methods, efficiencies and modern optimizations. *Renew. Sustain. Energy Rev.* **2018**, *97*, 414–427. [CrossRef]
5. Müller, G.; Kauppert, K. Performance characteristics of water wheels. *J. Hydraul. Res.* **2004**, *42*, 451–460. [CrossRef]
6. Quaranta, E. Stream water wheels as renewable energy supply in flowing water: Theoretical considerations, performance assessment and design recommendations. *Energy Sustain. Dev.* **2018**, *45*, 96–109. [CrossRef]
7. Quaranta, E.; Revelli, R. Output power and power losses estimation for an overshot water wheel. *Renew. Energy* **2015**, *83*, 979–987. [CrossRef]
8. Quaranta, E.; Revelli, R. Performance characteristics, power losses and mechanical power estimation for a breastshot water wheel. *Energy* **2015**, *87*, 315–325. [CrossRef]
9. Quaranta, E.; Revelli, R. Optimization of breastshot water wheels performance using different inflow configurations. *Renew. Energy* **2016**, *97*, 243–251. [CrossRef]
10. Quaranta, E.; Revelli, R. Hydraulic Behavior and Performance of Breastshot Water Wheels for Different Numbers of Blades. *J. Hydraul. Res.* **2017**, *143*, 04016072. [CrossRef]
11. Vidali, C.; Fontan, S.; Quaranta, E.; Cavagnero, P.; Revelli, R. Experimental and dimensional analysis of a breastshot water wheel. *J. Hydraul. Res.* **2016**, *54*, 473–479. [CrossRef]
12. Quaranta, E.; Revelli, R. CFD simulations to optimize the blades design of water wheels. *Drink. Water Eng. Sci. Discuss.* **2017**, *10*, 27-32. [CrossRef]
13. Quaranta, E.; Müller, G. Sagebien and Zuppinger water wheels for very low head hydropower applications. *J. Hydraul. Res.* **2018**, *56*, 526–536. [CrossRef]
14. Senior, J.; Saenger, N.; Müller, G. New hydropower converters for very low-head differences. *J. Hydraul. Res.* **2010**, *48*, 703–714. [CrossRef]
15. Belaud, G.; Cassan, L.; Baume, J.P. Calculation of Contraction Coefficient under Sluice Gates and Application to Discharge Measurement. *J. Hydraul. Eng.* **2009**, *135*, 1086–1091. [CrossRef]
16. Belaud, G.; Cassan, L.; Baume, J.P. Contraction and Correction Coefficients for Energy-Momentum Balance under Sluice Gates. In Proceedings of the World Environmental and Water Resources Congress 2012, Albuquerque, NM, USA, 20–24 May 2012. [CrossRef]
17. Guiot, L.; Cassan, L.; Belaud, G. Modeling the Hydromechanical Solution for Maintaining Fish Migration Continuity at Coastal Structures. *J. Irrig. Drain. Eng.* **2020**, *146*, 04020036. [CrossRef]
18. Guiot, L.; Cassan, L.; Dorchies, D.; Sagnes, P.; Belaud, G. Hydraulic management of coastal freshwater marsh to conciliate local water needs and fish passage. *J. Ecohydraul.* **2020**, in press. [CrossRef]
19. UNCTAD. *The Least Developed Countries Report 2017: Transformational Energy Access*; United Nations Conference on Trade and Development: Geneva, Switzerland, 2017.
20. Baudoin, J.M.; Burgun, V.; Chanseau, M.; Larinier, M.; Ovidio, M.; Sremski, W.; Steinbach, P.; Voegtle, B. *The ICE Protocol for Ecological Continuity—Assessing the Passage of Obstacles by Fish: Concepts, Design and Application*; Onema: Vincennes, France, 2014.
21. Müller, G.; Kauppert, K. Old watermills—Britain's new source of energy? *Civ. Eng.* **2002**, *150*, 178–186. [CrossRef]

sustainability

Article

Improved Operation Strategy of the Pumping System Implemented in Timisoara Municipal Water Treatment Station

Ionel Aurel Drăghici [1], Ionuț-Daniel Rus [2], Adrian Cococeanu [1,2] and Sebastian Muntean [2,3,*]

1 AQUATIM S.A., 300081 Timisoara, Romania; ionel.draghici@aquatim.ro (I.A.D.); adrian.cococeanu@aquatim.ro (A.C.)
2 Faculty of Mechanical Engineering, University Politehnica Timisoara, Blvd. Mihai Viteazu, No. 1, 300222 Timisoara, Romania; ionut.rus@student.upt.ro
3 Center for Fundamental and Advanced Technical Research, Romanian Academy—Timisoara Branch, Blvd. Mihai Viteazu, No. 24, 300223 Timisoara, Romania
* Correspondence: sebastian.muntean@academiatm.ro; Tel.: +40-256-403692

Abstract: Water treatment stations (WTSs) provide drinking water to the community and are critical infrastructures of any village or city. The main energy consumption of WTSs is associated with the operation of the pumping units installed in the water supply stations (WSSs). The parameters of the pumping units installed in the WSSs are continuously adjusted during service to meet the requirements of the customers. Therefore, variable-speed pumping units (VSPUs) are feasible technical solutions implemented in WSSs. Several strategies combining VSPUs and constant-speed pumping units (CSPUs) have been developed to operate in WSSs. A technical solution with four pumping units (two VSPUs and two CSPUs) is implemented in Timisoara pumping station No. 1 (TPS1) in the Bega municipal water treatment station (MWTS). The layout of TPS1 is detailed, and its energy consumption from the budget of the Bega MWTS is quantified. The operation strategy with four pumping units selected in TPS1 is investigated. The number of hours in service of each pumping unit and the total operating time of all pumping units in the last six years are examined. The specific power consumption associated with the operation of the pumping units installed in TSP1 is detailed. The failure incidents of the pumping units counted in service are enumerated and correlated with the operating conditions of the pumping units. A new strategy developed for the operation of the pumping units installed in TPS1 is proposed to better adapt to the operating conditions, improving the specific power consumption as well as diminishing the failure incidents. The new operation strategy is presented and assessed based on the data acquired from TPS1 over one year. The conclusions and the lessons learned in this case study are drawn in the last section.

Keywords: improved operation strategy; pumping system; municipal water treatment station

Citation: Drăghici, I.A.; Rus, I.-D.; Cococeanu, A.; Muntean, S. Improved Operation Strategy of the Pumping System Implemented in Timisoara Municipal Water Treatment Station. *Sustainability* **2022**, *14*, 9130. https://doi.org/10.3390/su14159130

Academic Editor: Ioannis Katsoyiannis

Received: 23 May 2022
Accepted: 21 July 2022
Published: 25 July 2022

Publisher's Note: MDPI stays neutral with regard to jurisdictional claims in published maps and institutional affiliations.

1. Introduction

The annual energy consumption of all pumps (constant- and variable-flow applications) was 225 TWh in 2015 [1]. Drinking-water processes and wastewater activities account for 3–4% of energy consumption in the U.S. [2]. The energy consumption of the water sector is about 3.5% of the total EU budget [3]. The estimated total energy consumption of the water sector is projected to be 253 TWh/year in 2025 [1], with an increase of 12.5% over a decade.

The above data clearly show that the activities associated with the water treatment processes and drinking-water supply activities require high energy consumption. Based on market research and data from the industry for constant- and variable-flow applications, it has been determined that a significant part of clean-water-pump application is for variable flow. However, the number of pumps installed in the water sector with variable-speed drive (VSD) is still limited [1]. Therefore, there would be a large energy-saving potential if VSD pumps were implemented in applications with variable flow.

Drinking-water supply systems (DWSSs) are adapted to the geographic conditions and the requirements of each city. Therefore, DWSSs are uniquely developed and implemented in each city. Commonly, DWSSs include water treatment stations (WTSs), booster pumping stations, storage tanks/reservoirs, and drinking-water distribution systems (e.g., tunnels, aqueducts or pipelines, siphons, valves) [4]. A few researchers have developed methods for optimizing municipal drinking-water distribution networks considering different criteria, but they applied them to hypothetical cases [5]. A literature survey revealed several investigations that have analyzed drinking-water distribution systems, taking into account the specific features of each case [6–8]. The application of multiple criteria (e.g., the financial constraints associated with operational and maintenance costs, the minimization of water residence time, maintaining positive water pressure to avoid contamination, and controlling the pressure in the water distribution network to maintain the hydraulic integrity) for the optimization of municipal drinking-water distribution networks requires taking into account the specific constraints for each case [9,10]. Mathematical models and numerical simulations deliver effective and feasible results to investigate the hydraulic integrity and water quality conditions of DWSSs [11–13]. These predictive capabilities are useful for detecting the integrity loss of DWSSs [4].

WTSs deliver drinking water to the community through supply systems. The major energy consumption of WTSs is associated with the operation of pumping units in the water supply system (WSS). The pumping units installed in the WSS consume up to 90% of the total energy [14,15]. The energy consumption depends upon the pumping-unit type (e.g., constant-speed pumping units (CSPUs) and variable-speed pumping units (VSPUs)) and the selected operation strategy [16]. A significant reduction in energy consumption can be achieved by combining variable-speed pumping units (VSPUs) and constant-speed pumping units (CSPUs) in the operation strategy of WSSs. A remarkable reduction in energy consumption through the pumping units installed in the WSS of Milan city has been reported by Castro-Gama et al. (2017) [17].

The purpose of the investigation conducted in our research study is to select a strategy with reduced operating and maintenance costs for installed pumping units in the water sector. An improved operation strategy for pumping units is developed based on an in-depth analysis of the available infrastructure and water treatment processes conducted in 2019 and then implemented in September 2020 in an infrastructure with four pumping units (two VSPUs and two CSPUs) available in Timisoara pumping station No. 1 (TPS1) in the Bega municipal water treatment station (MWTS). The analyzed data collected in 2021 from the pumping units installed in TPS1 highlight a decrease in operating and maintenance costs.

An overview of the drinking-water supply system and the water treatment stations available in Timisoara city is detailed in Section 2. The layout of TPS1 is presented, highlighting its contribution to the Bega MWTS. The specific energy consumption associated with the operation of the pumping units installed in TSP1 is detailed. Two specific energy consumption indicators (ECIs) are determined to assess the performance of TSP1. Both old and new operating strategies for the four pumping units selected in TPS1 are examined in Section 3. The old operation scenario of the pumping units installed in TSP1 is investigated, revealing its limitations. Then, a new strategy is developed and implemented for the operation of the pumping units installed in TPS1 to better adapt to the operating conditions, improving the specific energy consumption as well as diminishing the failure incidents. The new operation strategy is assessed based on the data acquired from TPS1. The number of hours in service in the last six years for each pumping unit and the total time for all pumping units are analyzed in Section 4. The failure incidents of the pumping units counted in service are enumerated and correlated with the operating conditions of the pumping units. The conclusions and the lessons learned in this case study are drawn in the last section.

2. Drinking-Water Supply System (DWSS) of Timisoara City

At the beginning of the 20th century, the water source for Timisoara city was groundwater captured from deep drilled wells. The water supply infrastructure of Timisoara city was designed and developed by the technical service of the mayor's office under the coordination of engineer Stan Vidrighin. The groundwater source was supplemented with treated water from the Bega River starting with the industrial development of Timisoara city in the second part of the last century [18]. The first development of the surface water treatment station (denoted WTS2), which took water from the Bega River with a designed capacity of 0.12 m^3/s, was carried out at the end of 1959. Several extensions of the WTS2 capacity have been successively implemented: 0.3 m^3/s in 1965 and 0.48 m^3/s in 1976. A new surface-water treatment station (denoted WTS4) with a capacity of 0.9 m^3/s was put into operation in 1982 in the same location as WTS2 [18]. The WTS4 capacity was expanded to 1.24 m^3/s in 1994. As a result, a maximum capacity for both surface-water treatment plants (WTS2 + WTS4) of 1.72 m^3/s was reached.

Nowadays, the drinking-water supply system in Timisoara city is a combination of two sources: WTS2 and WTS4 deliver up to 72% of surface water from the Bega River and the rest from groundwater is provided by WTS1 (called Urseni WTS) and WTS5 (called Ronaț WTS), as shown in Figure 1. Thus, the groundwater is an alternative and complementary source to the primary water surface source for Timisoara city. The water supply infrastructure of Timisoara city has been managed by the municipal company AQUATIM S.A. since 1991 [18].

Figure 1. Drinking-water supply system (DWSS) of Timisoara city together with locations of the water treatment stations (WTS): WTS2 and WTS4 (Bega MWTS) for surface water from the Bega River and WTS1 (Urseni WTS) and WTS5 (Ronaț WTS) for groundwater.

The layout of the Bega MWTS (WTS2 + WTS4) of Timisoara city for surface water from the Bega River is given in Figure 2.

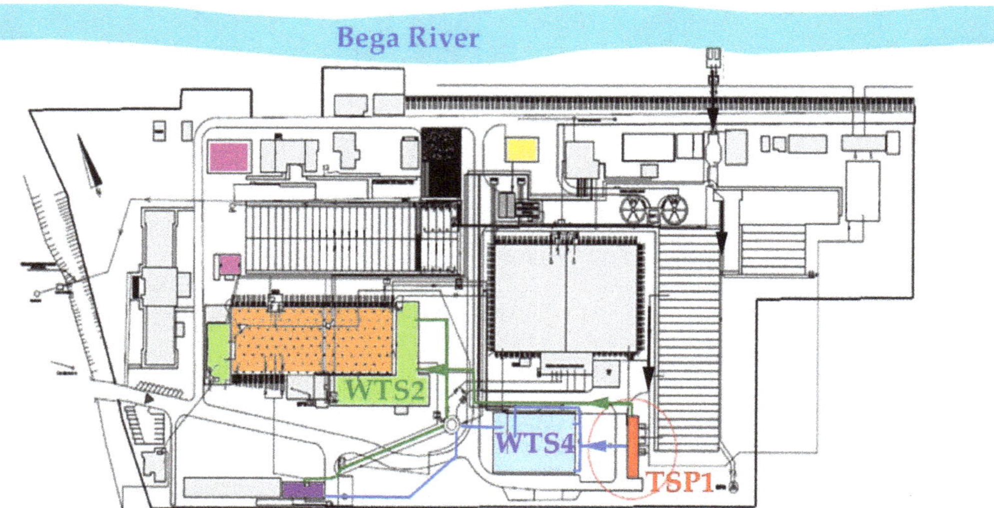

Figure 2. Bega MWTS (WTS2 and WTS4) layout of Timisoara city for surface water collected from Bega River.

The water flow into the Bega MWTS is gravitational along the passage from the river intake to the TSP1 pumping station because it is installed below the Bega River water level (see Figure 2). The water settling stage takes place along this passage to the underground suction tank of the hydraulic pumping units installed in TPS1. Four hydraulic pumping units installed in TPS1 deliver water to the WTS2 and WTS4. Then, the water processed in WTSs is pumped into the DWSS of Timisoara city, as presented in Figure 1.

The energy consumption in the water treatment process of Bega MWTS (WTS2 and WTS4) for three years is shown in Figure 3. One can remark in this figure that around 50% of the energy is consumed in the water distribution pumping process and 33% of energy consumption is associated with the pumping process in TSP1. The energy consumption from 4% to 7% is accounted for by the sludge treatment process, while around 11% is related to other processes (e.g., household reagents, chlorination station, thermal power plant, workshop, and so on). In contrast, the energy consumption in the filtering processes is less than 2% of the total budget.

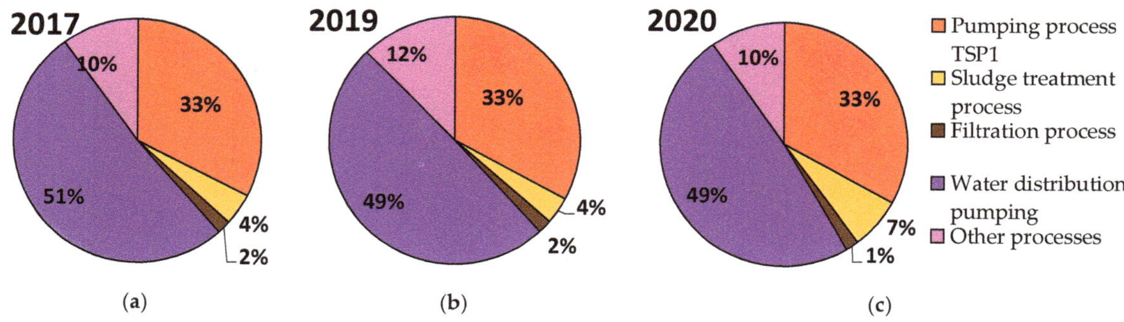

Figure 3. Energy consumption in the water treatment process of Bega MWTS (WTS2 + WTS4) for three years: (**a**) 2017; (**b**) 2019; (**c**) 2020.

The volume of water pumped by TSP1 and treated by WTS2 and WTS4 from 2016 to 2021 is shown in Figure 4a. There was an increase in the volume of water pumped by

TSP1 from 25.45 million m^3 in 2016 to 28.18 million m^3 in 2019. We would like to point out that this increase equates to 10% of the volume of pumped water. Moreover, the volume of water pumped by TSP1 in 2020 and 2021 was practically the same as in 2019. The energy consumption of the pumping units installed in TSP1 for the treatment of water volumes by WTS2 and WTS4 from 2016 to 2021 is shown in Figure 4b. As expected, the energy consumption of the pumping units installed in TSP1 increased with the volume of water processed. The energy consumption of the pumping units increased from 1.3 million kWh in 2016 to 1.51 million kWh in 2019, an increase of 16.3%. In 2020 and 2021 there was a 1% and 2% decrease in energy consumption compared to 2019.

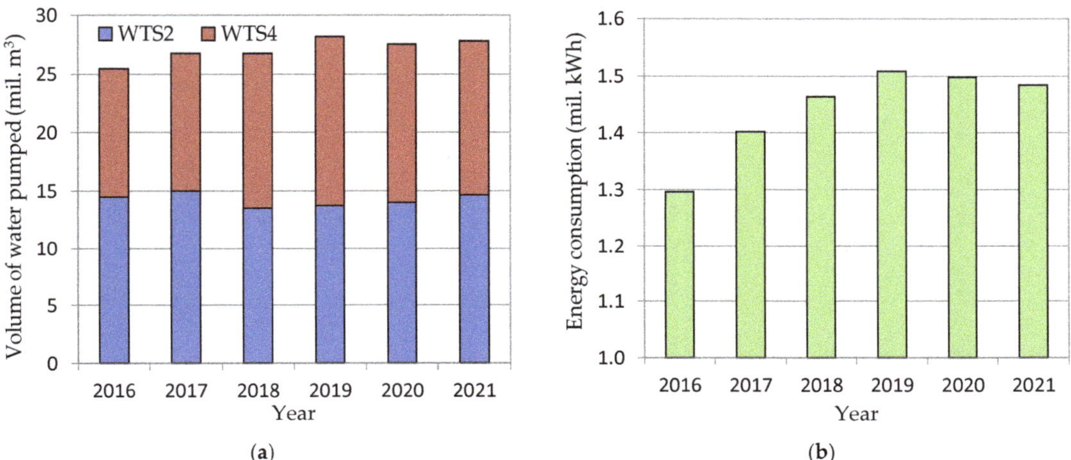

(a)

(b)

Figure 4. (**a**) Annual volume of water pumped by TSP1 and treated by WTS2 and WTS4 from 2016 to 2021; (**b**) annual energy consumption of the pumping units installed in TSP1 from 2016 to 2021.

The annual population equivalents (PE) for Bega MWTS (WST2 + WST4) are given in Figure 5 from 2016 to 2021. There is a monotonous increase in the population equivalents that is served by Bega MWTS every year. An average value of 290,000 population equivalents (PE) can be estimated for Bega MWTS.

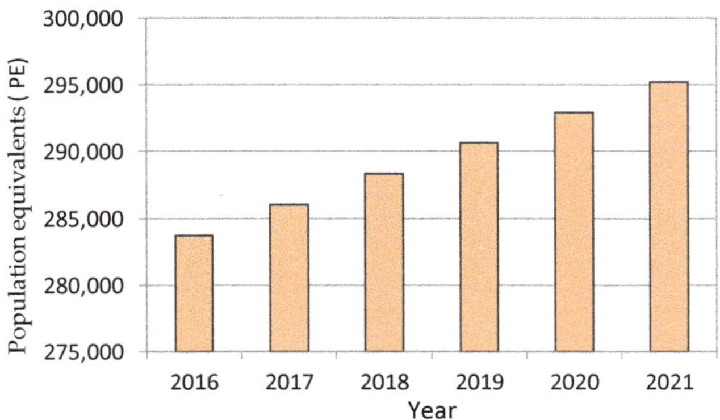

Figure 5. The evolution of the population equivalents (PE) from 2016 to 2021 for Bega MWTS (WST2 + WST4).

Two specific energy consumption indicators (ECIs) are determined to assess the performances of TSP1. The first indicator is the specific energy consumption in relation to the volume of water pumped. The indicator ECI_{m3} is defined as the ratio between the annual energy consumption and the annual volume pumped by units installed in TSP1.

$$ECI_{m3}(kWh/m^3) = \frac{Energy\ consumption\ (kWh/year)}{Volume\ of\ water\ pumped\ (m^3/year)} \tag{1}$$

The second indicator ECI_{PE} is defined as the ratio between the annual energy consumption and population equivalents (PE) served by pumping units installed in TSP1 [19]:

$$ECI_{PE}(kWh/PE/year) = \frac{Energy\ consumption\ (kWh/year)}{Population\ equivalents\ (PE)} \tag{2}$$

The ECI_{m3} indicator was calculated for each year from 2016 to 2021 using Equation (1) with the data included in Figure 4. The data obtained for the pumping units installed in TSP1 for the ECI_{m3} indicator are plotted in Figure 6. The minimum value of 0.0509 kWh/m^3 of the ECI_{m3} indicator was reached in 2016, while the maximum value of 0.0548 kWh/m^3 was reached in 2018. We noticed an improvement in the ECI_{m3} indicator of the specific energy consumption of the pumping process in TSP1 after 2018 by decreasing the indicator to 0.0535 kWh/m^3 in 2019 and 2021, respectively to 0.0544 kWh/m^3 in 2020. There was an improvement of up to 2% in specific energy consumption.

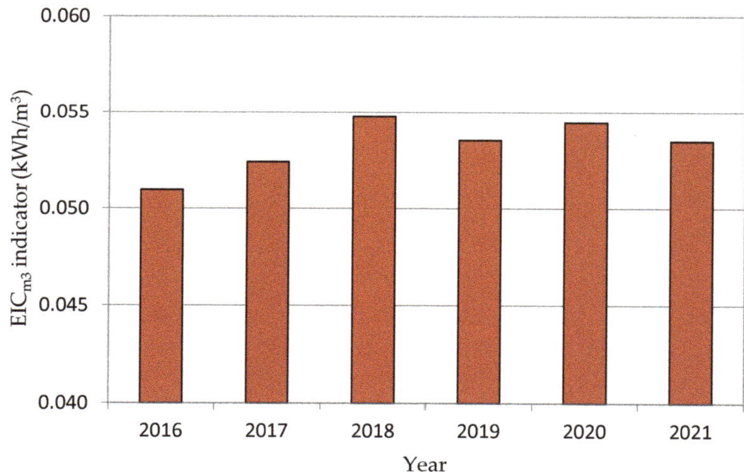

Figure 6. The specific energy consumption indicator ECI_{m3} (kWh/m^3) from 2016 to 2021 for pumping units installed in TSP1.

It is shown in Figure 3 that the energy consumption in TSP1 represents 33% of the energy consumption of the full process of Bega MWTS (WTS2 + WTS4). As a result, a range from 0.154 kWh/m^3 to 0.166 kWh/m^3 was obtained for the six years taken into account for the ECI_{m3} specific energy consumption indicator of the full water treatment process in the Bega MWTS. Benchmarking the energy consumption undertaken for selected regions in Europe shows that electricity consumption for water supply is in the range of 0.5–0.7 kWh/m^3 in Germany, 0.2–0.6 kWh/m^3 in Denmark, and 0.7–0.93 kWh/m^3 in Sweden, with a median value of 0.76 kWh/m^3 [20]. The range of the specific energy consumption indicator ECI_{m3} obtained for the Bega MWTS (WST2 + WST4) is near to the lower limit of the range 0.14 ÷ 0.71 kWh/m^3 indicated by Vaccari et al. (2018) [19] for water treatment plants available in Italy with population equivalents larger than 100,000.

The ECI_{PE} indicator is calculated using Equation (2) with the data included in Figures 4b and 5 from 2016 to 2021. The data for the ECI_{PE} indicator corresponding to the pumping units installed in TSP1 is shown in Figure 7. The minimum value of 4.57 kWh/PE/year of the ECI_{PE} indicator was reached in 2016 while the maximum value of 5.19 kWh/PE/year was reached in 2019. We noticed an improvement in the ECI_{PE} indicator of the specific energy consumption of the pumping process in TSP1 after 2019 with a decrease in the indicator to 5.11 kWh/PE/year in 2020 and 5.03 kWh/PE/year in 2021. There was an improvement in this specific energy consumption indicator in 2021 of over 3% compared to 2019.

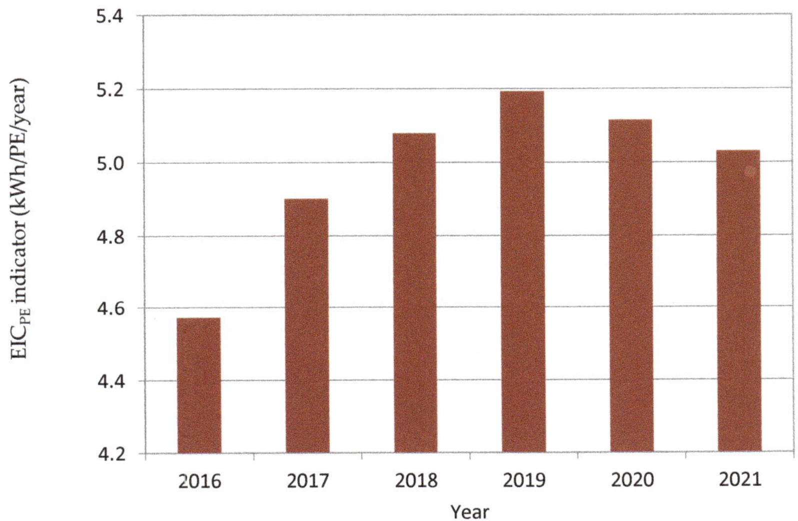

Figure 7. Specific energy consumption indicator ECI_{PE} (kWh/PE/year) from 2016 to 2021 for pumping units installed in TSP1.

The specific energy consumption indicator ECI_{PE} of the full water treatment process in the Bega MWTS (WTS2 + WTS4) covers a range from 13.85 kWh/PE/year to 15.73 kWh/PE/year over the six years taken into account in our investigation. The range limits of the ECI_{PE} indicator are determined based on the fact that the energy consumption in TSP1 represents 33% of the energy consumption of the whole process (see Figure 3). A literature survey led to a reference value for the ECI_{PE} indicator corresponding to large water treatment stations (>100,000 PE), as follows: 18 kWh/PE/year for German water treatment stations [21], 20 ÷ 22.5 kWh/PE/year for northwest European water treatment stations [22], and 23 kWh/PE/year for Italian water treatment stations [19]. The maximum value of 15.73 kWh/PE/year determined for the full water treatment process in Bega MWTS (WTS2 + WTS4) was lower than the reference values specified in the literature. It can be concluded based on the specific indicators determined above that the energy consumption in Bega MWTS is in line with the international recommendations for large WTSs.

An in-depth analysis of the pumping process in TSP1 performed in 2019 led to the identification of limitations in the operation of the pumping units. A new operating strategy for TP1 pumping units was implemented in September 2020. The improvements in the reduced operating and maintenance costs due to the new operating strategy were quantified in 2021. The next section presents the two operating strategies (old and new) of the pumping units installed in TSP1.

3. Timisoara Pumping Station No. 1 (TPS1)

Layout of the TPS1

The water flow along the passage from the river intake to TSP is gravitational. The water settling stage takes place along this passage to the underground suction tank of the hydraulic pumping units installed in TPS1. This underground suction tank is made of concrete with a volume of around 160 m^3 and the bottom is located at -3.55 m below ground level. Four DN 600 suction pipes are installed in the underground suction tank of the TPS1. These four pipes supply four hydraulic pump units (from P1 to P4) installed in TPS1 which deliver water to the WTS2 and WTS4 filters located approximately 10–12 m above ground level. Two DN600 valves are installed upstream and downstream of each hydraulic pump to allow for its maintenance operations. A DN800 butterfly valve installed on the discharge pipe ensures the interconnected or separate operation of the two WTSs. An electromagnetic flow meter and a butterfly valve are mounted on each water transport pipe of DN800 to WTS2 and WTS4 to monitor and adjust the flow rate delivered to the filters. From each discharge pipe the necessary water is taken for the turbidimeter and the residual chlorine analyzer is located in the automation panel in order to monitor water quality.

The value of the flow rate delivered by the pumping units in operation in TPS1 depends on the volumetric flow rate captured from the Bega River. The water level in the suction basin of the TPS1 is monitored by an ultrasonic level transducer. This water level is constantly compared with the water level recorded by the transducer level located on the Bega River. A warning signal is sent to the operator's room when a level of 50 cm is exceeded. The pumps installed in TSP1 are switched off automatically for safety reasons when the minimum level in the suction tank drops to the level where there is a danger of sucking air and/or the risk of developing cavitation phenomena.

Two groups of two pumping units installed in TPS1 are supplied with electricity by two dry transformers of 1000 kVA 20/0.4 kV installed in two transformer cells located in the 20 kV power station. Each pumping unit is driven by a 90 kW/0.4 kV electrical motor. Each group of two hydraulic pumping units is controlled by a frequency converter that provides variable speed for one hydraulic pumping unit (VSPU) and a constant speed for the second one (CSPU). As a result, a technical solution with four pumping units (two VSPUs and two CSPUs) has been implemented in the Timisoara pumping station no. 1 (TPS1) of the Bega municipal water treatment station (MWTS). A SCADA system is installed to monitor and control the operation of the pumping units and the parameters of the water treatment process. The pumping units installed in TSP1 fulfill the EU regulation 2006/42/EC.

The P1 and P2 pumping units (marked with blue) are included in the group G1 while the P3 and P4 pumping units (colored with green) belong to the group G2, as shown in Figure 8a. Each pumping group (G1 and G2) is serviced by a frequency converter to ensure the variable speed operation of one of two available pumping units. Therefore, only one pumping unit in each group can be controlled with variable speed at the same time, Figure 8b. This means that the operation in each pumping group can be performed with a single pumping unit at a variable speed while the other pumping unit is stopped or operating at a maximum speed of 730 rpm. The pumping group (G1 or G2) supplies water towards two water treatment stations (denoted WTS2 and WTS4) (see Figure 2). The volumetric flow rates required for each WTS are different from one to another and vary throughout the day.

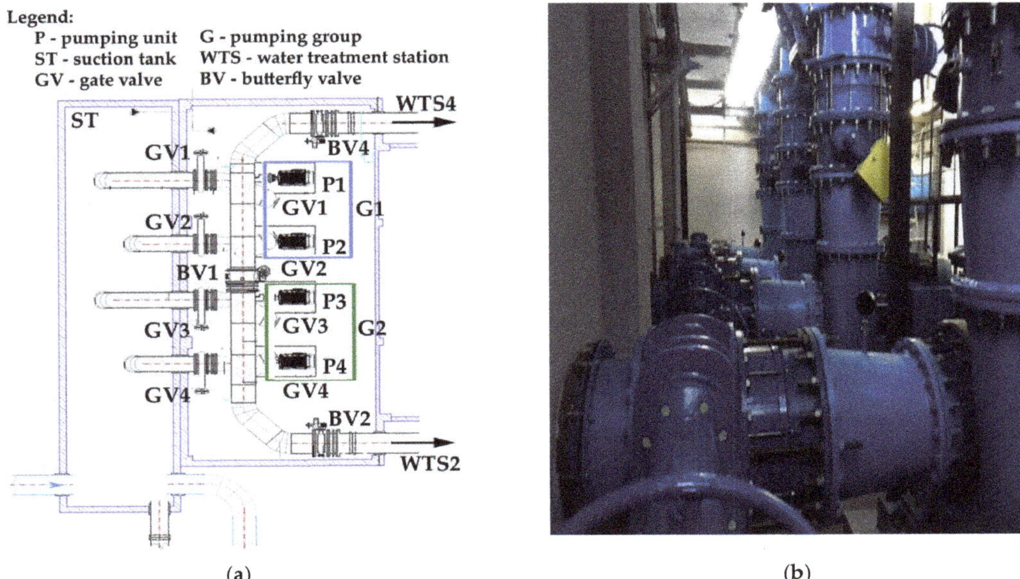

(**a**)   (**b**)

Figure 8. Timisoara Pumping Station no. 1 (TSP1) equipped with four pumping units: (**a**) layout; (**b**) photo.

Since commissioning in 2011, the TSP1 operating procedure has been selected to pump with only one single group (e.g., G1), while the second group is in reserve (G2). This selection of pumping operation in TPS1 implies that the butterfly valve (BV1) located between the pumping groups (G1 and G2) is always opened (see Figure 9) to supply water to both WTS2 and WTS4. This operation procedure in TSP1 is called scenario no. 1, or the old operation strategy.

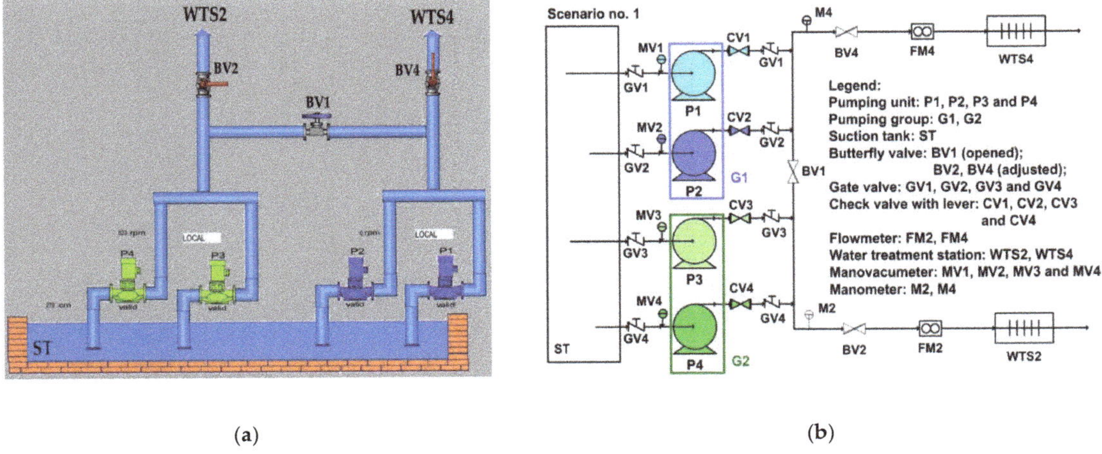

(**a**)   (**b**)

Figure 9. Layout of the TSP1 with two pumping units selected in one group (G1 or G2) in scenario no. 1. Butterfly valve BV1 is always opened while butterfly valves BV2 and BV4 are adjusted to control the volumetric flow rate required by WTS2 and WTS4: (**a**) panel view; (**b**) schematic view.

The performances (pumping head and power versus volumetric flow rate and speed) of the variable-speed pumping units (VSPUs) installed in TSP1 are plotted in Figure 10a while the main parameters are given in Table 1. The main parameters (e.g., speed, volumetric flow rate at best efficiency point, pumping head at best efficiency point and reference diameter) of the pumping units are selected from the manufacturer's documentation [23]. Both the characteristic speed value $n_q \approx 104$ and the dimensionless specific speed value $\nu = 1.112$ are determined according to IEC standard [24]. These values correspond to a mixed-type flow pump, Figure 10b.

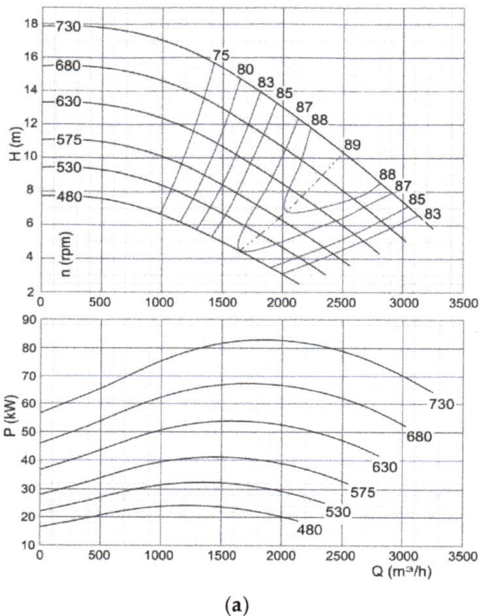

(a)   (b)

Figure 10. Variable-speed pumping unit (VSPU) installed in TSP1: (**a**) performance of the pumping units: pumping head and power vs. volumetric flow rate and speed; (**b**) photo of the pumping unit.

Table 1. Main parameters of the pumping units installed in TSP1.

Parameters of the Pumping Unit	Symbol (Unit)	Value
Maximum speed	n (rpm)	730
Angular velocity	ω (rad/s)	76.45
Volumetric flow rate at best efficiency point	Q_{BEP} (m^3/s)	0.694
Pumping head at best efficiency point	H_{BEP} (m)	10.5
Reference diameter	$D_{ref} = 2 \cdot R_{ref}$	0.478
Characteristic speed	$n_q = nQ_{BEP}^{\frac{1}{2}}/H_{BEP}^{\frac{3}{4}}$ (-)	104.3
Discharge coefficient	$\varphi = Q_{BEP}/\left(\pi\omega R_{ref}^3\right)$ (-)	0.212
Head coefficient	$\psi = gH/\left(\omega^2 R_{ref}^3\right)$ (-)	0.308
Dimensionless specific speed	$\nu = \varphi^{\frac{1}{2}}/\psi^{\frac{3}{4}}$ (-)	1.112

In this case, the hydraulic parameters delivered by the pumping group in operation to each WTS are directly correlated with the losses along the hydraulic passage. In the normal operation of the TPS1 station, the discharge value delivered to WTS4 is higher than the value delivered to WTS2 due to the fact that the hydraulic losses along the passage to WTS4 are smaller than that to WTS2. Therefore, the hydraulic parameters required by WTS2 are the ones provided by the pumping group while the hydraulic parameters for WTS4 are controlled by the BV4 butterfly valve. The operating scenario changes when the hydraulic parameters required by WTS4 are modified due to the filtration system. In these cases, the discharge value delivered to WTS2 is higher than the value delivered to WTS4. Therefore, the hydraulic parameters required by WTS4 are the ones to be provided by the pumping group while the hydraulic parameters for WTS2 are controlled by the BV2 butterfly valve. This selection of operation of pumps from the same group in TPS1 has two major drawbacks: (1) part of the energy supplied by the pumping group is lost in the butterfly valves which are controlled to ensure the hydraulic parameters required by each WTS; and (2) several events have been recorded which resulted in the failure of various mechanical components of the hydraulic pump due to the operating conditions beyond the limits prescribed by the pump designer (see Section 4.2).

A new operating strategy for the pumping units installed in TSP1 has been developed based on in situ investigations conducted in 2019. The new operating strategy was implemented in TSP1 in September 2020 to better adapt to the operating conditions by improving specific energy consumption and reducing the failure incidences. The newer operating strategy for TSP1 pumping units is called scenario no. 2 and involves operation with the butterfly valve (BV1) closed, as in Figure 11.

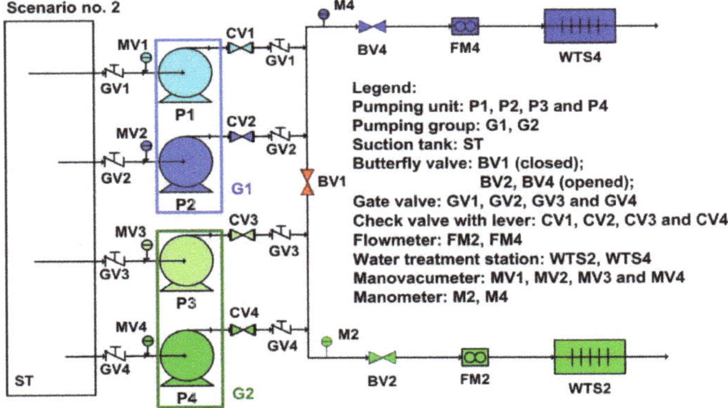

Figure 11. Layout of the TSP1 with two pumping units (one pumping unit from G1 and one pumping unit from G2) operating in scenario no. 2 implemented in September 2020.

In this scenario, the pumping group no. 1 serves the WTS4 and the pumping group no. 2 supplies water to the WTS2. In other words, one pumping unit (e.g., P1 or P2) corresponding to the G1 pumping group operates with variable speed to supply water to WTS4, and one pumping unit (e.g., P3 or P4) available to the G2 pumping group delivers water to WTS2. Each pumping unit independently adjusts its volumetric flow rate according to the requirements of each water treatment station. One pumping unit from each pumping group is in reserve in scenario no. 2. As a result, the energy losses from adjusting the butterfly valves in scenario no. 1 are eliminated in scenario no. 2. Part of this energy recovered through the implementation of scenario no. 2 in TSP1 can be found in the improved values of the specific energy consumption indicators (ECI_{m3} and ECI_{PE}) for 2021. It is certain that the operating and maintenance costs due to failure incidents in the operation of the pumping units have decreased, as no incidents were reported in 2021. An analysis of the failure incidents in the operation of the pumping units installed in TSP1 can be found in Section 4.3.

4. Analysis of the Pumping Units Installed in TPS1 in Both Scenarios

4.1. Time Operation Analysis of the Pumping Units Installed in TPS1

The operating time corresponding to the pumping units P1 (light blue), P2 (blue), P3 (light green) and P4 (blue) from 2016 to 2021 are presented in Figure 12. The cumulative operating time in service for all pumping units is 1200 ± 250 h per month from 2016 to 2019, except for December 2018. It can be noticed that all four pumping units were in operation during this time. This means that the pumping groups were in successive operation. Water pumped to both WTS2 and WTS4 was provided with one or two pumping units available in a single group (e.g., G1). Then, the pump(s) in the other group (e.g., G2) were only operational when the first pumping group (G1) was stopped.

There has been an increasing trend in drinking-water consumption since May 2020 due to the SARS-CoV-2 pandemic. In particular, one can notice a cumulative operating time of the pumping units installed in TSP1 of over 1350 h every month in 2021, except for February and November.

The cumulative operating time in service per year of each pumping unit installed in TSP1 is shown in Table 2 (column marked with grey). The cumulative operating time per year of all pumps installed in TSP1 increased monotonically from 14,448 h in 2016 to 17,230 h in 2021. There was an increase in the cumulative operating time in service for all four pumps installed at TSP1 of 2782 h over the six years. The cumulative operating time per year of all pumping units installed in TSP1 computed as a percentage of the total available time is determined in the last column of Table 2. It can be noticed that the cumulative operating time per year for all pumps computed as a percentage from the total available time increased from 41.2% to 49.2% in last six years. We emphasize that this 8% represents the increase in the number of operating hours in the last six years for the pumps installed in TSP1 from the total available hours. The mean operating time for the last six years of all pumping units installed in TSP1 is 44%. This percentage value of 44% is slightly lower than the 50% taken into account at the design stage.

The cumulative operating time for each pumping unit installed in TSP1 in the last six years is presented in Table 2 (row marked with grey). As expected, each pumping unit installed in TSP1 had a different number of operating hours. It can be seen that the cumulative operating time in the last six years was shorter for pumping units no. 4 (18,522 h) and no. 1 (22,593 h) and higher for pumping units no. 3 (25,820 h) and no. 2 (25,676 h). It is noted that, in each pumping group, there was a pumping unit with a large number of operating hours (e.g., P2 in G1 and P3 in G2) and a pumping unit with a small number of operating hours (e.g., P1 in G1 and P4 in G2). Both pumping units with a large number of operating hours had a "quiet" operation being the first option in the selection of the operator. The "quiet" operation of the two pumping units was due to their central location in TSP1 (see Figure 8a).

Table 2. Cumulative operating time in service for each pumping unit installed in TSP1.

Pumping Group	G1		G2			
Year (Y) \ Pumping Unit (P)	P1	P2	P3	P4	Total Time/Y (h)	Total Time/Y (%)
2016	4328	4265	2931	2924	14,448	41.1
2017	4165	4431	3243	2647	14,486	41.3
2018	2703	2697	4725	4827	14,952	42.7
2019	4453	4380	3311	3276	15,420	44.0
2020	4750	3488	3327	4510	16,075	45.8
2021	2194	6415	8283	338	17,230	49.2
Total time/P (h)	22,593	25,676	25,820	18,522	92,611	44.0
Total time/P (%)	42.9	48.8	49.1	35.2	44.0	

The cumulative operating time of each pumping unit installed in TSP1 computed as a percentage of the total available time is determined in the last row of Table 2. It can be noticed that the cumulative operating time values for the two pumping units computed as a percentage of the total available time are 48.8% and 49.1% for P2 and P3, respectively. These percentage values around 49% are slightly lower than the 50% taken into account at the design stage.

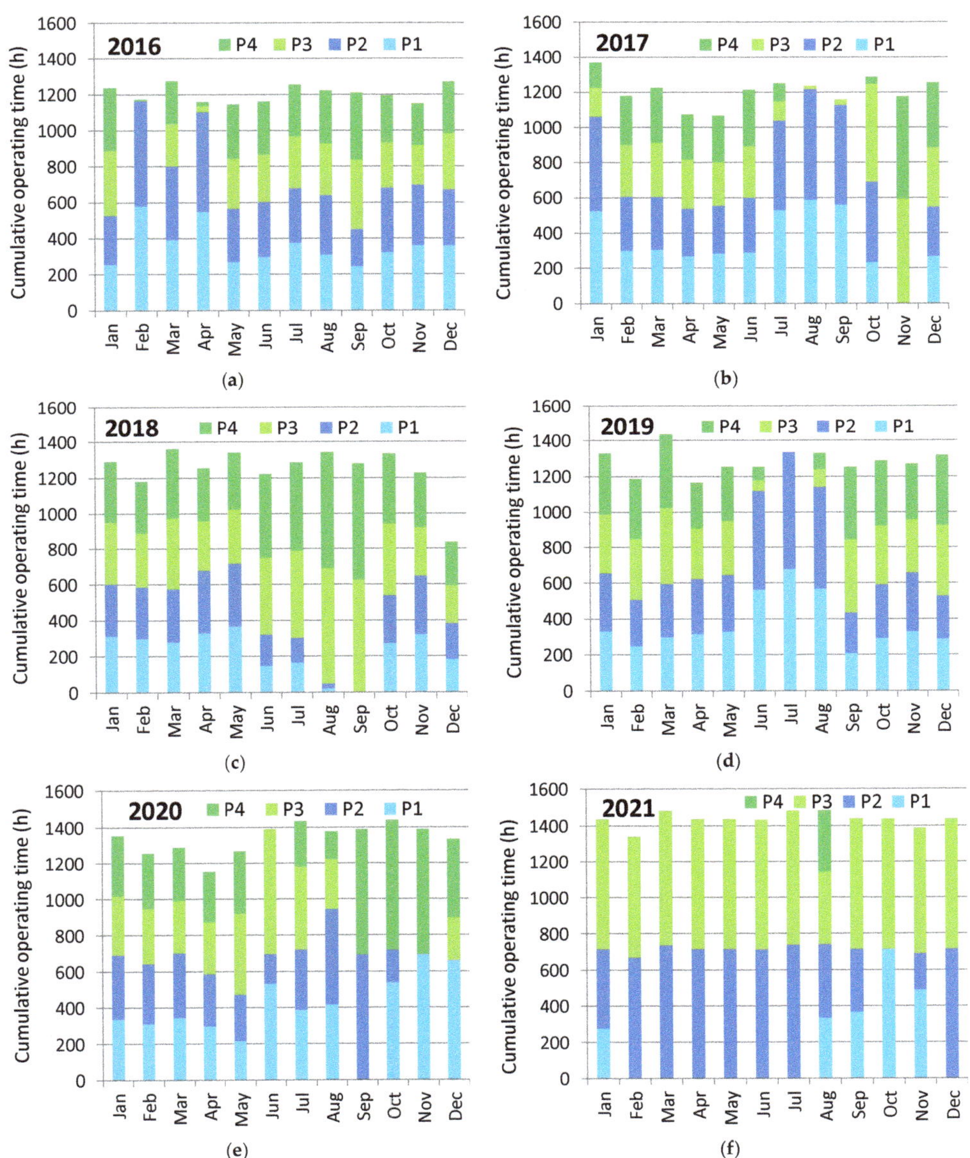

Figure 12. In situ operating time in hours for the pumping units installed in TSP1: (**a**) 2016; (**b**) 2017; (**c**) 2018; (**d**) 2019; (**e**) 2020; (**f**) 2021.

4.2. Operating Conditions of the Pumping Units Installed in TSP1

Figure 8a shows that two pumping groups (G1 and G2) make up TSP1. It has already been underlined that a pumping group is equipped with two pumping units (e.g., P1 and

P2 belong to G1, while P3 and P4 are part of G2). The pumping units installed in TSP1 are operated in two ways: (1) with a variable-speed pumping unit (VSPU) and (2) with a constant-speed pumping unit (CSPU), respectively.

The operation scenario no. 1 in TSP1 involves the variable-speed operation of one pumping unit within a pumping group while the butterfly valve (BV1) is open (see Figure 9b). This means that one single pumping group supplies water for both WTS2 and WTS4. The increase in water requirement for WTS2 and WTS4 over the volumetric flow rate provided by the pumping unit in operation is ensured by starting the second pumping unit available in the same pumping group. Operation with two pumping units installed in the same pumping group means operation with a variable-speed pump unit (VSPU) and a constant-speed pump unit (CSPU) because there is only one frequency converter for each pumping group. In this scenario, the volumetric flow rate balance between WTS2 and WTS4 is achieved by adjusting the butterfly valves (BV2 and BV4) installed on the pipes that supply them. In this scenario, the second pumping group is in reserve. Two short-comings are associated with the operation of the pumping units in TSP1 in scenario no. 1: (1) the energy losses from adjusting the butterfly valves; and (2) the failure incidents in the operation of the pumping units that increase operating and maintenance costs.

A careful analysis of the operation of variable-speed pumping units (VSPUs) installed in TSP1 has been conducted. The pumping head curves H(Q) for variable-speed pumping units (VSPUs) installed in TSP1 are given in Figure 13.

This VSPU operation corresponds with both scenarios no. 1 and no. 2. The solid lines of the pumping head for six speed values (480 rpm (orange), 530 rpm (light blue), 575 rpm (magenta), 630 rpm (light green), 680 rpm (dark red), and the maximum value of 730 rpm (blue) recommended by the manufacturer are plotted based on the experimental data provided by the manufacturer for VSPU. The red stripe marked on Figure 13 corresponds to the VSPU maximum speed of the 780 rpm setup in the frequency converter to start the second pumping unit available in the same pumping group when operating in scenario no. 1. The red stripe marked in Figure 13 is no longer valid for scenario no. 2 due to the second pumping unit not starting. This is one of the main advantages of scenario no. 2.

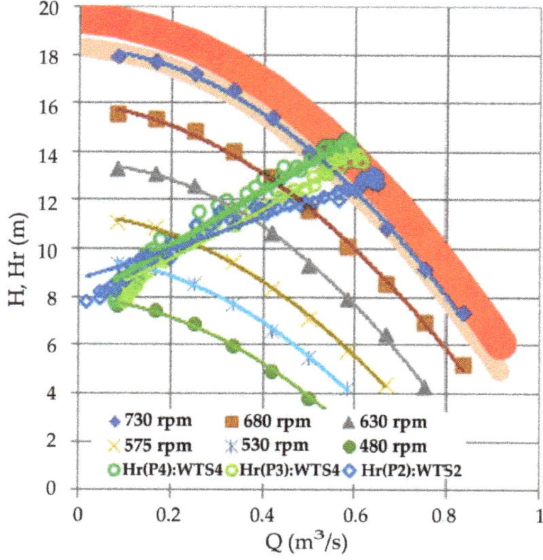

Figure 13. Pumping head curves H(Q) for variable-speed pumping unit (VSPU) installed in TSP1 and the hydraulic losses Hr(Q) corresponding to the hydraulic passages of WTS2 (green circles) and WTS4 (blue diamond).

In situ experimental investigations performed on three pumping units installed in TSP1 are marked with circle symbols in Figure 13. The curve for the hydraulic losses Hr(Q) corresponding to the hydraulic passage of each water treatment station (WTS) is determined with the butterfly valve closed (see BV1 in Figure 9b). Then, the hydraulic losses Hr(Q) associated with the hydraulic passage of WTS2 are determined with the pumping unit no. 3 (light green circle symbol ◯ in Figure 13) and the pumping unit no. 4 (green circle symbol ◯ in Figure 13), respectively. On the other hand, the hydraulic losses Hr(Q) corresponding to the hydraulic passage of WTS4 are experimentally obtained with the pumping unit no. 2 (blue diamond symbol ◇ in Figure 13) in the same operating conditions with the butterfly valve closed. Unfortunately, the determination of the hydraulic losses Hr(Q) corresponding to the hydraulic passage of WTS4 with the pumping unit no. 1 was not possible due to suction pressure sensor failure.

Each pumping unit (P3 or P4) in variable-speed operation delivers a volumetric flow rate value up to 0.55 m^3/s at a maximum speed of 730 rpm for WST2. The P2 variable-speed pumping unit operation provides a volumetric flow value up to 0.6 m^3/s at a maximum speed of 730 rpm for WST4. It is important to note that the same pump units deliver a 10% higher flow rate for WTS4 than WTS2. This is because the hydraulic passage associated with WTS4 is shorter than WTS2. As a result, the operation in scenario no. 1 with one pumping unit with variable speed (VSPU) or two pumping units available in the same pumping group (one pumping unit with constant speed (CSPU) and one pumping unit with variable speed (VSPU)), and the butterfly valve (BV1) in the open position, leads to the dissipation of part of the hydraulic energy by adjusting the butterfly valves (BV2/BV4) on the hydraulic passages of the water treatment stations (WTS2 and/or WTS4). In addition, the operation of pumping units at speed values higher than 730 rpm leads to issues that increase operating and maintenance costs.

Next, a deep analysis of the operation of variable-speed pumping units (VSPUs) together with constant-speed pumping units (CSPUs) is given. This combined operation of VSPUs together with CSPUs is specific to scenario no. 1. The pumping head curves H(Q) for VSPU + CSPU combined operation are given in Figure 14. The black solid line corresponds to CSPU operation at 730 rpm. The solid lines of the VSPU pumping head for six speed values (480 rpm (orange), 530 rpm (light blue), 575 rpm (magenta), 630 rpm (light green), 680 rpm (dark red), and a maximum value of 730 rpm (blue)) are plotted in combination with CSPU.

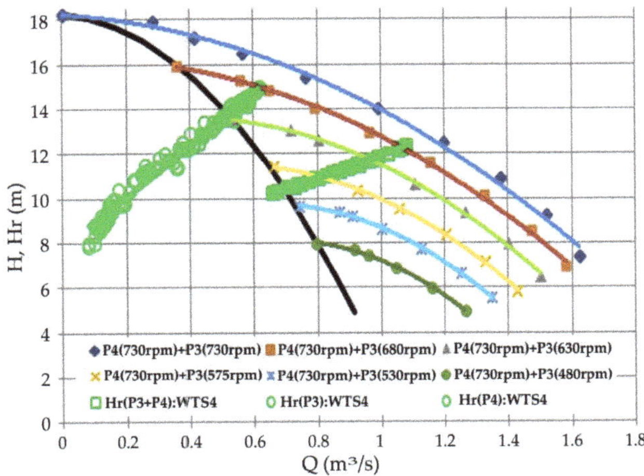

Figure 14. Pumping head curves H(Q) for combined variable-speed pumping unit (VSPU) together with a constant-speed pumping unit (CSPU) installed in TSP1 and the hydraulic losses Hr(Q) corresponding to the hydraulic passages of WTS4.

In situ experimental investigations conducted on combined operation P3 (VSPU) + P4 (CSPU) are marked with square symbols □ in Figure 14. The curve for the hydraulic losses Hr(Q) corresponding to the hydraulic passage of WTS2 is determined with the butterfly valve closed (see BV1 in Figure 9b). In situ experimental data determined for P3 and P4 under the VSPU operation plotted in Figure 13 is added with circle symbols ○ in Figure 14. Unfortunately, the determination of the hydraulic losses Hr(Q) corresponding to the hydraulic passage of WTS4 in P1 (VSPU) + P2 (CSPU) combined operation was not possible due to pressure sensor failure on the suction of pumping unit no. 1 (P1).

The parallel operation of P3 (VSPU) + P4 (CSPU) pumping units cover the volumetric flow rate range from 0.7 m^3/s to 1.2 m^3/s required for WST2 + WST4 operation in scenario no. 1. The volumetric flow rate range specified above is ensured by the operation of the pumping unit no. 3 (P3) on a speed range between 550 rpm and 730 rpm and the pumping unit no. 4 (P4) operates at a constant speed of 730 rpm.

4.3. Major Incidents Recorded in the Operation of the Pumping Units Installed in TPS1

The operation of the pumping units in TSP1 has been affected over time by incidents. Listed below are ten major incidents that affected the pumping units installed in TSP1 during the period 2016–2021 that have been taken into account in our analysis.

The distribution of major incidents on the pumping units installed in TSP1 is shown in Figure 15. The following distribution of major incidents is noted: P1—one incident in 2017; P2—two incidents in 2018 and 2020; P3—four incidents in 2016; and P4—three incidents in 2017, 2019 and 2020.

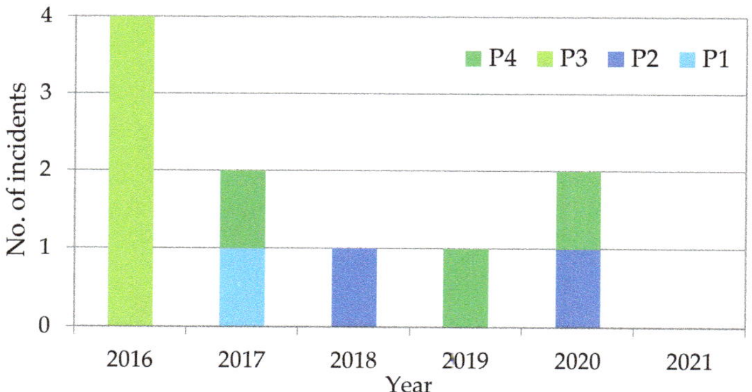

Figure 15. Major incidents encountered in service during 2016–2021 for the pumping units installed in TPS1.

Three major incidents of single-stage pumping units installed in TSP1 are detailed below. First, the major failure of the mixed-flow-type impeller of the pumping unit no. 3 registered on 9 March 2016 is shown in Figure 16. This major failure corresponds to incident no. 3 in Table 3.

The impeller failure of pumping unit no. 3 recorded on 9 March 2016 was found in the vicinity of the trailing edge of the missing part of the blade (see Figure 16). We note that the impeller of these pump units was manufactured with stainless steel 316SS. The impeller was repaired but the cause of the incident has not yet been identified.

Figure 16. Impeller failure of the pumping unit no. 3 on 9 March 2016.

Table 3. Major incidents encountered in the service of the pumping unit installed in TSP1.

No.	Pumping Unit	Date	Pumping Unit Incident
1	P3	15 January 2016	Sealing leaks
2	P3	23 February 2016	Sealing leaks
3	P3	9 March 2016	Impeller failure, sealing leaks
4	P3	4 April 2016	Ball bearings failure
5	P4	11 July 2017	Ball bearings failure
6	P1	9 October 2017	Shaft failure
7	P2	9 October 2018	Ball bearings failure
8	P4	21 June 2019	Bearing heating
9	P2	27 May 2020	Bearing heating
10	P4	9 October 2020	Sealing leaks

There have been several failures of ball bearings in the operation of the pumping units at TSP1 (see Figure 17). Wear and even breakage of the bearing balls was found when inspecting the bearing box after noises were heard in the operation of the pumping units. Operating at speeds higher than those recommended by the designer, eccentricities and shaft alignment deviations caused the pump bearings to overload. The implementation of scenario no. 2 in September 2020 eliminated the operation of pumping units with speeds higher than those recommended by the designer.

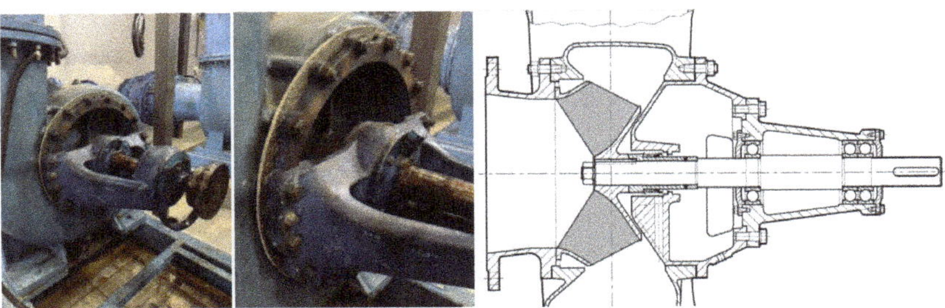

Figure 17. Ball bearing failure of the pumping unit no. 2 on 9 October 2018.

A small clearance (0.5–1 mm) between the shroud front of the impeller and the suction housing is recommended by the manufacturer to obtain minimum internal leakage from the high-pressure area (discharge flow at volute) of the pump to the low-pressure area (eye of the impeller). This clearance is controlled by adjusting the number of seals or the thickness of the sealed box gasket. The lateral loading of the impeller and the axial movement of the rotating assembly of the pump lead to wear on the front surfaces of the shroud impeller and the pump housing (see Figure 18). A new sealing kit is needed each time to fix problems of this type.

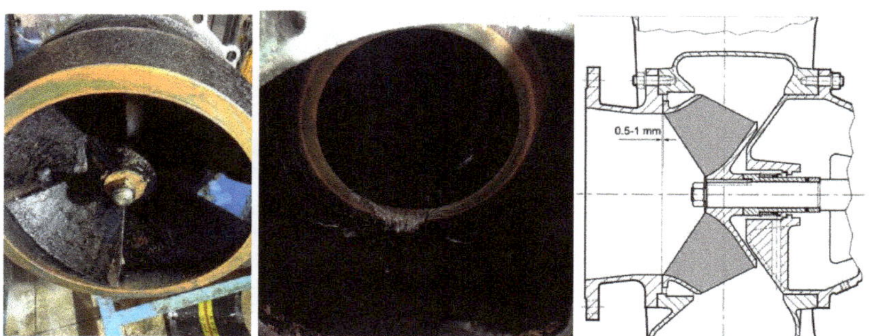

Figure 18. Sealing failure of the pumping unit no. 4 on 9 October 2020.

It is important to note that in 2021 no incidents were recorded in the operation of the pumping units. All the information and data included in this case study provide lessons regarding the service and maintenance activities for the pumping units, and offer valuable support for other cases in the water sector.

5. Conclusions

The paper focuses on the issues associated with the drinking-water infrastructure of large cities. The key elements of the water infrastructure corresponding to the drinking-water supply system and water treatment stations (MWTSs) of Timisoara city are presented. The Bega municipal water treatment station (MWTS) is considered as a case study for our investigation. The main components of the water treatment process included in the Bega MWTS are shown.

Two specific energy consumption indicators (ECIs) associated with water treatment processes from the Bega MWTS are determined. Both the ECI_{m3} and ECI_{PE} indicators are determined for the Bega MWTS from 2016 to 2021, covering the pandemic period. The ECI_{m3} indicator corresponded to a range of 0.154 kWh/m^3 to 0.166 kWh/m^3 over the six years taken into account for the full process of the Bega MWTS. The maximum value of the ECI_{m3} specific energy consumption indicator for Bega MWTS corresponded to the reference value for water treatment plants with population equivalents larger than 100,000 available in Europe (e.g., Italy, Germany, Scandinavian countries).

It has been identified, based on recorded data from the Bega MWTS, that 83% of the energy consumption corresponds to the pumping activities (50% of the energy is consumed in the water distribution pumping process and 33% of the energy consumption is associated with the pumping process in TSP1). As a result, the ECI_{m3} indicator associated with the energy consumption in the pumping process in TSP1 is improved up to 2% in the last three years. The ECI_{PE} indicator corresponding to the pumping units installed in TSP1 showed an improvement in 2021 of over 3% compared to 2019. This slight reduction in energy consumption was also due to the implementation of the new operating strategy of the pumping units installed in TSP1.

The number of hours in service for each pumping unit and the total time for all pumping units were examined for the last six years. It was determined that the cumulative operating time per year for all pumping units, computed as a percentage from total available time, has increased from 41.2% to 49.2%. It is emphasized that this 8% represents the increase in the number of operating hours in the last six years of the pumps installed in TSP1. The mean operating time over the last six years for all pumping units installed in

TSP1 is 44%. This percentage value of 44% is slightly lower than the 50% taken into account at the design stage. It is noticed that there has been an increasing trend in drinking-water consumption since May 2020 due to the SARS-CoV-2 pandemic.

The layout of the TPS1 has been detailed, and its contribution to the Bega MWTS has been highlighted. A technical solution with four pumping units (two VSPUs and two CSPUs) is implemented in TPS1. This old operating strategy, labeled scenario no. 1, has been implemented in TSP1 since 2011. This operating procedure involves only one group in service, while the second group is in reserve. The selection of pumping operation in TPS1 implies that the butterfly valve (BV1) located between pumping groups G1 and G2 is always open and both water treatment plants (WTS2 + WTS4) are supplied by a single pumping group (with one or two pumping units). The failure incidents of the pumping units counted in service were enumerated and correlated with operating conditions. The main drawback of this operating scenario, identified over time, is the high maintenance and repair cost of the pumping units. It was identified that the high maintenance and repair costs of the pump units in this scenario are associated with their operation beyond the limit set by the manufacturer.

A new operating strategy, labeled scenario no. 2, was developed based on in situ investigations conducted in 2019. The new operating strategy was implemented in TSP1 in September 2020 to better adapt to in situ operating conditions, improving the specific power consumption as well as diminishing the failure incidents. The new operating strategy for TSP1 pumping units involves operation with the butterfly valve (BV1) closed. As a result, the pumping group G1 serves the WTS4 and the pumping group G2 supplies water to the WTS2. One pumping unit corresponding to the G1 pumping group operates with variable-speed drive to supply water to WTS4 and one pumping unit with variable-speed drive is available in the G2 pumping group to deliver water to WTS2. Each pumping unit independently adjusts its volumetric flow rate according to the requirements of each water treatment station it supplies. One pumping unit from each pumping group is in reserve in scenario no. 2. The energy losses from adjusting the butterfly valves in scenario no. 1 are not found in scenario no. 2. Therefore, it is expected that part of this energy recovered through the implementation of scenario no. 2 in TSP1 is found in the improved values of the specific energy consumption indicators (ECI_{m3} and ECI_{PE}) for 2021. It is certain that the operating and maintenance costs due to failure incidents in the operation of the pumping units were decreased, as no incidents were reported in 2021. The lessons learned in this case study are useful for other cases in the water sector.

The AQUATIM Company operates a fleet of hundreds of hydraulic pumping units that cover a quite wide range of powers, from a few kilowatts to megawatts, installed in dozens of water treatment stations. Further work will focus on investigating the operating conditions of pumping systems available in the water treatment plants administrated by AQUATIM in order to reduce operating and maintenance costs using the lessons learned in this case study.

Author Contributions: Conceptualization, S.M.; methodology, S.M.; validation, I.A.D., I.-D.R., A.C. and S.M.; investigation, I.A.D., I.-D.R., A.C. and S.M.; resources, I.A.D., A.C. and S.M.; data curation, I.A.D., A.C. and S.M.; writing—original draft preparation, I.A.D., I.-D.R., A.C. and S.M.; writing— review and editing, I.A.D., A.C. and S.M.; supervision, S.M.; project administration, I.A.D. and S.M. All authors have read and agreed to the published version of the manuscript.

Funding: This research received no external funding.

Institutional Review Board Statement: Not applicable.

Informed Consent Statement: Not applicable.

Data Availability Statement: Not applicable.

Acknowledgments: The author affiliated with Romanian Academy—Timisoara Branch has been supported by the research program of Hydrodynamic and Cavitation Laboratory from Center for Fundamental and Advanced Technical Research.

Conflicts of Interest: The authors declare no conflict of interest.

References

1. Maagøe, V.; Van Holsteijn en Kemna, B.V. *Ecodesign Pump Review Study of Commission Regulation (EU) No. 547/2012 (Ecodesign Requirements for Water Pumps) Extended Report—Final Version*; European Commission: Brussels, Belgium, 2018.
2. U.S. Environmental Protection Agency. Strategies for Saving Energy at Public Water Systems, EPA report 816-F-13-004, EPA Office of Water. 2013. Available online: http://water.epa.gov/type/drink/pws/smallsystems/upload/epa816f13004.pdf (accessed on 20 July 2022).
3. EURACTIV: How Europe Can Make Its Water Sector Energy Neutral. 2018. Available online: https://www.euractiv.com/section/energy-environment/opinion/how-europe-can-make-its-water-sector-energy-neutral/ (accessed on 20 July 2022).
4. National Academies of Sciences, Engineering, and Medicine. *Drinking Water Distribution Systems: Assessing and Reducing Risks*; The National Academies Press: Washington, DC, USA, 2006. [CrossRef]
5. McCormick, G.; Powell, R.S. Optimal Pump Scheduling in Water Supply Systems with Maximum Demand Charges. *J. Water Resour. Plann. Manage.* **2003**, *129*, 372–379. [CrossRef]
6. Georgescu, A.-M.; Georgescu, S.-C.; Petrovici, T.; Culcea, M. Pumping stations operating parameters upon a variable demand, determined numerically for the water distribution network of Oradea. *U.P.B. Sci. Bull. Ser. C* **2007**, *69*, 643–650.
7. Georgescu, S.-C. HBMOA applied to design a water distribution network for a town of 50000 inhabitants. *U.P.B. Sci. Bull. Ser. D* **2012**, *74*, 91–102.
8. Georgescu, S.-C.; Georgescu, A.-M.; Madularea, R.A.; Piraianu, V.-F.; Anton, A ; Dunca, G. Numerical model of a medium-sized municipal water distribution system located in Romania. *Procedia Eng.* **2015**, *119*, 660–668. [CrossRef]
9. Gullick, R.W.; LeChevallier, M.W.; Svindland, R.C.; Friedman, M.J. Occurrence of transient low and negative pressures in distribution systems. *J. Amer. Water Work. Assoc.* **2004**, *96*, 52–66. [CrossRef]
10. Cromwell, J.; Nestel, G. ; Albani. R. *Financial and Economic Optimization of Water Main Replacement Programs*; AwwaRF: Denver, CO, USA, 2001.
11. Lansey, K.E.; Boulos, P.F. *Comprehensive Handbook on Water Quality Analysis for Distribution Systems*; MWH Soft Pub: Pasadena, CA, USA, 2005.
12. Panguluri, S.; Grayman, W.M.; Clark, R.M. *Distribution System Water Quality Report: A Guide to the Assessment and Management of Drinking Water Quality in Distribution Systems*; EPA Office of Research and Development: Cincinnati, OH, USA, 2005.
13. Boulos, P.F.; Lansey, K.E.; Karney, B.W. *Comprehensive Water Distribution Systems Analysis Handbook for Engineers and Planners*, 2nd ed.; MWH Soft Pub: Pasadena, CA, USA, 2006.
14. Electric Power Research Institute (EPRI). *Water & Sustainability (Volume 4): U.S. Electricity Consumption for Water Supply & Treatment—The Next Half Century*; Electric Power Research Institute (EPRI): Washington, DC, USA, 2002.
15. Magagna, D.; Hidalgo González, I.; Bidoglio, G.; Peteves, S.; Adamovic, M.; Bisselink, B.; De Felice, M.; De Roo, A.; Dorati, C.; Ganora, D.; et al. *Water—Energy Nexus in Europe*; Publications Office of the European Union: Luxembourg, 2019; ISBN 978-92-76-03385-1. [CrossRef]
16. Plappally, A.K.; Lienhard, V.J.H. Energy requirements for water production, treatment, end use, reclamation, and disposal. *Renew. Sust. Energ. Rev.* **2012**, *16*, 4818–4848. [CrossRef]
17. Castro-Gama, M.; Pan, Q.; Lanfranchi, E.A.; Jonoski, A.; Solomatine, D.P. Pump scheduling for a large water distribution network. Milan, Italy. *Procedia Eng.* **2017**, *186*, 436–443. [CrossRef]
18. Vlaicu, I.; Haţegan, I. *Timisoara's Water Supply System: History, Present and Perspectives*; Brumar Publishing House: Timisoara, Romania, 2012. (In Romanian)
19. Vaccari, M.; Foladori, P.; Nembrini, S.; Vitali, F. Benchmarking of energy consumption in municipal wastewater treatment plants—A survey of over 200 plants in Italy. *Water Sci. Technol.* **2018**, *77*, 2242–2252. [CrossRef] [PubMed]
20. Ronen, P.; Jacobsen, B.N. *Performance of Water Utilities beyond Compliance*; EEA Technical report No 5 (Appendix 2); EEA: Brussels, Belgium, 2014.
21. Baumann, P.; Roth, M. *Senkung des Stromverbrauchs auf Kläranlagen, Leitfaden für das Betriebspersonal (Reduction of the Energy Consumption of WWTPs—Manual for Operators)*; Heft 4 (In German); DWA Landesverband Baden-Württemberg: Stuttgart, Germany, 2014.
22. Torregrossa, D.; Schutz, G.; Cornelissen, A.; Hernández-Sancho, F.; Hansen, J Energy saving in WWTP: Daily benchmarking under uncertainty and data availability limitations. *Environ. Res.* **2016**, *148*, 330–337. [CrossRef] [PubMed]
23. Apex Fluid Engineering Ltd. *Apex AMF Mixed Flow Pumps*; Datasheet TCAMF-2001 Rev. 1; Apex Fluid Engineering Ltd.: Bristol, UK, 2003.
24. International Electrotechnical Commision (IEC). *Hydraulic Turbines, Storage Pumps and Pumps-Turbines Model Acceptance, International Standard IEC No. 60193*, 2nd ed.; International Electrotechnical Commision (IEC): Lausanne, Switzerland, 1999.

 *sustainability*

Article

Optimizing the Potential Impact of Energy Recovery and Pipe Replacement on Leakage Reduction in a Medium Sized District Metered Area

Gideon Johannes Bonthuys [1,*], **Marco van Dijk** [1] and **Giovanna Cavazzini** [2]

1   Department of Civil Engineering, University of Pretoria, 0002 Pretoria, South Africa; marco.vandijk@up.ac.za
2   Department of Industrial Engineering, School of Engineering, University of Padua, 35131 Padova, Italy; giovanna.cavazzini@unipd.it
*   Correspondence: gbonthuys@golder.com; Tel.: +27-76-412-4035

Abstract: The drive for sustainable societies with more resilient infrastructure networks has catalyzed interest in leakage reduction as a subsequent benefit to energy recovery in water distribution systems. Several researchers have conducted studies and piloted successful energy recovery installations in water distribution systems globally. Challenges remain in the determination of the number, location, and optimal control setting of energy recovery devices. The PERRL 2.0 procedure was developed, employing a genetic algorithm through extended period simulations, to identify and optimize the location and size of hydro-turbine installations for energy recovery. This procedure was applied to the water supply system of the town of Stellenbosch, South Africa. Several suitable locations for pressure reduction, with energy recovery installations between 600 and 800 kWh/day were identified, with the potential to also reduce leakage in the system by 2 to 4%. Coupling the energy recovery installations with a pipe replacement model showed a further reduction in leakage up to a total of above 6% when replacing 10% of the aged pipes within the network. Several solutions were identified on the main supply line and the addition of a basic water balance, to the analysis, was found valuable in preliminarily evaluation and identification of the more sustainable solutions.

Keywords: energy recovery; hydropower; water distribution network; optimization processes

Citation: Bonthuys, G.J.; van Dijk, M.; Cavazzini, G. Optimizing the Potential Impact of Energy Recovery and Pipe Replacement on Leakage Reduction in a Medium Sized District Metered Area. *Sustainability* **2021**, *13*, 12929. https://doi.org/10.3390/su132212929

Academic Editors: Mosè Rossi, Massimiliano Renzi, David Štefan and Sebastian Muntean

Received: 4 October 2021
Accepted: 15 November 2021
Published: 22 November 2021

Publisher's Note: MDPI stays neutral with regard to jurisdictional claims in published maps and institutional affiliations.

1. Introduction

Numerous studies have been conducted on the potential of energy recovery in water supply and distribution systems, with several successful installations globally [1–4]. Small-scale hydropower technology, applied in Water Supply Infrastructure, has been demonstrated to be a viable renewable energy to develop in South Africa [5,6]. Small-scale hydropower technology or energy recovery devices utilise excess energy, which is conventionally dissipated, in order to control the maximum admissible pressure in the system and to avoid pipe rupture [7]. In both the case of a conventional pressure reducing valve (PRV) and energy recovery devices (ERD) such as hydro turbines, a pressure drop across the component allows for downstream pressure control [8].

Since leakages are proportional to pressure within the system, a decrease in system pressure will reduce the rate of leakage from the system [9]. Extensive research has been conducted on leakage reduction through pressure management in water supply and distribution systems. Challenges faced with pressure management are related to the determination of the number, location, and optimal control settings of pressure management devices such as conventional PRVs or ERDs such as hydro turbines or pumps-as-turbines (PAT). The last decade has seen many researchers investigating the optimization of pressure management in water supply/distribution networks employing optimization techniques including linear programming (LP), nonlinear program (NLP) algorithms, mixed integrated nonlinear programming (MINLP), mathematical programming with complimentary constraints (MPCC), and genetic algorithms (GAs) [9,10]. Interest in leakage reduction as a benefit of energy recovery has rapidly gained popularity along with the drive for more sustainable societies and resilient infrastructure networks [11,12]. Several researchers shifted focus to recovering excess pressure from water supply networks while reducing water losses at the same time [13–15].

Creaco and Haidar [16] developed a methodology proposing a hybrid algorithm that attempts to find an installation of control valves for pressure management in District Metered Areas (DMAs) based on the total installation cost, daily leakage volume, and demand uniformity.

Analysis of the previous research [1–16] catalysed the 2020 research study by Bonthuys et al. [17], which developed a procedure employing a genetic algorithm (GA) to identify and optimize the location and sizes of hydro-turbine installations or conduit hydropower installations for energy recovery. The procedure, named PERRL [17], differed from the Creaco and Haidar [16] methodology by specifically focusing on energy recovery rather than pressure management. Bonthuys et al. [18] improved the developed procedure to include extended period simulations, enabling the analysis of demand patterns and demand uniformity similar to Creaco and Haidar [16]. The improved procedure was termed PERRL 2.0.

A comparison of the PERRL 2.0 procedure and recent related research studies in energy recovery and leakage reduction was undertaken. The comparison focused on the study methodology and the conclusions with regards to both energy recovery and leakage reduction. A summary of the recent research studies related to energy recovery and leakage reduction from water supply and distribution networks is shown in Table 1.

Table 1. Recent research studies related to energy recovery and leakage reduction from water supply and distribution networks.

Study	Ref.	Abridged Methodology	Conclusions	
			Energy Recovery	**Pressure Management/Leakage Reduction**
Modelling Irrigation Networks for the Quantification of Potential Energy Recovering	[19]	• Calculates the theoretical recoverable energy by utilizing an energy balance	• Discriminate the energy needed for irrigation, friction head losses, non-recoverable energy, and potentially recoverable energy in any line in a network based on flows	• N/A
Simulated Annealing in Optimization of Energy Production in a Water Supply Network	[20]	• Proposed optimization algorithm to provide a selection of optimal locations for the installation of turbines in a distribution network. • A simulated annealing process was developed to optimize the location of the turbines	• Rapid convergence for the cases of installing up to three turbines • Best solutions were not the combinations of the branches with highest energetic potential, indicating the need for a detailed analysis in terms of daily variation in flow and turbine efficiency.	• N/A
Energy Recovery Using Micro-Hydropower Technology in Water Supply Systems	[21]	• Critical network points identified to assess excess pressure • Algorithm developed to access the potential for energy recovery from identified excess pressure	• The optimization maximizes the economic value of the installation • Sensitivity analysis of the demand proposed to be considered to verify the impact of energy recovery and in the behaviour of the network.	• Replacement of PRV with ERD evaluated • Effect of ERD on leakage reduction not considered

Table 1. *Cont.*

Study	Ref.	Abridged Methodology	Conclusions	
			Energy Recovery	**Pressure Management/Leakage Reduction**
Pressure Management by Combining Pressure Reducing Valves and Pumps as Turbines for Water Loss Reduction and Energy Recovery	[15]	• Critical network points identified to assess excess pressure • Hydraulic modelling (EPANET) used to evaluate the pressure situation and pressure management strategies	• Hydraulic modelling is an essential tool for understanding the relationship between flow and pressure in water distribution systems and to evaluate energy recovery potential	• Leakage modelled and the effect of different pressure management strategies using energy recovery was analysed.
Recent Innovations and Trends in In-conduit Hydropower Technologies and Their Applications in Water Distribution Systems	[22]	• In-depth review of the currently available hydro-turbine technologies that are suitable for various in-conduit applications such as energy recovery	• Several conventional and new turbine technologies identified for possible application within water supply and distribution networks.	• N/A
Framework Development for the Evaluation of Conduit Hydropower within Water Distribution Systems	[23]	• Development of a generic framework to quantify energy recovery potential and to identify potential sites in bulk water supply systems when limited data poses a challenge	• Evaluation frameworks were developed for nine hydropower types (incl. energy recovery in WDS) to assist in the evaluation and quantification of hydropower potential.	• N/A
Recovering energy by hydro-turbines application in water transmission pipelines	[24]	• Quantification of the embedded power within a WDS from hydraulic analysis of the network simulated in WaterCAD • Identification of energy recovery potential hotspots • Redesign of network links (maximum velocity) to allow conveying of required design flow rate • Selection of the best fitting hydro-turbine type • Financial and environmental evaluation of the simulated scenarios with respect to the original design	• Redesign of network links in terms of maximum allowable velocity increased the residual pressure and consequently the potential energy recovery value. • From an economic perspective, installing traditional turbines in water transmission systems concluded to be feasible for energy recovery	• N/A
Methodology for Determining the Maximum Potentially Recoverable Energy in Water Distribution Networks	[25]	• Determination of the potentially recoverable energy (PRE) • A new energy balance is proposed • Network resilience index (Recoverable Energy Index (PREI)) is proposed	• New energy balance gives detail on excess energy in WDNs • Developed PREI as an indicator of the % of excess energy that can be recovered	• Includes energy consumed by apparent losses in leakages • Does not include analysis of leakage reduction from energy recovery

Energy Recovery in Pressurized Hydraulic Networks	[26]	• Evaluate the possibility of using non-utilized hydraulic energy in urban water distribution systems in terms of:
		  ○ feasibility and efficiency of installing turbines or PATs
		  ○ required hydraulic capacity of the turbines or PATs
		  ○ analysis of installation points in the pipeline network
		  ○ evaluation of possible energy recovery
			• Recommended that energy recovery devices be installed in a parallel position with the valve that regulates the distribution of the water flow to each of the intakes of the urban areas or DMAs
			• Payback period alone does not support the decision to promote energy recovery; socio-environmental analyses proposed in addition to current studies.
				• N/A

From Table 1, it can be seen that recent studies on energy recovery predominantly focusses on identifying the optimal installation locations within a network based on hydraulic capacity and economic feasibility. The addition of energy balances to the latest research studies improved the evaluation of excess energy in water distribution systems.

Research conducted on the leakage reduction as an effect of pressure management was not included in the comparison in Table 1, and only studies which focused on leakage reduction as a benefit or value-add from energy recovery were considered. As can be seen from Table 1, several studies included analyses of the effect of energy recovery on pressure management and leakage reduction, however, none included the benefits (economic, environmental, social) of leakage reduction in the optimisation procedure employed.

Within the South African context, the series of rolling blackouts, known as loadshedding, that started in 2008 when the total electricity demand within the country started encroaching on the supply capacity [27] sparked various studies on alternative energy supply. Additionally, studies on leakage detection and reduction methods have been sparked by recent severe droughts that have plagued the KwaZulu Natal and Western Cape provinces following poor rainfall during the 2014 to 2017 rainy seasons.

In May 2018, storage reservoirs in Cape Town, in the Western Cape province of South Africa, dropped to about 20%. This resulted in the implementation of strict water-usage restrictions to delay water levels from reaching 13.5%, at which point total failure of the Cape Town Metropolitan Municipal Water Supply System would have occurred [28]. Above average rainfall in the remainder of 2018 allowed the City of Cape Town to avoid a complete water supply system failure. The effects of the droughts were worsened by poor water-management practices and infrastructure deficiencies [29,30]. Although it remains a priority to understand the probability of similar meteorological droughts in future it is imperative to start facilitating the improvement of water-management practices as well as the infrastructure necessary to develop a more resilient system [28].

The Stellenbosch Local Municipality, situated to the northeast of the City of Cape Town, receives two-thirds of its municipal water from Cape Town via the Wemmershoek and Theewaterskloof dams, resulting in Stellenbosch experiencing similar water distress during the drought period. In 2018, the Stellenbosch municipality imposed strict water restrictions, limiting users to 6000 L a month and banning several uses of potable water, such as its application for gardening. Restrictions were also placed on use from groundwater sources [31].

The current study was conducted to evaluate the Stellenbosch Local Municipality's Water Supply System's suitability for energy recovery installations, and to quantify the potential leakage reduction emanating from such energy recovery installations. The PERRL 2.0 procedure developed by Bonthuys et al. [18] was incorporated in the analysis of the Stellenbosch Local Municipality Water Supply System.

2. Methodology

The objective of the study is to test the application of the PERRL 2.0 procedure [18] and assess the advantage thereof in terms of the operation of a District Metered Area (DMA). The PERRL 2.0 procedure is an enhanced optimization procedure which incorporates user-defined weighted importance of specific objectives and extended period simulations into a genetic algorithm (GA) that identifies the optimum size and location of potential installations for energy recovery and leakage reduction [18].

The methodology of the study can be outlined as shown in Figure 1. To meet the study objective, the PERRL 2.0 procedure was used to evaluate the energy recovery and leakage reduction potential within the Stellenbosch Water Distribution Network, and the subsequent effects the installation of the recommended energy recovery devices would have on the operation of the network.

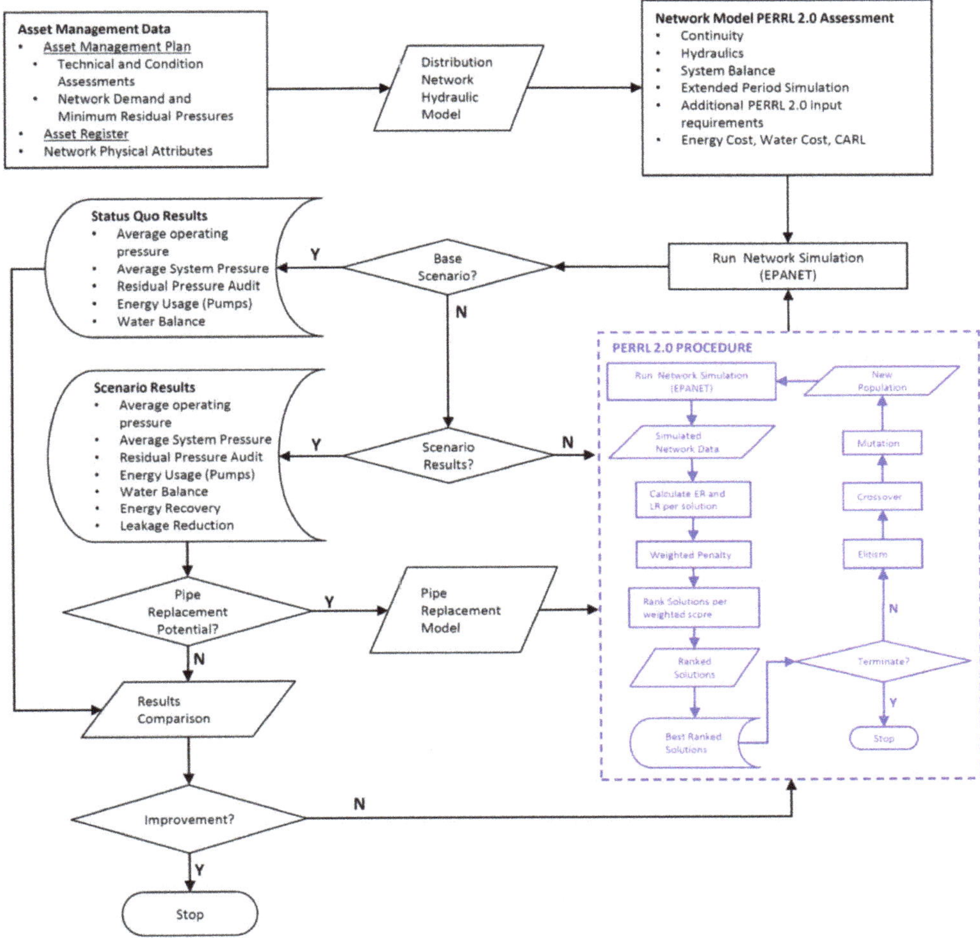

Figure 1. Study methodology schematic summary.

The paper first evaluates the suitability of the current Stellenbosch Water Distribution Network Model for use within the PERRL 2.0 procedure to obtain a simplified, stable, and closely representative model of reality. The current Stellenbosch Water Distribution Network Model as obtained from the Asset Management Operations of the Stellenbosch Local Municipality. All changes made to the current model, to make it adaptable for use within the PERRL 2.0, procedure have been documented and the effect of these changes on the baseline hydraulic analysis has been evaluated.

The energy recovery installations identified using the PERRL 2.0 procedure, were modelled within EPANET to obtain hydraulic analyses comparable to the baseline hydraulic analysis. The comparison served as a basis for evaluating the energy recovery installations identified by the PERRL 2.0, hereby assessing the efficacy and applicability of the procedure to the Stellenbosch Water Distribution Network.

Finally, the comparison of the energy recovery hydraulic analyses and the baseline hydraulic analysis, was used to identify additional areas where infrastructure changes, such as pipe replacement programs, increase the potential for energy recovery within the system.

3. Stellenbosch Water Distribution Network

The Stellenbosch Local Municipality (LM) is responsible for the governance of the towns of Stellenbosch, Franschhoek, and Pniel, as well as a number of rural towns, in the Western Cape Province of South Africa. The Stellenbosch LM covers approximately 900 km^2 and supplies water to approximately 199,800 people, including a fairly extensive industrial area. Water supply to the consumers in the Stellenbosch LM is achieved through the following water supply systems [32]:

- Stellenbosch (Jonkershoek and Theewaterskloof tunnel);
- Franschhoek;
- Wemmershoek (treated water imported from the City of Cape Town);
- Blackheath (treated water imported from the City of Cape Town);
- Faure (treated water imported from the City of Cape Town);
- Other own sources (Boreholes).

The total water supply system is comprised of 56 reservoirs, holding tanks, and water towers, 36 water pump stations, 35 pressure-reducing valve installations, 667 km of pipeline, and 79 water supply zones, with a bulk water input of 8,015,027 kL in 2019/20 or an annual average daily demand (AADD) value of 21.9 ML/day. A total of 69% of the AADD is supplied by the three water treatment works managed by the LM, namely, Ida's Valley, Paradyskloof, and Franschhoek.

Based on the municipality's IWA Water Balance sheet for 2019/20 shown in Table 2, the municipality recorded 20.5% for non-revenue water (NRW) and 6.5% for Real Network Losses which is below the best practice value of 15% [32]. The Stellenbosch LM Long Term Water Conservation and Demand Management (WC/WDM) Strategy was approved by Council in February 2014 and updated in 2019. The strategy implicitly includes measures to further reduce real water losses in the water network.

Table 2. Stellenbosch LM 2019/20 IWA water balance (kL) [32].

Authorized consumption 88.5%	Revenue Water 79.5%	Billed metered	4,360,974	54.4%
		Billed metered (indigent)	1,625,305	20.3%
		Billed unmetered	388,195	4.8%
		Unbilled metered	240,451	3.0%
Unaccounted for Water (UAW) 11.5%	Non-revenue Water 20.5%	Unbilled unmetered	16,030	0.2%
		Informal areas not metered	462,601	5.8%
		Losses in bulk supply system	0	0.0%
		Apparent losses	400,751	5.0%
		Real network losses	520,720	6.5%

The Stellenbosch LM water supply networks can be classified into several DMAs that. For the purpose of the study, only the Stellenbosch Town DMA was modelled and analysed. The Stellenbosch Town DMA demand comprises 65% of the water supply of the entire Stellenbosch LM, which amounts to an AADD of 14.8 ML/day. Figure 2 shows the locality of the Stellenbosch Town DMA and the configuration of the Water Distribution Network.

The estimated Current Annual Real Losses (CARL) in the Stellenbosch Town DMA, based on the municipality's 6.9% for Real Network Losses, is roughly 350,000 kL.

The two main sources of raw water supply for the Stellenbosch Town are as follows:

- Eerste River—Kleinplaas Dam (7.224 Mm^3/a registered abstraction);
- Western Cape Water Supply System (WCWSS) (3 Mm^3/a registered abstraction)—via Theewaterskloof Tunnel.

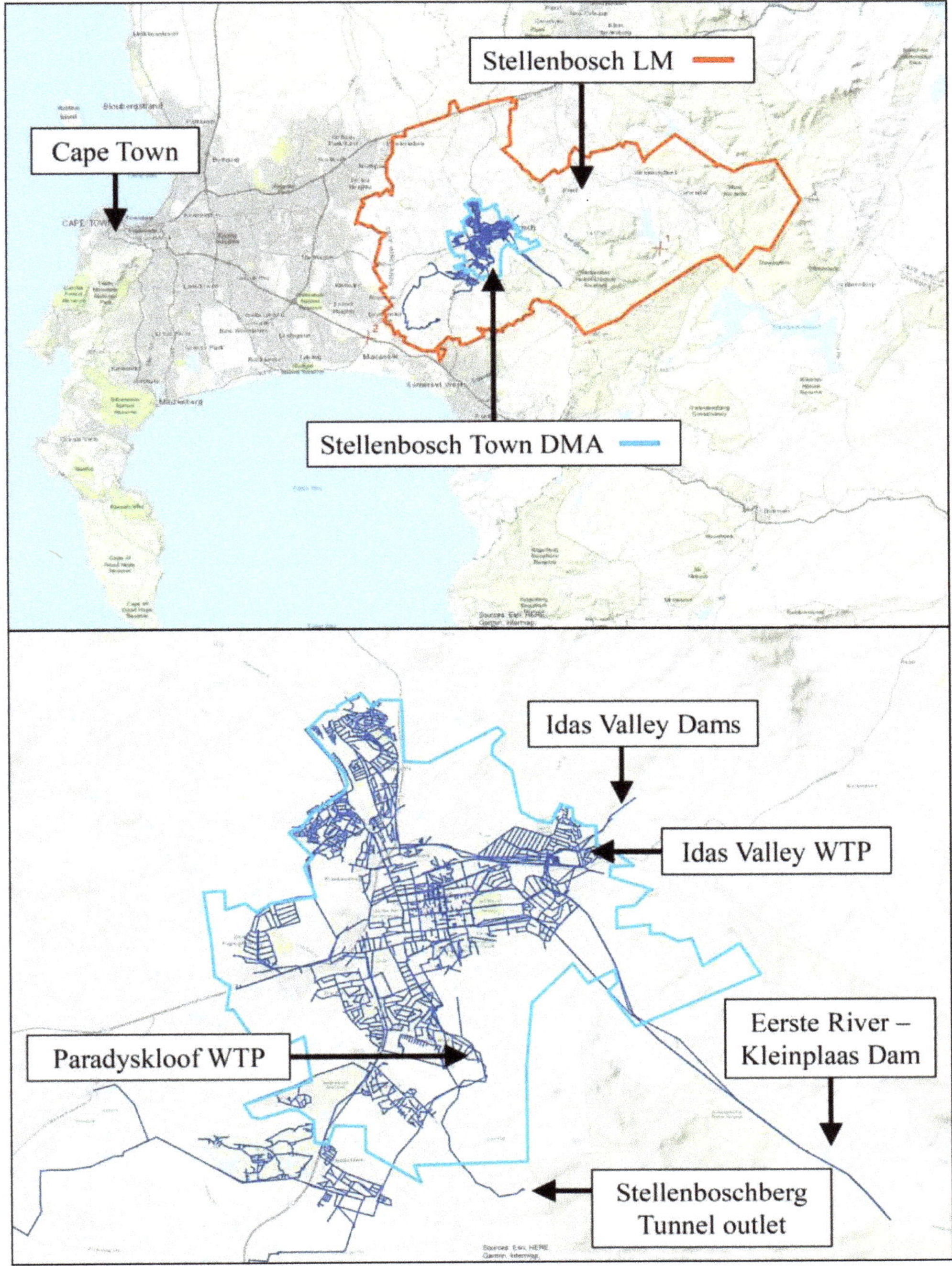

Figure 2. Stellenbosch Town DMA and water distribution network configuration.

The Jonkershoek Weir diverts water from the Eerste River in the Jonkershoek Valley at Kleinplaas Dam and conveys it to two off-channel storage dams in Idas Valley through a gravity pipeline. The Jonkershoek Weir combined with the two Idas Valley dams is the most important source of water for Stellenbosch Town. The treatment capacity of the Idas Valley Water Treatment Plant (WTP) is 28 ML/day. The WCWSS supplies water to the Paradyskloof WTP through a pipeline leading from the Stellenboschberg Tunnel outlet of the Riviersonderend GWS tunnel system. Under normal operation, a volume of 3 Mm^3/a is available from this source. The treatment capacity of the Paradyskloof WTP is 10 ML/day [33].

The Stellenbosch Town DMA has a varying topography with the main water sources of the Jonkershoek Weir and Stellenboschberg Tunnel outlet having elevations of 310 m and 246 m, respectively. The average elevation of consumptive nodes in the system is 134 m, with the lowest consumptive node elevation equal to 78 m and the standard deviation of node elevation equal to 34.7 m, indicating a preliminary potential for energy recovery.

As suggested in Bonthuys et al. [34], the study leverages asset management data, contained within municipal infrastructure asset registers and asset management plans, to identify energy recovery and leakage reduction potential in municipal water distribution systems. In this particular case study, the Stellenbosch LM provided access to their IMQS data, representing the majority of data outlined in the Bonthuys et al. [34] study. The IMQS data is contained in a GIS-centric, web-based software for Infrastructure Asset Management. IMQS data is conventionally used for the maintenance and operational planning of physical water infrastructure assets by integrating with specialist hydraulic software packages to offer a geographically linked, infrastructure-lifecycle-focused representation of a municipality's water reticulation network [35].

In addition to the Stellenbosch LM IMQS asset management data, the latest Wadiso planning model for the Stellenbosch LM was also received. Wadiso is a comprehensive application for the analysis and optimal design of water distribution systems through its inclusion of a seamless interface to the public domain EPANET program module which is used in the PERRL 2.0 application [36]. However, the Wadiso model data received was for a steady state analysis on the peak operation of the LM, whereas the PERRL 2.0 procedure uses an extended period simulation. The differences in analyses or simulation types coupled with the several other issues related to the interface between Wadiso, EPANET and the PERRL 2.0 procedure, necessitated certain amendments to the current numerical Stellenbosch Water Distribution Network Model.

4. Amendments to Stellenbosch Water Distribution Network Model

Several amendments to the current Stellenbosch LM Water Distribution Network Model were required to optimally run the PERRL 2.0 procedure and analyse and evaluate the energy recovery and leakage reduction potential in the network through the installation of energy recovery devices. The most prominent change was the introduction of a demand pattern to enable the extended period simulation employed by the GA within the PERRL 2.0 procedure. Standard residential demand patterns, for the applicable levels of service, were used for the extended period simulation similar to the demand pattern employed by Bonthuys et al. [34] (Figure 3).

Implementing the demand pattern and changing the model from steady state peak analysis to extended period simulation resulted in model and procedural instabilities. These instabilities centred around the synergy between the PERRL 2.0 procedure and the EPANET model outputs. The PERRL 2.0 procedure was developed to use a 24 h, hourly timestep Extended Period Simulation (EPS). The hydraulic time step in EPANET was also set by the user to 1 h. However, in EPANET, time steps shorter than normal occur automatically in cases of one of the following events [37]:

- the next output reporting time period occurs;
- the next time pattern period occurs;
- a tank becomes empty or full;
- a simple control or rule-based control is activated.

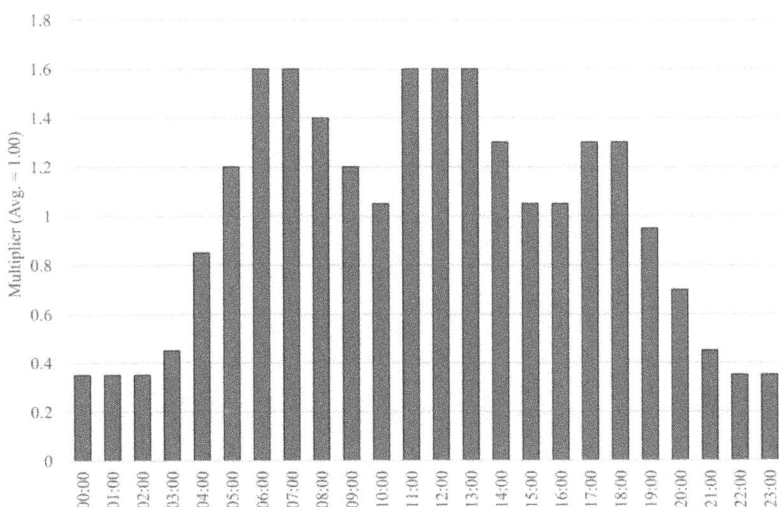

Figure 3. Stellenbosch Town DMA—adapted demand pattern.

Further analyses of hourly flows in and out of tanks in conjunction with tank storage volumes indicated that the filling and emptying of tanks occurred at intervals shorter than the set hydraulic timestep. This resulted in an unbalanced system, runtime errors, and reporting periods shorter than the 1-h intervals required by the PERRL 2.0 procedure. The issue was overcome through a reduction in the set hydraulic time steps and the amendment of several tank storage sizes, to reflect the actual current operations more accurately and to overcome the errors due to tanks filling or emptying in shorter time steps than the set hydraulic timestep.

In addition to these amendments, an analysis on the roughness coefficients of the current network scenario was conducted in order to better estimate pressure losses. This was conducted by identifying the pipes with higher-than-average roughness coefficients for known materials. Table 3 shows a comparison of the roughness coefficients for the different pipe materials in the modelled data with the recommended ranges from literature. The percentage of modelled pipe concluded as "aged" or in bad condition, due to low Hazen Williams C-Factors in the source data, are shaded in Table 3.

All pipe segments were also evaluated in terms of head loss (m), pipe length (m), and unit head loss (m/km). The worst hydraulically performing pipe segments, in terms of head loss and unit head loss were isolated and are shown in Table 4.

Table 3. Modelled roughness coefficients per pipe length.

Pipe Material	Modelled Length (km)	% of Modelled Pipe Length per Hazen Williams C-Factor							Recommended Range [38,39]		
		90	100	110	115	120	130	150	New	25 Years	50 Years
Asbestos Cement (AC)	192	24	31	9	28	7	1	0	150	130	120
Cast Iron (CI)	0.2	0	100	0	0	0	0	0	130	110	90
HDPE	16	5	26	11	56	2	0	1	150	140	140
uPVC	108	2	34	11	46	1	6	0	150	140	140
Steel [1]	0.2	0	100	0	0	0	0	0	150	130	100
Unknown	49	10	42	2	44	3	0	0			

[1] for welded and seamless steel, the recommended C-Factor can reduce to 100 [40].

Table 4. Ten worst performing pipe segments at peak flow—Base Scenario.

Pipe ID	Length (m)	Diameter (mm)	Material	C-Value	Unit Headloss (m/km)	Head Loss (m)
3516	5.23	200.00	n/a	115	508.50	2.66
4155	7.67	37.00	uPVC	100	433.96	3.33
4335	18.55	37.00	uPVC	100	433.96	8.05
3526	25.55	200.00	n/a	115	508.50	12.99
3819	34.60	19.00	Steel	100	783.38	27.11
4240	11.30	37.00	uPVC	100	122.63	1.39
4339	18.06	37.00	uPVC	100	122.63	2.22
4033	21.2	54.00	uPVC	100	112.80	2.39
3735	47.68	19.00	uPVC	100	808.03	38.53
3258	12.18	125.00	n/a	100	97.82	1.19

From the above analysis, two different scenarios were proposed for the analysis of energy recovery potential within the Stellenbosch Town DMA network.

1. Scenario 1—Base scenario analysis with the amended Stellenbosch Water Distribution Network Model as the status quo, excluding energy recovery installations;
2. Scenario 2—Analysis of the status quo using PERRL 2.0;
3. Scenario 3—Analysis of energy recovery potential with the following amendments to the network:

    a. Replacement of "aged" pipe segments for a defined length upstream of energy recovery locations identified in Scenario 1;
    b. Replacement of defined percentage of "aged" pipes and re-application of the PERRL 2.0 procedure.

5. Scenario 1—Base Scenario Analysis

To evaluate the starting network performance, a base scenario was run using the EPANET software.

Figure 4 shows a comparison of the average operating pressure of all consumptive nodes and the consumed flow within a 24-h modelled operation of the network.

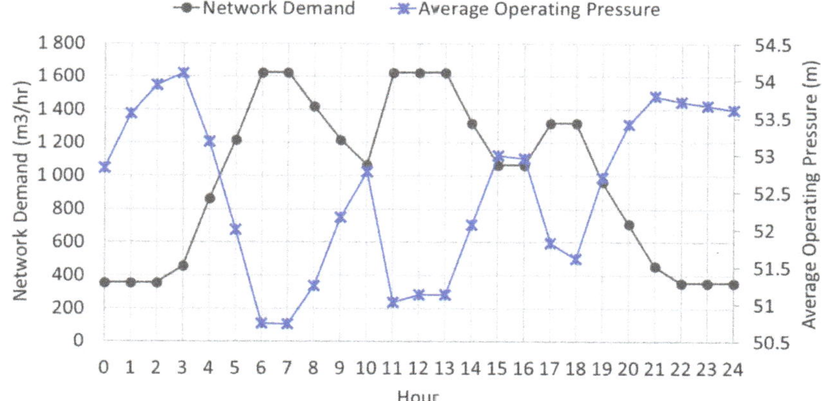

Figure 4. Average operating pressure and consumed flow—Stellenbosch Town DMA—base scenario.

As it can be seen, the hourly averaged operating pressure, at the consumptive nodes in the base scenario operation of the Stellenbosch Town DMA, fluctuates between 50 and 55 m of pressure with minimum values obtained during the hours where the network demand is at the maximum. The Council for Scientific and Industrial Research [41] specifies the minimum residual pressure within the South African municipal water services environment as 24 m (2.4 bar) for house connections under instantaneous peak demand. So, it is clear that pressures within the system are above the minimum acceptable limit, and the potential excess pressure in the system could be exploited through ERD.

Figure 5 shows the distribution of the hourly operating pressure at consumptive nodes in the system during both peak (06h00—07h00) and off-peak (22h00—02h00) times. From Figure 5 the median of the peak and off-peak operating pressures can be calculated as 50.8 m and 53.3 m, respectively. In both instances, more than 90% of the consumptive nodes are operating at pressures above the acceptable minimum residual pressure, and more than 50% are operating above double the acceptable minimum residual pressure, further indicating the potential for energy recovery within the system. Figure 6 shows the spatial distribution of different operating pressure ranges for the base scenario of the Stellenbosch Town DMA during peak and off-peak operations. As it can be seen, there is a significant percentage of the consumptive nodes characterized by operating pressure values exceeding double the inferior limit and a number of them with operating pressure values exceeding the triple of the inferior limit.

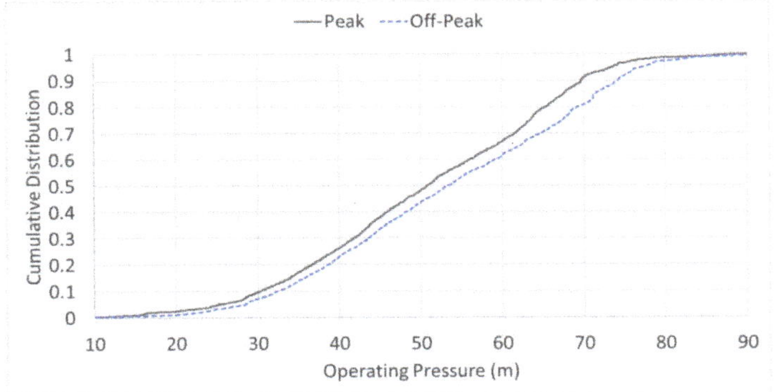

Figure 5. Consumptive node operating pressure cumulative distribution.

The following metrics (Table 5) were calculated from the results of the base scenario run (Scenario 1) and used in the evaluation of solutions from Scenario 2 and 3 discussed in the following sections.

Table 5. Base scenario—performance metrics.

Average operating pressure		52.47 m
Modelled % consumptive nodes below minimum residual pressure		2.8%
Modelled % demand below minimum residual pressure		3.7%
Pumps	kWh/m^3 (avg)	0.3
	Average kW	121.95
Water Balance	Demand	24.3 ML
	Supply	25.3 ML
	Network Start Storage	90.9 ML
	Network End Storage	92.3 ML
	Change in storage	1.4 ML (1.6%)
	% Imbalance	1.9%

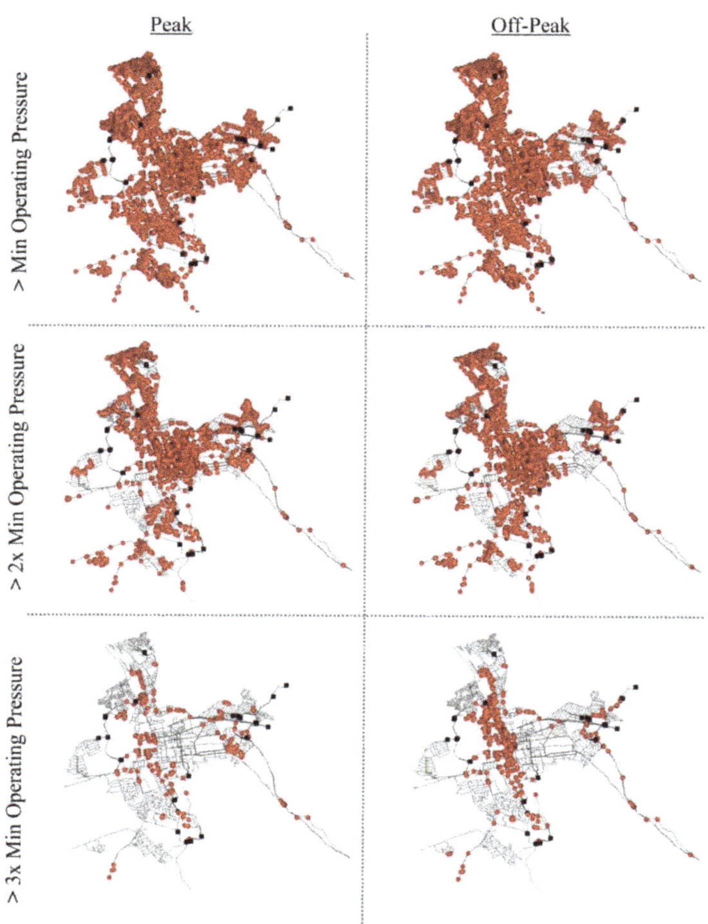

Figure 6. Consumptive node operating pressure spatial distribution.

6. Results and Discussion

The base scenario analysis of the Stellenbosch Town DMA showed potential for energy recovery. The mean operating pressures calculated at the consumptive nodes, for both the peak and off-peak hours, are considerably more than the minimum residual pressure prescribed by the Council for Scientific and Industrial Research [41].

The results of both Scenarios 2 and 3 are shown in the following sections and discussed in Section 7.

6.1. Scenario 2

Figure 7 shows the three energy recovery locations for the five top ranked solutions for energy recovery and leakage reduction in the Stellenbosch Town DMA for the Scenario 2 analysis. The main performance metrics of these five solutions are summarized in Table 6, which also reports the total energy recovery and leakage reduction potential. In the Table, the proposed solutions are compared with the base case scenario discussed in Section 5.

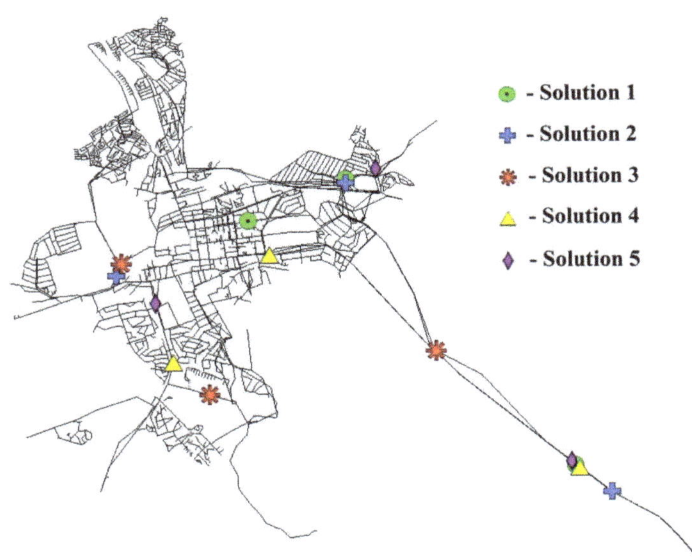

Figure 7. Scenario 2—Top proposed energy recovery locations.

Table 6. Scenario 2—Solution comparison. The three individual energy recovery locations (sites) are indicated in terms of the numerical sequence assigned to the pipes at which the installations are proposed.

		Base	Solution 1	Solution 2	Solution 3	Solution 4	Solution 5
Average operating pressure at consumptive nodes (m)		52.47	52.49	52.10	51.94	52.02	51.86
Average system pressure		53.53	52.64	52.29	52.50	52.58	52.53
Modelled % consumptive nodes below minimum residual pressure		2.8%	2.8%	2.8%	2.8%	2.8%	3.0%
Modelled % demand below minimum residual pressure		3.7%	3.7%	3.8%	3.8%	3.8%	4.0%
Pumps	kWh/m^3 (avg.)	0.30	0.30	0.30	0.29	0.29	0.30
	Average kW	121.95	161.59	160.43	122.19	122.15	122.25
Water Balance	Demand (ML)	24.3	24.3	24.3	24.3	24.3	24.3
	Supply (ML)	25.3	30.7	30.9	26.3	26.3	26.3
	Network Start Storage (ML)	90.9	90.9	90.9	90.9	90.9	90.9
	Network End Storage (ML)	92.3	97.0	97.0	93.2	93.1	93.3
	Change in storage	1.4 ML (1.6%)	6.2 ML (6.8%)	6.1 ML (6.7%)	2.3 ML (2.6%)	2.2 ML (2.5%)	2.4 ML (2.7%)
	% Imbalance	1.9%	0.8%	1.5%	1.5%	1.2%	1.6%
Energy Recovery	kWh/day		812	692	665	667	649
Leakage Reduction	%		2.47%	3.65%	2.98%	2.82%	2.95%
Site1	Link		3 060	500	3 582	2 717	4 445
Site2	Link		3 673	3 668	2 044	1 750	1 242
Site3	Link		4 126	3 064	689	3 673	3 673

6.2. Scenario 3

Table 7 shows the pipe sections upstream of the energy recovery locations (Solution 3—Scenario 3), which has been replaced by new uPVC pipes. Link 3 582 is on one of the 450-mm diameter main supply lines from the Jonkershoek Weir. The entire length of pipe from Link 3 582 back to the supply from the Jonkershoek Weir was modelled to be replaced.

Table 7. Pipe replacement details.

Energy Recovery Location (Model Link No.)		3582	2044	689
Pipe Replacement	Start Node	3138	3945	416
	End Node	3055	1749	579
	Pipe Diameter (mm)	400	450	450
	Pipe Length (m)	61	838	485
	Start Node	3933		
	End Node	3138		
	Pipe Diameter (mm)	450		
	Pipe Length (m)	6380		
Total Pipe Replacement Length (m)		6441	838	485

Scenario 3a was run with the preferred proposed energy recovery solution as per Scenario 2 (Solution 3) for the base case hydraulics and roughness parameters, as well as the amended roughness parameter corresponding to the proposed pipe replacement (Table 8).

Table 8. Top proposed energy recovery installation—Hydraulic effect of pipe replacement.

	Roughness	**Base Case**	**Pipe Replacement**
	Energy Recovery Location Induced Headloss	Scenario 2 Results	Scenario 2 Results
	Average operating pressure at consumptive nodes (m)	51.94	51.89
	Average system pressure (m)	52.50	52.58
	Modelled % consumptive nodes below minimum residual pressure	2.8%	2.8%
	Modelled % demand below minimum residual pressure	3.8%	3.8%
Pumps	kWh/m^3 (avg.)	0.29	0.30
	Average kW	122.1	122.3
Water Balance	Demand (ML)	24.3	24.3
	Supply (ML)	26.3	26.4
	Network Start Storage (ML)	90.9	90.9
	Network End Storage (ML)	93.2	93.2
	Change in storage	2.3 ML (2.6%)	2.3 ML (2.6%)
	% Imbalance	1.5%	1.2%
Site 1	Link	3582	3582
	Loss Coefficient	203.6	203.6
	Induced Head loss (Peak) (m)	25.6	30.0
	Flow (Peak) (LPS)	196.9	214
	Energy Recovery Potential (Peak) (kW)	34.6	44.1
	Energy Recovery Potential (Daily Avg.) (kWh)	448	573

Table 8. *Cont.*

	Roughness	Base Case	Pipe Replacement
	Link	2044	2044
	Loss Coefficient	833.4	833.4
	Induced Head loss (Peak) (m)	19.6	19.6
Site 2	Flow (Peak) (LPS)	108.2	108.5
	Energy Recovery Potential (Peak) (kW)	14.6	14.64
	Energy Recovery Potential (Daily Avg.) (kWh)	119	120
	Link	689	689
	Loss Coefficient	948.1	948.1
	Induced Head loss (Peak) (m)	15.2	15.2
Site 3	Flow (Peak) (LPS)	88.9	89.3
	Energy Recovery Potential (Peak) (kW)	9.3	9.32
	Energy Recovery Potential (Daily Avg.) (kWh)	98	99
Total Energy Recovery Potential (Daily Avg.) (kWh)		665	792
Total Leakage Reduction Potential		2.98%	2.73%

For Scenario 3b, 10%, 25%, and 50% of all the "aged" pipes identified in Table 3 were replaced by uPVC pipes with roughness parameters within the recommended range for new pipes. For Scenario 3b, the effect of pipe replacement on the CARL within the system was also incorporated by reducing the CARL proportionally to the sizes and lengths of the pipes replaced.

Figure 8 shows the modelled leakage reduction and energy recovery for the system post pipe replacement for the top 20 solutions of each pipe replacement scenario. The large markers were used for each pipe replacement scenario to indicate the average modelled leakage reduction and energy recovery.

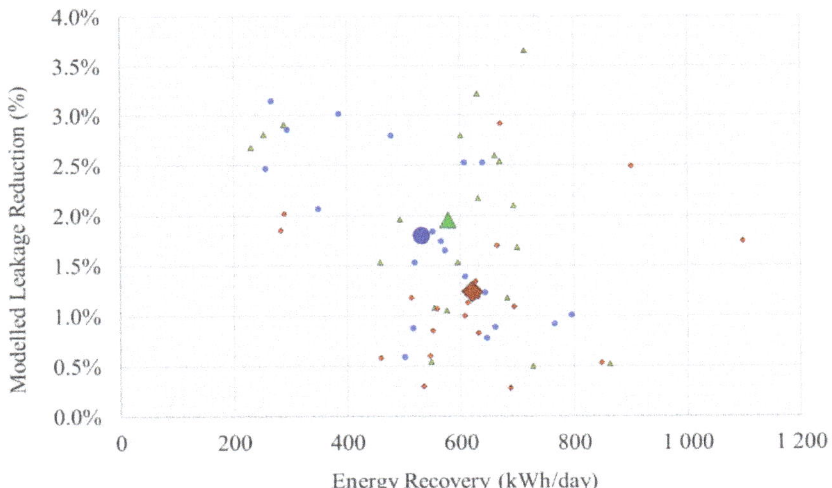

Figure 8. Scenario 3b—Energy recovery and leakage reduction potential—top 20 solutions of each pipe replacement scenario.

Figure 9 shows the location of the three energy recovery locations (sites) for the top five proposed energy recovery solutions for Scenario 3b with 10%, 25%, and 50% of all the "aged" pipes, whereas Table 9 shows the comparison between the solutions for Scenario 3b and the base case scenario. Some of the top five proposed energy recovery solutions identified the same or very similar locations for more than one solution. For the sake of clarity, in Table 9, only the top proposed energy recovery solution was reported.

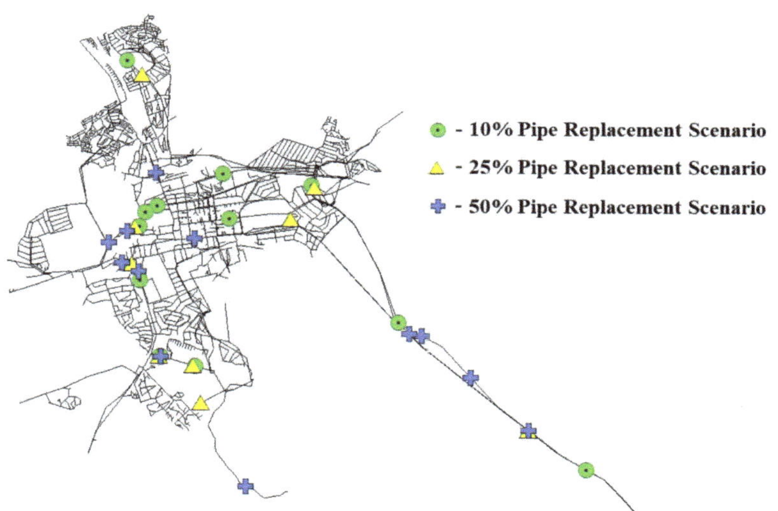

Figure 9. Scenario 3b—Top proposed energy recovery locations.

Table 9. Scenario 3b—Solution comparison with different pipe replacement models.

Pipe Replacement		Base	Solution 1		
		Base Case	10%	25%	50%
Average operating pressure	Pipe Replacement Only	52.47 m	53.02 m	53.30 m	53.35 m
	Incl. Energy Recovery		52.53 m	52.40 m	52.67 m
Average "aged" piped system pressure	Pipe Replacement Only	53.53 m	56.45 m	55.59 m	53.87 m
	Incl. Energy Recovery		56.47 m	55.62 m	53.15 m
Modelled % consumptive nodes below minimum residual pressure		2.8%	3.4%	4.1%	2.5%
Modelled % demand below minimum residual pressure		3.7%	4.3%	5.0%	3.4%
Pumps	kWh/m^3 (avg.)	0.3	0.29	0.29	0.29
	Average kW	121.95	131.98	131.66	132.34
Water Balance	Demand	24.3 ML	24.3 ML	24.3 ML	24.3 ML
	Supply	25.3 ML	27.0 ML	28.3 ML	22.7 ML
	Network Start Storage	90.9 ML	90.9 ML	90.9 ML	90.9 ML
	Network End Storage	92.3 ML	94.0 ML	93.7 ML	89.6 ML
	Change in storage	1.4 ML (1.6%)	3.1 ML (3.4%)	2.8 ML (3.1%)	−1.2 ML (−1.3%)
	% Imbalance	1.9%	1.8%	3.8%	2.0%
Energy Recovery	kWh/day		606	713	902
Leakage Reduction	Pipe Replacement Only		3.8%	10.1%	21.6%
	Incl. Energy Recovery		6.2%	13.4%	23.5%
Site 1	Link		4137	1597	1597
Site 2	Link		2044	4454	3675
Site 3	Link		4428	4319	3677

7. Discussion

The Stellenbosch Town DMA was analysed using the PERRL 2.0 procedure to identify potential for energy recovery and to optimize the size and location of the energy recovery installations, based on several factors including the potential for leakage reduction. In Scenario 2 the DMA is analysed with the current network hydraulics as described in Sections 3 and 4. The varying topography and clear potential for energy recovery highlighted in Figures 5 and 6, coupled with the dynamic nature of a network supplied by numerous sources and several consumptive nodes with operating pressures above the minimum residual pressure, made the PERRL 2.0 procedure an ideal analysis method for energy recovery optimization.

The desktop analysis as well as the base case scenario run of the Stellenbosch Town DMA indicated a large degree of "aged" pipes in the network with low roughness coefficients causing high friction losses and less available potential head for energy recovery. High friction losses, however, lower the average operating pressure in the system and in turn cause lower losses due to leakages. However, the high friction losses are from "aged" pipework, which is more susceptible to damage and subsequent leakage. Pipe replacement lowers the friction losses in replaced pipe segments and potentially increases the available head for energy recovery. Higher available head, however, also results in higher leakages from the system, although pipe replacement by nature decreases the number of damaged pipes and in turn the leakage from the system. Scenario 3 was run to analyse the effect of selective pipe replacement on energy recovery and leakage reduction within the Stellenbosch Town DMA.

The results shown in Section 6 are discussed in more detail in the following sections.

7.1. Scenario 2

Scenario 2 was run using the base case hydraulic characteristics of the Stellenbosch Town DMA, and applying the PERRL 2.0 optimization procedure, which incorporates the user-defined weighted importance of specific objectives and extended-period simulations into a genetic algorithm, to identify the optimum size and location of potential installations for energy recovery and leakage reduction [18].

Within the PERRL 2.0 process, 10 independent simulation runs, with an initial population of 100 and an energy-recovery location number set to 3, were performed. The PERLL 2.0 process is a multi-objective optimization procedure, and the independent simulation runs were used to define a Pareto frontier. Similar to previous research [18], the objective functions within the PERRL 2.0 optimization procedure included maximizing the energy recovered and minimizing the cost of water lost through leakage by an overall leakage reduction in the network. The potential yearly revenue along with the current annual real losses (CARL) of the system and the nominal cost of water were used within the PERRL 2.0 procedure to calculate a weighted score to rank the potential energy recovery solutions. The monetary value of energy recovered was calculated per unit rate using an average electrical cost and the water cost per kilolitre was defined as all cost savings from the leakage reduction and included the surface water abstraction and treatment electricity consumption cost, the water distribution electricity consumption cost, and the average water rate (loss of sales) [34]. The CARL of the Stellenbosch Town DMA was calculated as 350 ML.

All solutions highlighted the potential of reducing the leakage up to 3.65% (Solution 2) and of recovering energy up to 812 kWh/day (Solution 1). However, the maximum values of both these goals are not achieved by a single solution. In particular Solution 1 allows for significantly more energy recovery than the other solutions (812 kWh/day vs. 649–692 kWh/day) at the cost of the minimum reduction in leakage (2.47%). Looking at the locations in Figure 7, all solutions had proposed installations on the main supply line from the Jonkershoek Weir. In particular, Solution 1 has two locations in common with Solution 2 (Sites 1 and 2 vs. Sites 2 and 3), and hence it is the third site that makes the difference between the achievement of the maximum energy recovery (Solution 1) and the maximum leakage reduction (Solution 2). Site 3 of Solution 1 is located in the centre of the considered DMA with the operating conditions allowing for more energy recovery than all the other sites, however, the corresponding pressure drop deriving from the ERD installation negatively affects the pressure distribution in the other zones of the DMA. In contrast, the third site of the Solution 2 (Site 1) is located on the periphery of the DMA (blue cross on the left of Figure 7) with less energy recovered but also less impact on the average pressure of the system.

To identify the best solution among the top five, the water balance in the district was also analysed since it affects the pumping power needs of the district. In this case, Solution 3 has a water balance more comparable to the base case scenario (121.95 kW vs. 122.19 kW of average pump power) and, hence, is preferable. Despite their higher weighted score in the PERRL 2.0 procedure, Solution 1 and 2 both have considerably higher average kW pump usages than the base case scenario (161.59 and 160.43 kW).

Solution 5 performed well in terms of the average pump energy use and storage volume changes but showed an increased percentage of demand below the minimum residual pressure. Solution 5 was shown to have the lowest operating pressure at consumptive nodes but a higher average system pressure than both Solution 2 and 3, indicative of the lower leakage reduction potential and, therefore, was not selected as the preferred option.

Solutions 3 and 4 compared well to the status quo operations of the DMA with similar results for the consumptive node operating pressure analysis, water balance analysis and pump pressure usage. These solutions also have very similar average operating pressures at consumptive nodes; however, Solution 3 has a lower average system pressure, which results in higher leakage reduction potential and a higher overall weighted score in the PERRL 2.0 procedure. Thus, for the above-mentioned reasons, Solution 3 was chosen as the preferred solution for Scenario 2.

7.2. Scenario 3

The third scenario that was run aimed at showing the effect of "aged" pipes in the Stellenbosch Town DMA on the energy recovery and leakage reduction potential. Hydraulically poor-performing or "aged" pipes with relatively high friction losses over short pipe lengths reduce the available energy in the system for energy recovery in gravity-fed systems. In systems where additional head is added through pumps, these hydraulically poor-performing pipe segments increase the overall pumping energy consumption in the system.

For the first option of Scenario 3, Scenario 3a, only the pipe segments associated with the immediate upstream pipe network of the top proposed energy recovery solutions were considered for pipe replacement with new pipes of similar diameters. The pipe replacement changes the hydraulic characteristics of the network by changing the roughness coefficients of the replaced pipe to fall within the recommended range of factors, as indicated in Table 3. The change in roughness theoretically increases the head available for energy recovery but also has the potential to increase the average operating pressure within the system, which will in turn increase overall leakage. Pipe sections immediately upstream of the proposed installation were replaced back to a point where the supply line convergences with another line or back to the supply reservoir. This was done to ensure the pipe replacement benefits the proposed energy recovery installation. The proposed pipe replacement details are shown in Table 7.

Scenario 3a was run with the preferred proposed energy recovery solution of Scenario 2 for both the base case hydraulics and roughness parameters, as well as the amended roughness parameter corresponding to the proposed pipe replacement. The hydraulic performance of the two simulation runs were compared and analysed in terms of flow, head loss and energy recovery potential, as well as the effect on the average operating pressure and leakage reduction within the system. This was done in order to indicate the effect of the pipe replacement on the energy recovery and leakage reduction potential in the system.

As expected, the pipe replacement in Scenario 3a increased the energy recovery potential in all three identified sites by decreasing the roughness of the pipes upstream of the potential locations. At all points, this increased the flow through the section of pipe due to the lower friction losses. The lower friction losses also account for higher available pressure head above the minimum residual pressure. For Sites 2 and 3, the increases were slight, which can be attributed to the short section of pipes being theoretically replaced upstream of the installation sites.

For Link 3 582, over 6 km of pipe was modelled to be theoretically replaced. This increased the daily potential energy recovery at the site by 28%. This does, however, also increase the average operating pressure head downstream of Link 3 582. Figure 10 shows the operating pressure at Node 3 055, immediately downstream of Link 3 582, before the implementation of energy recovery, with energy recovery and with energy recovery following pipe replacement. From Figure 10 it can be noted that the pipe replacement scenario increases the daily operating pressures. Table 8 indicates an increase in the overall system pressure, but due to the dynamic nature of the network, it still reports a slight decrease in average operating pressure at consumptive nodes. With the operating pressure at Node 3 055, with pipe replacement being higher than the pressure with energy recovery alone, there still exists excess pressure available for energy recovery. Utilizing this excess energy will undoubtably further reduce the average operating pressure at consumptive nodes, resulting in a lesser weighted score through the PERRL 2.0 procedure. The overall increase in system pressure also explains the decrease in Total Leakage Reduction Potential between the scenario with base case roughness and the scenario with pipe replacement roughness implemented, notwithstanding the increase in Total Energy Recovery Potential

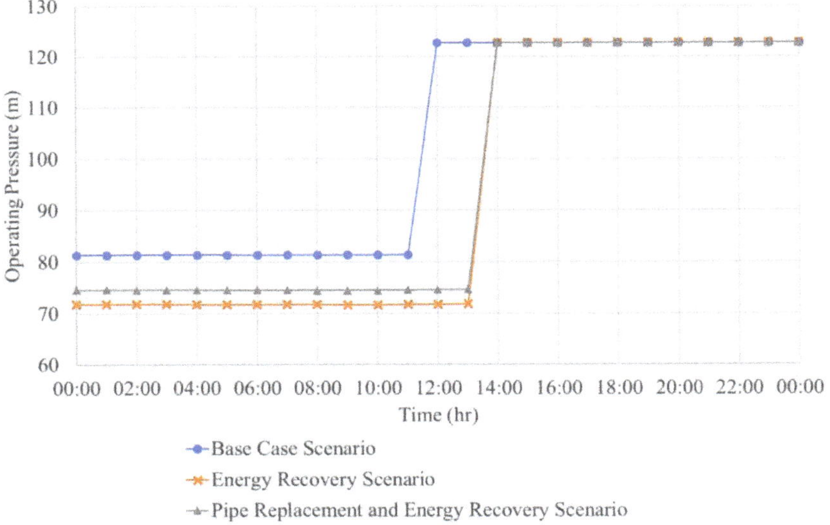

Figure 10. Node 3 055 daily operating pressure—scenario analysis.

The Scenario 3b analysis replaced the "aged" pipes as discussed previously and calculated a new CARL based on the modelled system, with the assumption that the base CARL was equally distributed in size and frequency along the length of the pipe network. Replaced pipes were modelled not to have any leakage and thus effectively reduces leakage prior to the installation of energy recovery devices.

Following the pipe replacement, the PERRL 2.0 procedure was applied to identify and optimise potential energy recovery installations. The leakage reduction calculation in the PERRL 2.0 procedure now incorporated the effective reduced CARL calculated in the previous step. The PERRL 2.0 procedure was applied to the amended networks and new proposed energy recovery solutions were identified.

As expected, the average operating pressure for the 10%, 25%, and 50% pipe replacement scenarios, without energy recovery implemented, increased from 52.47 m to 53.02 m, 53.30 m and 53.35 m, respectively. Nevertheless, the replacement of the "aged" pipes allows for the reduction of leakages with respect to the base case scenario from 3.8% up to 21.6% and also with respect to the solutions of the Scenario 2 (only energy recovery). This leakage reduction is further increased when the energy recovery is considered in the simulations and reaches a maximum of 23.5%.

As regards the energy recovery, it is interesting to notice that the amount of energy recovery increases starting from the 25% of pipe replacement. In case of 10%, the contribution of the pipe replacement is not significant. A possible reason of this negligible contribution could be the locations of the replaced pipes in comparison with the locations of the energy recovery devices: The replaced pipes are not sufficient for positively affecting all the potential ERD installations.

The objective of the study was to test the application of the PERRL 2.0 procedure and assess the advantage thereof in terms of the operation of the Stellenbosch Water Distribution Network by evaluating the energy recovery and leakage reduction potential within the network and the subsequent effects that the installation of energy recovery devices has on the operation of the network. Through the application of the PERRL 2.0 procedure the "status quo" of the Stellenbosch Town DMA was analysed and compared with several scenarios with varied degrees of pipe replacement. The PERRL procedure identified several potential locations for energy recovery within the current Stellenbosch Town DMA (status quo) based on a weighted score of energy recovered, leakage reduction, and estimated capital installation cost. Conclusions and recommendations based on the analysed scenarios are discussed in the following section.

8. Conclusions and Recommendations

The PERRL 2.0 procedure, similar to recent research compared in Table 1, optimizes the installation of energy recovery devices within a water distribution system. The procedure differs from current applied methods in that it incorporates the effect and benefits of leakage reduction within the optimisation process, as illustrated by the current study.

The post-processing of the discussed results were compared the top proposed solutions/installations based on several factors from consumptive node pressures, overall water balance and correlation to the status quo, average pump energy usage, as well as network storage changes.

The following overall conclusions were made from the initial analysis:

- The hydraulic operations of the energy recovery solutions need to mimic the status quo operations as closely as possible, barring the increased pressure reduction at proposed installations. Varying too far from the current operations will not result in sustainable energy recovery installations without a change in system operations;
- The addition of a water balance analysis to the energy recovery and leakage reduction analysis is valuable in identifying sustainable solutions;
- All proposed installations had considerably high average system operating pressures post energy recovery, which indicated additional potential in the network not accessible through current system operation or configuration;
- Applying additional DMAs within the Stellenbosch Town and limiting the effect of energy recovery installations on all consumptive nodes could potentially further increase the total energy recovery potential in the system. The existence of several network water storages between sources and end users indicates potential for such network operations and needs to be further investigated.

The third analysed Scenario incorporated pipe replacement, firstly only on pipe segments upstream of proposed energy recovery solution (3a) and secondly at set pipe replacement levels of 10%, 25%, and 50% of all the "aged" pipes identified in the district (3b).

Scenario 3a indicated a higher total energy recovery potential with a lower total leakage reduction potential. This is attributed to the slightly higher overall average system pressure indicated in Table 8 and the fact that Scenario a did not account for reduced CARL due to pipe replacement.

Scenario 3b was run to show the effect of varying degrees of pipe replacement within the system. Scenario 3b differs from 3a, since it re-applies the PERRL 2.0 procedure to evaluate the energy recovery and leakage reduction potential after pipe replacement considering the dynamic changes to the system operation following the hydraulic changes to the system. The following observations and conclusions were made from the Scenario 3b modelling:

- Leakage reduction in the system for Scenario 3b was first calculated as leakage reduction from the pipe replacement only, and the reduced CARL was then used to calculate the additional leakage reduction as a benefit from energy recovery through the PERRL 2.0 process;
- The leakage reduction increased as pipe replacement increased for 10% and 25% pipe replacement. For 50% pipe replacement, the modelled leakage reduction reduces. With 50% of the network pipe replaced and in effect not being affected by energy recovery in terms of leakage reduction, this occurrence is validated;
- The less pipes are replaced, the higher the average operating pressure in the replaced pipes are (both for the pipe replacement only and energy recovery scenarios). Increasing pipe replacement percentage, however, still increases the average operating pressure in the entire system, which indicates a very localized effect of pipe replacement when conducted on a small scale;
- The average operating pressure in the "aged" pipes move closer to the average operating pressure of the entire network as the percentage of pipe replacement increases. This is largely due to the fact that with an increased pipe replacement percentage, the "aged" pipe network length becomes closer to the entire network length. It can also be attributed to the fact that the more pipes are replaced, the closer the network operations become to the base case scenario. This is indicated by the closer correlation of the water balance figures for 50% pipe replacement;
- The previous observation is reinforced by Figure 9, which shows the average distribution of solutions for the 50% pipe replacement scenario similar to the optimum solution for the base case scenario;
- In all scenarios, several solutions are contained in the main supply lines from the Jonkershoek Weir. These supply lines carry the largest flow volumes resulting in large energy recovery potential changes with small induced headloss.
- In general, the following conclusions were made from the study:
- The PERRL 2.0 procedure performed well in conducting a baseline assessment and scenario modelling of the energy recovery and leakage reduction potential within the Stellenbosch Town DMA.
- Water distribution and water supply systems are dynamic in nature and require a holistic approach to energy recovery modelling.
- Due to the dynamic nature of water supply systems, energy recovery potential installation evaluation should include investigations on the effect of the installation on the overall systems operations to indicate the sustainability of the potential installations. The incorporation of a basic water balance in the evaluation procedure provided valuable insight into the effect of energy recovery on the network operations.
- The replacement of aged infrastructure increases the potential for both energy recovery and leakage reduction within a system. Pipe replacement, however, changes the hydraulic operations of the system and could to some extent require revision of the system operating procedures.

9. Disclaimer

The output of the research conducted in this article was generated from developed hydraulic models of the Stellenbosch Town Water Supply Infrastructure. These models incorporate assumptions informed by demand modelling and have not been calibrated to any specific time, date, or scenario of measured data from the Stellenbosch Water Supply Networks. This research does not reflect or constitute the views of the Stellenbosch LM or any individuals affiliated with the Stellenbosch LM.

Author Contributions: Conceptualization, G.J.B. and M.v.D.; methodology, G.J.B. and M.v.D.; software, G.J.B. and M.v.D.; validation, G.J.B., M.v.D. and G.C.; formal analysis, G.J.B. investigation, G.J.B.; resources, G.J.B., M.v.D. and G.C.; data curation, G.J.B.; writing—original draft preparation, G.J.B.; writing—review and editing, G.J.B., M.v.D. and G.C.; visualization, G.J.B., M.v.D. and G.C.; supervision, M.v.D. and G.C. All authors have read and agreed to the published version of the manuscript.

Funding: This research received no external funding.

Institutional Review Board Statement: Not applicable.

Informed Consent Statement: Not applicable.

Data Availability Statement: Restrictions apply to the availability of these data. Data was obtained from the Stellenbosch Local Municipality and are available from the authors with the permission of the Stellenbosch Local Municipality.

Acknowledgments: The authors would like to acknowledge the Stellenbosch Local Municipality for their assistance and willingness to cooperate and share their knowledge as it pertains to the research conducted in this study.

Conflicts of Interest: The authors declare no conflict of interest.

References

1. Gono, M.; Kyncl, M.; Gono, R. Hydropower Stations in Czech Water Supply System. *AASRI Procedia* **2012**, *2*, 81–86. [CrossRef]
2. Güttinger, M. *Water Resource Management Programme (WARM-P)—Drinking Water and Hydropower: Feasibility Study for Inline Hydropower in Rural NEPAL*; Helvetas Swiss Intercooperation: Surkhet, Nepal, 2012.
3. Perez-Sanchez, M.; Sanchez-Romero, F.J.; Ramos, H.M. Energy recovery in existing water networks: Towards greater sustainability. *Water* **2017**, *9*, 97. [CrossRef]
4. Van Dijk, M.; Kgwale, M.; Bhagwan, J.; Loots, I. Bloemwater conduit hydropower plant launched. *Civ. Eng. Siv. Ing.* **2015**, *23*, 42–46.
5. Loots, I.; Van Dijk, M.; Van Vuuren, S.J.; Bhagwan, J.N.; Kurtz, A. Conduit-hydropower potential in the City of Tshwane water distribution system: A discussion of potential applications, financial and other benefits. *J. S. Afr. Inst. Civ. Eng.* **2014**, *56*, 2–13.
6. Van Dijk, M.; Scharfetter, B.; Bonthuys, G.; Bhagwan, J. Growing hydropower opportunities. *Civ. Eng. Sivil. Ing.* **2017**, *25*, 18–21.
7. Van Dijk, M.; Cavazzini, G.; Bonthuys, G.J.; Santolin, A.; Van Delft, J. Integration of water supply, conduit hydropower generation and electricity demand. *Proceedings* **2018**, *2*, 689. [CrossRef]
8. Ramos, H.M.; Mello, M.; De, P.K. Clean power in water supply systems as a sustainable solution: From planning to practical implementation. *Water Sci. Technol. Water Supply* **2010**, *10*, 39–49. [CrossRef]
9. Gupta, A.; Bokde, N.; Marathe, D.; Kulat, K. Leakage Reduction in Water Distribution Systems with Efficient Placement and Control of Pressure Reducing Valves Using Soft Computing Techniques. *Eng. Technol. Appl. Sci. Res.* **2017**, *7*, 1528–1534. [CrossRef]
10. Nicolini, M.; Zovatto, L. Optimal location and control of pressure reducing valves in water networks. *J. Water Resour. Plann. Manag.* **2009**, *135*, 178–187. [CrossRef]
11. Fecarotta, O.; Aricò, C.; Carravetta, A.; Martino, R.; Ramos, H.M. Hydropower potential in water distribution networks: Pressure control by PATs. *Water Resour. Manag.* **2015**, *29*, 699–714. [CrossRef]
12. Butera, I.; Balestra, R. Estimation of the hydropower potential of irrigation networks. *Renew. Sustain. Energy Rev.* **2015**, *48*, 140–151. [CrossRef]
13. Colombo, A.; Kleiner, Y. Energy recovery in water distribution systems using microturbines. In Proceedings of the Probabilistic Methodologies in Water and Wastewater Engineering 2011, Toronto, ON, Canada, 23 September 2011; pp. 23–27.
14. Tricarico, C.; Morley, M.S.; Gargano, R.; Kapelan, Z.; de Marinis, G.; Savi´c, D.; Granata, F. Optimal water supply system management by leakage reduction and energy recovery. *Procedia Eng.* **2014**, *89*, 573–580. [CrossRef]
15. Parra, S.; Krause, S. Pressure Management by Combining Pressure Reducing Valves and Pumps as Turbines for Water Loss Reduction and Energy Recovery. *Int. J. Sustain. Dev. Plan.* **2017**, *12*, 89–97. [CrossRef]

16. Creaco, E.; Haidar, H. Multiobjective Optimization of Control Valve Installation and DMA Creation for Reducing Leakage in Water Distribution Networks. *J. Water Resour. Plan. Manag.* **2019**, *145*, 04019046. [CrossRef]
17. Bonthuys, G.J.; Van Dijk, M.; Cavazzini, G. Energy Recovery and Leakage-Reduction Optimization of Water Distribution Systems Using Hydro Turbines. *J. Water Resour. Plan. Manag.* **2020**, *146*, 04020026. [CrossRef]
18. Bonthuys, G.J.; Van Dijk, M.; Cavazzini, G. The Optimization of Energy Recovery Device Sizes and Locations in Municipal Water Distribution Systems during Extended-Period Simulation. *Water* **2020**, *12*, 2447. [CrossRef]
19. Pérez-Sánchez, M.; Sánchez-Romero, F.; Ramos, H.; López-Jiménez, P. Modelling Irrigation Networks for the Quantification of Potential Energy Recovering: A Case Study. *Water* **2016**, *8*, 234. [CrossRef]
20. Samora, I.; Franca, M.; Schleiss, A.; Ramos, H.M. Simulated Annealing in Optimization of Energy Production in a Water Supply Network. *Water Resour. Manag.* **2016**, *30*, 1533–1547. [CrossRef]
21. Samora, I.; Manso, P.; Franca, M.J.; Schleiss, A.J.; Ramos, H.M. Energy Recovery Using Micro-Hydropower Technology in Water Supply Systems: The Case Study of the City of Fribourg. *Water* **2016**, *8*, 344. [CrossRef]
22. Sari, M.A.; Badruzzaman, M.; Cherchi, C.; Swindle, M.; Ajami, N.; Jacangelo, J.G. Recent Innovations and Trends in In-conduit Hydropower Technologies and Their Applications in Water Distribution Systems. *J. Environ. Manag.* **2018**, *228*, 416–428. [CrossRef]
23. Bekker, A.; Van Dijk, M.; Niebuhr, C.M.; Hansen, C.D. Framework Development for the Evaluation of Conduit Hydropower within Water Distribution Systems: A South African Case Study. *J. Clean. Prod.* **2020**, *283*, 125326. [CrossRef]
24. Itani, Y.; Soliman, M.R.; Kahil, M. Recovering energy by hydro-turbines application in water transmission pipelines: A case study west of Saudi Arabia. *Energy* **2020**, *211*, 118613. [CrossRef]
25. Cubides-Castro, E.D.; López-Aburto, C.S.; Iglesias-Rey, P.L.; Martínez-Solano, F.J.; Mora-Meliá, D.; Iglesias-Castelló, M. Methodology for Determining the Maximum Potentially Recoverable Energy in Water Distribution Networks. *Water* **2021**, *13*, 464. [CrossRef]
26. Rodríguez-Pérez, A.M.; Pérez-Calañas, C.; Pulido-Calvo, I. Energy Recovery in Pressurized Hydraulic Networks. *Water Resour. Manag.* **2021**, *35*, 1977–1990. [CrossRef]
27. South African Government. Energy Challenge. Available online: https://www.gov.za/issues/energy-challenge/ (accessed on 22 April 2021).
28. Pascale, S.; Kapnick, S.B.; DElworth, T.L.; Cooke, W.F. Increasing risk of another Cape Town "Day Zero" drought in the 21st century. *Proc. Natl. Acad. Sci. USA* **2020**, *117*, 29495–29503. [CrossRef]
29. Muller, M. Cape Town's drought: Don't blame climate change. *Nature* **2018**, *559*, 174–176. [CrossRef]
30. Bischoff-Mattson, Z.; Maree, G.; Vogel, C.; Lynch, A.; Olivier, D.; Terblanche, D. Shape of a water crisis: Practitioner perspectives on urban water scarcity and 'Day Zero' in South Africa. *Water Policy* **2020**, *22*, 193–210. [CrossRef]
31. Sunday Times. Available online: https://www.timeslive.co.za/news/south-africa/2018-02-19-stellenbosch-imposes-stricter-water-restrictions/ (accessed on 22 July 2021).
32. Stellenbosch Local Municipality. *Fourth Review of the Fourth Generation Integrated Development Plan for 2017–2022*; Stellenbosch LM: Stellenbosch, South Africa, 2021.
33. Hatch. *Stellenbosch Municipality Drought Response Plan*. 2017. Available online: https://fsmountain.org/f369/water/DroughtResponsePlan9June2017.pdf (accessed on 22 April 2021).
34. Bonthuys, G.J.; Van Dijk, M.; Cavazzini, G. Leveraging Water Infrastructure Asset Management for Energy Recovery and Leakage Reduction. *Sustain. Cities Soc.* **2019**, *46*, 101434. [CrossRef]
35. IMQS. Available online: https://www.imqs.co.za/ (accessed on 22 April 2021).
36. Du Plessis, P.C.; (GLS, Stellenbosch, Western Cape, South Africa). Personal communication, 2021.
37. Rossman, L.A. *EPANET 2, User Manual*; United States Environmental Protection Agency: Cincinnati, OH, USA, 2001.
38. Connell, D. Hazen-Williams C-Factor Assessment in an Operational Irrigation Pipeline. Master's Thesis, McGill University, Montreal, Quebec, Canada, 2001.
39. Rahman, N.A.; Muhammad, N.S.; Abdullah, J.; Mohtar, W.H.M.W. Model Performance Indicator of Aging Pipes in a Domestic Water Supply Distribution Network. *Water* **2019**, *11*, 2378. [CrossRef]
40. Malaysian Water Association. *MWA Design Guidelines for Water Supply Systems*; The Malaysian Water Association: Kuala Lumpur, Malaysia, 1994.
41. CSIR. *Guidelines for Human Settlement Panning and Design*; Capture Press: Pretoria, South Africa, 2005; Volume 2.

MDPI
St. Alban-Anlage 66
4052 Basel
Switzerland
Tel. +41 61 683 77 34
Fax +41 61 302 89 18
www.mdpi.com

Sustainability Editorial Office
E-mail: sustainability@mdpi.com
www.mdpi.com/journal/sustainability